WITHDRAWN

Elements of
Thermal Stress Analysis

Elements
of
Thermal Stress Analysis

David Burgreen
Professor of Mechanical Engineering, Polytechnic Institute of Brooklyn

First Edition

Published by: C.P. Press, 80-60 190th St., Jamaica, N.Y., 11423

Copyright - 1971, by David Burgreen. All rights reserved. No part of this book may be reproduced in any form without permission in writing by the author.

TA
418.58
B87

Published by: C.P. Press
80-60 190th Street
Jamaica, N.Y. 11423

Preface

The subject of thermal stress analysis is of broad scope. It encompasses a full spectrum of structural systems and components, with applications in power generating equipment, in nuclear energy components, in aircraft and rocket structures and propulsion systems, in process engineering plants, and, to a lesser degree, in civil engineering structures. From the point of view of structural element classification, thermal stress analyses are performed on bars, beams, rings, frames, trusses, plates and shells; and from the analytical point of view the thermal stress problems fall into the categories of extensional deformation, plane stress, plane strain, bending, axially symmetrical deformation, etc.

Because of the large variety of topics that need to be included in a comprehensive treatment of this subject, it is expedient to break up the material into several volumes. The present text serves as an introduction to thermal stress analysis. It contains detailed treatments of the simpler types of thermal stress problems; and these are generally the problems that are most commonly encountered. Included are one-dimensional problems of extensional and bending deformation of bars, beams, and rings, two-dimensional plane stress and plane strain problems, and problems of extensional deformations of cylindrical and spherical shells. Special attention is given to temperature distributions that do not produce stresses, and to strain energy methods that are particularly applicable to thermal stress analysis. The analytical method which utilizes "equivalent thermal loads" is described fully and is utilized throughout the text. It is a useful analytical tool, applicable to thermal stress analysis alone. Some topics which are unique to this text are: thermal moment area analysis; Castigliano's theorem application to thermal stress problems; continuous thermal beams; thermal beams on elastic foundations; bimetallic and layered beams and plates.

It is anticipated that the text will be used by design-oriented students and by practicing designers and engineers who perform thermal stress analyses. The text can serve as a reference source since it contains detailed treatments of a great variety of problems, and many illustrative examples. The material is presented as simply as possible. Those readers with considerable mathematical proficiency may find the treatment to be unnecessarily detailed. On the other hand, engineers who have been out of school for some period of time, will appreciate the added mathematical detail.

The text is an outgrowth of accumulated solutions of thermal stress problems, developed by the author during the years spent in industrial engineering design, and of notes accumulated during the last five years while teaching thermal stress analysis at the Polytechnic Institute of Brooklyn.

David Burgreen

Polytechnic Institute of Brooklyn
August 1, 1971

Contents

Chapter 1 — FUNDAMENTALS

1.1	Classical Equations	1
1.2	Equivalent Thermal Loads	5
1.3	Two-Dimensional Equivalent Loads	11
1.4	One-Dimensional Equivalent Loads	13
1.5	Equivalent Load Solution Procedure	15
1.6	Modified Elastic Constants	18
1.7	Pure Extension and Pure Bending	21
1.8	One-Dimensional Solutions	23
1.9	Equilibrium of Axial Thermal Loads	34
1.10	Transverse Thermal Loads	36

Chapter 2 — EXTENSIONAL DEFORMATIONS OF BARS AND RINGS

2.1	Plane Displacement of a Bar Assembly	44

2.2	Plane Strain in Bar Assembly	47
2.3	Shear Stress in Parallel Element Attachments	48
2.4	Biaxial Stresses in Bars and Rings	49
2.5	Bars with Transverse and Axial Temperature Variation	54
2.6	Bars with Three-Dimensional Temperature Variation	59
2.7	Layered Bars	61
2.8	Shear Stresses in Homogeneous Bars	63
2.9	Interfacial Shear Stress in Layered Bars	67
2.10	End Shear Stresses	69
2.11	Circular Ring with Axial Temperature Variation	73
2.12	Axially Layered Circular Ring	76
2.13	Circular Ring with Radial Temperature Variation	80
2.14	Concentric Rings with Radial Temperature Variation	84
2.15	Ring with Three-Dimensional Temperature Variation	87
2.16	Elements in Series	89
2.17	Bar with Non-Uniform Cross Section	100
2.18	Statically Indeterminate Bar Assembly	104

Chapter 3 — PLANE STRESS AND PLANE STRAIN

3.1	Plane Stress	127
3.2	Plane Strain	136
3.3	Stress-Free Temperature Distributions	141
3.4	Plane Stresses in Plate Strips	155
3.5	Rectangular Plates and Bars	161
3.6	Uniform Hot Area in Large Plate	169
3.7	Plate with Symmetrical Transverse Temperature Variation	171
3.8	Extensional Deformation of Layered Plate	172
3.9	Cylindrical Shell with Radial Temperature Variation	173
3.10	Concentric Cylindrical Shells	177
3.11	Spherical Shell with Radial Temperature Variation	181
3.12	Concentric Spherical Shells	182
3.13	Assemblies with Axial Symmetry	185
3.14	Maximum Stresses in Parallel Elements	190
3.15	Thermal Shock	192

Chapter 4 — RADIAL TEMPERATURE VARIATION

4.1	Concentric Circular Plates	204
4.2	Circular Plate with Radial Temperature Variation	212

4.3	Axially Symmetrical Temperature Variation	227
4.4	Concentric Circular Cylinders	239
4.5	Circular Rod with Radial Temperature Variation	246
4.6	Concentric Hollow Spheres	256
4.7	Radial Temperature Variation in Spheres	261

Chapter 5 — ENERGY METHODS

5.1	Energy Methods of Thermal Stress Analysis	279
5.2	Complementary Energy	280
5.3	Castigliano's Theorem	289
5.4	Thermoelastic Reciprocal Theorem	306
5.5	Rayleigh-Ritz Method	319

Chapter 6 — BEAMS

6.1	Beam Thermal Stress Analysis	333
6.2	Thermal Beam Relationships	337
6.3	Stress and Deformation Equations	340
6.4	Shear Stresses	342
6.5	Solutions of Thermal Beam Equations	347
6.6	Two-Plane Bending	363
6.7	Boundary Load Superposition	365

6.8	Equivalent Load and Moment Area Analysis	371
6.9	Complementary Energy Solutions	381
6.10	Castigliano's Theorem Solutions	386
6.11	Reciprocal Theorem Solutions	392
6.12	Continuous Beams	397
6.13	Elastic Foundations	405
6.14	Layered Beams	412
6.15	Pure Bending of Plates	426

INDEX 455

1 Fundamentals

1.1 CLASSICAL EQUATIONS

A problem of thermal stress analysis differs analytically from an isothermal stress problem (one in which there is no temperature variation throughout the body) only by the presence of the term αT, in the classical stress-strain relationships shown in Eqs. (1.1-1). All of the other equations of elasticity, as well as the boundary conditions, are alike in both thermal and isothermal stress analysis. The formulation of a problem of thermal stress analysis requires consideration of the following:

(a) stress-strain relationships,

(b) strain-displacement relationships,

(c) equilibrium of forces acting on an element,

(d) equilibrium of boundary loads with local internal stresses, and (or) specification of boundary displacements.

These relationships can be expressed either in the classical manner or, as described in Section 1.2, in terms of "equivalent thermal loads" and "displacement stresses." We start with the classical equations in rectangular coordinates.

Rectangular Coordinates

In the classical treatment of thermal stress problems the relationships between stresses and strains are given as

$$\epsilon_x = \frac{1}{E}(\sigma_x - \nu\sigma_y - \nu\sigma_z) + \alpha T$$

$$\epsilon_y = \frac{1}{E}(\sigma_y - \nu\sigma_z - \nu\sigma_x) + \alpha T$$

$$\epsilon_z = \frac{1}{E}(\sigma_z - \nu\sigma_x - \nu\sigma_y) + \alpha T$$

$$\gamma_{xy} = \tau_{xy}/G \qquad (1.1\text{-}1)$$

$$\gamma_{xz} = \tau_{xz}/G$$

$$\gamma_{yz} = \tau_{yz}/G$$

where ϵ and γ are the observable strains or true strains, σ and τ are the stresses due to temperature effects, and αT is the free strain or the unit expansion that an element would experience if it were to expand freely. The normal strains are thus composed of two parts; the free strain, $\epsilon_f = \alpha T$ and the normal thermal strains ϵ_{xt}, ϵ_{yt}, ϵ_{zt}, which are defined as

$$\epsilon_{xt} = \frac{1}{E}(\sigma_x - \nu\sigma_y - \nu\sigma_z)$$

$$\epsilon_{yt} = \frac{1}{E}(\sigma_y - \nu\sigma_z - \nu\sigma_x) \qquad (1.1\text{-}2)$$

$$\epsilon_{zt} = \frac{1}{E}(\sigma_z - \nu\sigma_x - \nu\sigma_y)$$

If the individual elements of a structure were free to expand without restraint, in conformance with the local temperatures, the resulting local deformations would generally yield individual elements which do not fit together. In order to make these deformations compatible (that is, to maintain material continuity) it is necessary to superimpose a stress distribution such that the deformations produced by the superimposed stress distribution (the thermal stresses) together with the free expansions, αT, due to the local temperatures, produce deformations and strains that are compatible. The strains are expressed in terms of the three displacements, u, v, and w, as

$$\epsilon_x = \frac{\partial u}{\partial x} \qquad \gamma_{xy} = \frac{\partial u}{\partial y} + \frac{\partial v}{\partial x}$$
$$\epsilon_y = \frac{\partial v}{\partial y} \qquad \gamma_{xz} = \frac{\partial u}{\partial z} + \frac{\partial w}{\partial x} \qquad (1.1\text{-}3)$$
$$\epsilon_z = \frac{\partial w}{\partial z} \qquad \gamma_{yz} = \frac{\partial v}{\partial z} + \frac{\partial w}{\partial y}$$

The distributed stresses satisfy the equilibrium equations,

$$\frac{\partial \sigma_x}{\partial x} + \frac{\partial \tau_{xy}}{\partial y} + \frac{\partial \tau_{xz}}{\partial z} = 0$$
$$\frac{\partial \sigma_y}{\partial y} + \frac{\partial \tau_{yz}}{\partial z} + \frac{\partial \tau_{xy}}{\partial x} = 0 \qquad (1.1\text{-}4)$$
$$\frac{\partial \sigma_z}{\partial z} + \frac{\partial \tau_{xz}}{\partial x} + \frac{\partial \tau_{yz}}{\partial y} = 0$$

and the condition that the boundary stresses are in equilibrium with the surface loads, is expressed as

$$\overline{X} = \sigma_{xs}\ell + \tau_{xys}m + \tau_{xzs}n$$
$$\overline{Y} = \sigma_{ys}m + \tau_{yzs}n + \tau_{xys}\ell \qquad (1.1\text{-}5)$$
$$\overline{Z} = \sigma_{zs}n + \tau_{xzs}\ell + \tau_{yzs}m$$

where $\overline{X}$, $\overline{Y}$, and $\overline{Z}$, are the vector components of the surface tractions; σ_{xs}, σ_{ys}, σ_{zs}, τ_{xys}, τ_{xzs} and τ_{yzs} are the stresses at the surface; and ℓ, m, and n, are the direction cosines.

The foregoing equations are the relationships that are used in the determination of the stresses and strains in problems of thermal stress analysis. As a rule, simplifications and approximations are needed in order to obtain a problem that is analytically tractable.

Cylindrical Coordinates

The foregoing equations express the basic relationships between the stresses, strains, displacements, and loading in rectangular coordinates. In cylindrical coordinates the stress-strain relationships are

$$\epsilon_r = \frac{1}{E}(\sigma_r - \nu\sigma_\varphi - \nu\sigma_z) + \alpha T$$

$$\epsilon_\varphi = \frac{1}{E}(\sigma_\varphi - \nu\sigma_z - \nu\sigma_r) + \alpha T \qquad (1.1\text{-}6)$$

$$\epsilon_z = \frac{1}{E}(\sigma_z - \nu\sigma_r - \nu\sigma_\varphi) + \alpha T$$

$$\gamma_{r\varphi} = \tau_{r\varphi}/G \qquad \gamma_{rz} = \tau_{rz}/G \qquad \gamma_{\varphi z} = \tau_{\varphi z}/G$$

The normal strains and shear strains are expressed in cylindrical coordinates as

$$\epsilon_r = \frac{\partial u}{\partial r} \qquad \epsilon_\varphi = \frac{u}{r} + \frac{\partial v}{r\,\partial\varphi} \qquad \epsilon_z = \frac{\partial w}{\partial z} \qquad (1.1\text{-}7)$$

$$\gamma_{r\varphi} = \frac{\partial u}{r\,\partial\varphi} + \frac{\partial v}{\partial r} - \frac{v}{r} \qquad \gamma_{rz} = \frac{\partial u}{\partial z} + \frac{\partial w}{\partial r} \qquad \gamma_{\varphi z} = \frac{\partial v}{\partial z} + \frac{\partial w}{r\,\partial\varphi}$$

$$e = \epsilon_r + \epsilon_\varphi + \epsilon_z \qquad \text{(dilatation)}$$

with u the displacement in the r direction, v the displacement in the φ direction, and w the displacement in the z direction. The angular rotations of the differential elements are

$$\omega_r = \frac{1}{2}\left(\frac{\partial w}{r\,\partial\varphi} - \frac{\partial v}{\partial z}\right) \qquad \omega_\varphi = \frac{1}{2}\left(\frac{\partial u}{\partial z} - \frac{\partial w}{\partial r}\right)$$

$$\qquad (1.1\text{-}8)$$

$$\omega_z = \frac{1}{2}\left(\frac{\partial(rv)}{r\,\partial r} - \frac{\partial u}{r\,\partial\varphi}\right)$$

When the equilibrium equations are expressed in terms of displacements it is convenient to use the rotation components of Eq. (1.1-8). Problem 1-3 (at the end of Chapter 1) shows the form of the equilibrium equations in terms of the dilatation and angular rotations. In terms of stress components the equilibrium equations are

$$\frac{\partial \sigma_r}{\partial r} + \frac{\partial \tau_{r\varphi}}{r \partial \varphi} + \frac{\partial \tau_{rz}}{\partial z} + \frac{\sigma_r - \sigma_\varphi}{r} = 0$$

$$\frac{\partial \sigma_\varphi}{r \partial \varphi} + \frac{\partial \tau_{\varphi z}}{\partial z} + \frac{\partial \tau_{r\varphi}}{\partial r} + \frac{2 \tau_{r\varphi}}{r} = 0 \quad (1.1-9)$$

$$\frac{\partial \sigma_z}{\partial z} + \frac{\partial \tau_{zr}}{\partial r} + \frac{\partial \tau_{\varphi z}}{r \partial \varphi} + \frac{\tau_{rz}}{r} = 0$$

At the boundary, the vector components $\overline{R}$, $\overline{\Phi}$, and $\overline{Z}$ of the surface traction $\overline{N}$, are in equilibrium with the local normal stresses and shear stresses. Thus

$$\overline{R} = \sigma_{rs}\ell + \tau_{r\varphi s} m + \tau_{rzs} n$$

$$\overline{\Phi} = \sigma_{\varphi s} m + \tau_{\varphi zs} n + \tau_{r\varphi s} \ell \quad (1.1-10)$$

$$\overline{Z} = \sigma_{zs} n + \tau_{rzs} \ell + \tau_{\varphi zs} m$$

where ℓ, m, and n are the cosines of the angles between the coordinate axes, x, y, and z, respectively, and the normal to the surface.

1.2 EQUIVALENT THERMAL LOADS

The equivalent thermal load method of solution of thermal stress problems is based on the substitution for the normal stresses, σ_x, σ_y and σ_z, another set of stresses σ_x', σ_y' and σ_z', termed displacement stresses. The stress transformation in rectangular coordinates, from the real stresses to the displacement stresses, is performed as follows:

$$\sigma_x = \sigma'_x - \frac{E\alpha T}{1-2\nu}$$

$$\sigma_y = \sigma'_y - \frac{E\alpha T}{1-2\nu} \tag{1.2-1}$$

$$\sigma_z = \sigma'_z - \frac{E\alpha T}{1-2\nu}$$

$$\tau'_{xy} = \tau_{xy} \qquad \tau'_{xz} = \tau_{xz} \qquad \tau'_{yz} = \tau_{yz} \tag{1.2-2}$$

As noted from these equations, the shear stresses are not modified. The quantity $-\frac{E\alpha T}{1-2\nu}$ in the foregoing transformations, has a special significance. It is the local uniform pressure that counterbalances the free expansion strain, αT. This can be shown by computing the uniform compressive stress, $\sigma''_x = \sigma''_y = \sigma''_z = \sigma''$, which is required to develop a compressive counter-balancing strain $\epsilon_c = -\alpha T$. From the isothermal stress-strain relationships we have

$$\epsilon_c = -\alpha T = \frac{1}{E}(\sigma''_x - \nu\sigma''_y - \nu\sigma''_z) \tag{1.2-3}$$

and as $\sigma''_x = \sigma''_y = \sigma''_z = \sigma''$, we have

$$\sigma'' = -\frac{E\alpha T}{1-2\nu} \tag{1.2-4}$$

This part of the thermal stress eliminates the free expansion strain, and brings the individual body elements back to their original size. It follows then that the displacements and the strains which accompany the thermal stresses are due to the stresses σ' alone. For this reason they are called displacement stresses.

When Eqs. (1.2-1) and (1.2-2) are set into the thermal stress-strain equations, Eqs. (1.1-1), we obtain the displacement stress-strain relationships

$$\epsilon_x = \frac{1}{E}(\sigma'_x - \nu\sigma'_y - \nu\sigma'_z)$$

$$\epsilon_y = \frac{1}{E}(\sigma'_y - \nu\sigma'_z - \nu\sigma'_x)$$

$$\epsilon_z = \frac{1}{E}(\sigma'_z - \nu\sigma'_x - \nu\sigma'_y)$$

$$\gamma_{xy} = \frac{\tau_{xy}}{G} = \frac{\tau'_{xy}}{G} \quad (1.2\text{-}5)$$

$$\gamma_{xz} = \frac{\tau_{xz}}{G} = \frac{\tau'_{xz}}{G}$$

$$\gamma_{yz} = \frac{\tau_{yz}}{G} = \frac{\tau'_{yz}}{G}$$

It is seen then that the stresses σ', which alone determine the normal strains, are related to the strains in the same manner as in the isothermal stress-strain relationships. The stress transformations, Eq. (1.2-1) and (1.2-2) are substituted into the equilibrium relationships, Eqs. (1.1-4), which become

$$\frac{\partial \sigma'_x}{\partial x} + \frac{\partial \tau'_{xy}}{\partial y} + \frac{\partial \tau'_{xz}}{\partial z} - \frac{E\alpha}{1-2\nu}\frac{\partial T}{\partial x} = 0$$

$$\frac{\partial \sigma'_y}{\partial y} + \frac{\partial \tau'_{yz}}{\partial z} + \frac{\partial \tau'_{xy}}{\partial x} - \frac{E\alpha}{1-2\nu}\frac{\partial T}{\partial y} = 0 \quad (1.2\text{-}6)$$

$$\frac{\partial \sigma'_z}{\partial z} + \frac{\partial \tau'_{xz}}{\partial x} + \frac{\partial \tau'_{yz}}{\partial y} - \frac{E\alpha}{1-2\nu}\frac{\partial T}{\partial z} = 0$$

These equations show that when equilibrium is expressed in terms of stresses σ'_x, σ'_y, σ'_z, τ'_{xy}, τ'_{xz}, and τ'_{yz}, it is required that there be included apparent body forces, X', Y', and Z', defined as

$$X' = -\frac{E\alpha}{1-2\nu}\frac{\partial T}{\partial x}$$

$$Y' = -\frac{E\alpha}{1-2\nu}\frac{\partial T}{\partial y} \quad (1.2\text{-}7)$$

$$Z' = -\frac{E\alpha}{1-2\nu}\frac{\partial T}{\partial z}$$

in the equations of equilibrium.

When the stresses σ_i and τ_{ij}, from Eqs. (1.2-1) and (1.2-2) are substituted into the boundary relationships, Eqs. (1.1-5) we obtain

$$\overline{X} + \frac{E\alpha T}{1-2\nu} \ell = \sigma'_{xs} \ell + \tau'_{xys} m + \tau'_{xzs} n$$

$$\overline{Y} + \frac{E\alpha T}{1-2\nu} m = \sigma'_{ys} m + \tau'_{yzs} n + \tau'_{xys} \ell \qquad (1.2\text{-}8)$$

$$\overline{Z} + \frac{E\alpha T}{1-2\nu} n = \sigma'_{zs} n + \tau'_{xzs} \ell + \tau'_{yzs} m$$

or

$$\overline{X} + \overline{X}' = \sigma'_{xs} \ell + \tau'_{xys} m + \tau'_{xzs} n$$

$$\overline{Y} + \overline{Y}' = \sigma'_{ys} m + \tau'_{yzs} n + \tau'_{xys} \ell \qquad (1.2\text{-}8)'$$

$$\overline{Z} + \overline{Z}' = \sigma'_{zs} n + \tau'_{xzs} \ell + \tau'_{yzs} m$$

The terms

$$\overline{X}' = \frac{E\alpha T}{1-2\nu} \ell = \overline{N}' \ell$$

$$\overline{Y}' = \frac{E\alpha T}{1-2\nu} m = \overline{N}' m \qquad (1.2\text{-}9)$$

$$\overline{Z}' = \frac{E\alpha T}{1-2\nu} n = \overline{N}' n$$

appear in the boundary equations in the form of additional surface tractions. Their vector sum is $\overline{N}' = \frac{E\alpha T}{1-2\nu}$, a surface traction acting normal to the surface.

Eqs. (1.2-6) and (1.2-8) show that equilibrium is established by the balance of the displacement stresses σ'_i and τ'_{ij} with the body forces $-\frac{E\alpha}{1-2\nu} \frac{\partial T}{\partial s_i}$ and surface tractions $\frac{E\alpha T}{1-2\nu}$. These apparent body forces and surface tractions are designated equivalent thermal loads, since they take the place of temperature effects and appear in the equilibrium equations and boundary equations in the form of loads. When the temperature variation is known, the equivalent body forces and surface tractions are also known. They can be treated as conventional loads, and the displacement stresses due to these loads determined in the same manner as stresses are found when structures are subjected to conventional loads. To obtain the net normal stresses, the internal pressure, $\sigma'' = -\frac{E\alpha T}{1-2\nu}$, is added to the normal displacement stresses σ'_x, σ'_y, and σ'_z in accordance with Eqs. (1.2-1). The displacement shear stresses and true shear stresses are the same. The strains and deformations are obtained directly from the displacement stresses using the relationships of Eqs. (1.2-5).

Cylindrical Coordinates

The transformation of the three-dimensional equations of elasticity in cylindrical coordinates, from the classical form, given in Section 1.1 to the equivalent loading form, in terms of displacement stresses, is obtained by means of the following substitutions.

$$\sigma_r = \sigma'_r - \frac{E\alpha T}{1-2\nu}$$

$$\sigma_\varphi = \sigma'_\varphi - \frac{E\alpha T}{1-2\nu}$$

$$\sigma_z = \sigma'_z - \frac{E\alpha T}{1-2\nu}$$

(1.2-10)

$$\tau_{r\varphi} = \tau'_{r\varphi} \qquad \tau_{rz} = \tau'_{rz} \qquad \tau_{\varphi z} = \tau'_{\varphi z}$$

When the foregoing equations are set into Eqs. (1.1-6), (1.1-9), and (1.1-10) we obtain the stress-strain equations, equilibrium equations, and boundary equations, in terms of the displacement stresses.

The displacement stress-strain equations are

EQUIVALENT THERMAL LOADS

$$\epsilon_r = \frac{1}{E}(\sigma'_r - \nu\sigma'_\varphi - \nu\sigma'_z)$$

$$\epsilon_\varphi = \frac{1}{E}(\sigma'_\varphi - \nu\sigma'_z - \nu\sigma'_r) \qquad (1.2\text{-}11)$$

$$\epsilon_z = \frac{1}{E}(\sigma'_z - \nu\sigma'_r - \nu\sigma'_\varphi)$$

The displacement stress equilibrium equations are

$$\frac{\partial \sigma'_r}{\partial r} + \frac{\partial \tau'_{r\varphi}}{r\partial \varphi} + \frac{\partial \tau'_{rz}}{\partial z} + \frac{\sigma'_r - \sigma'_\varphi}{r} - \frac{E\alpha}{1-2\nu}\frac{\partial T}{\partial r} = 0$$

$$\frac{\partial \sigma'_\varphi}{r\partial \varphi} + \frac{\partial \tau'_{\varphi z}}{\partial z} + \frac{\partial \tau'_{r\varphi}}{\partial r} + \frac{2\tau'_{r\varphi}}{r} - \frac{E\alpha}{1-2\nu}\frac{\partial T}{r\partial \varphi} = 0 \qquad (1.2\text{-}12)$$

$$\frac{\partial \sigma'_z}{\partial z} + \frac{\partial \tau'_{zr}}{\partial r} + \frac{\partial \tau'_{\varphi z}}{r\partial \varphi} + \frac{\tau'_{rz}}{r} - \frac{E\alpha}{1-2\nu}\frac{\partial T}{\partial z} = 0$$

The displacement stress boundary equations are

$$\bar{R} + \frac{E\alpha T}{1-2\nu}\ell = \sigma'_{rs}\ell + \tau'_{r\varphi s}m + \tau'_{rzs}n$$

$$\bar{\Phi} + \frac{E\alpha T}{1-2\nu}m = \sigma'_{\varphi s}m + \tau'_{\varphi zs}n + \tau'_{r\varphi s}\ell \qquad (1.2\text{-}13)$$

$$\bar{Z} + \frac{E\alpha T}{1-2\nu}n = \sigma'_{zs}n + \tau'_{rzs}\ell + \tau'_{\varphi zs}m$$

It is seen that when displacement stresses are utilized the stress-strain relationships have the form of the isothermal stress-strain equations. The equilibrium equations show that there are present equivalent body forces

$$R' = -\frac{E\alpha}{1-2\nu}\frac{\partial T}{\partial r} \qquad \Phi' = -\frac{E\alpha}{1-2\nu}\frac{\partial T}{r\partial \varphi} \qquad Z' = -\frac{E\alpha}{1-2\nu}\frac{\partial T}{\partial z} \qquad (1.2\text{-}14)$$

acting in the r, φ, and z directions respectively; and the boundary equations show that in the displacement stress formulation there are present apparent surface tractions of magnitude

$$\overline{N}' = \frac{E\alpha T}{1 - 2\nu} \qquad (1.2\text{-}15)$$

The strain-displacement relationships remain as given by Eq. (1.1-7), since the displacement strains are the true strains.

When a thermal stress problem is formulated in terms of the displacement stresses, the solution yields the displacement stresses. The true stresses are then obtained by means of the transformations, Eqs. (1.2-10).

1.3 TWO-DIMENSIONAL EQUIVALENT LOADS

The three-dimensional formulation of the thermal stress problem in terms of displacement stresses, is utilized in the solution of plane strain problems, and other common three-dimensional problems. A displacement stress formulation can also be derived for "two-dimensional" and "one-dimensional" problems. The meaning of "two-dimensional" and "one-dimensional," problems in equivalent load thermal stress analysis, differs from plane stress and plane strain problems, and from uniaxial stress problems. By plane stress we mean a two-dimensional state of stress, as in a flat plate, with the stress σ_x, σ_y and τ_{xy} present in the plane of the plate, and all other stresses equal to zero. In plane strain, we are concerned primarily with the state of stress in cross section planes in which σ_x, σ_y and τ_{xy} are not zero, with ϵ_z uniform, and σ_z also not equal to zero. The shear stresses τ_{zx} and τ_{zy}, in plane strain deformation, are zero.

Two-dimensional thermal stress on the other hand, is defined as the state of stress in which $\sigma_z = 0$. There are no other restrictions. In a two-dimensional problem the formulation of the elastic equations, in terms of displacement stresses, is based on the following stress transformation:

TWO-DIMENSIONAL EQUIVALENT LOADS

$$\sigma_x = \sigma'_x - \frac{E\alpha T}{1-\nu}$$

$$\sigma_y = \sigma'_y - \frac{E\alpha T}{1-\nu} \qquad (1.3-1)$$

$$\sigma_z = \sigma'_z - \frac{E\alpha T}{1-\nu} = 0$$

$$\tau_{xy} = \tau'_{xy} \qquad \tau_{xz} = \tau'_{xz} \qquad \tau_{yz} = \tau'_{yz}$$

Substitution of the above into Eqs. (1.1-1) yields

$$\epsilon_x = \frac{1}{E}(\sigma'_x - \nu \sigma'_y) \qquad \epsilon_z = -\nu(\sigma'_x + \sigma'_y) + \frac{1+\nu}{1-\nu}\alpha T$$

$$\epsilon_y = \frac{1}{E}(\sigma'_y - \nu \sigma'_x) \qquad (1.3-2)$$

$$\gamma_{xy} = \frac{\tau'_{xy}}{G} \qquad \gamma_{xz} = \frac{\tau'_{xz}}{G} \qquad \gamma_{yz} = \frac{\tau'_{yz}}{G}$$

The equations for ϵ_x and ϵ_y are similar in appearance to the isothermal plane stress equations. Shear stresses τ'_{xz} and τ'_{yz} and strains γ_{xz} and γ_{yz} may be present, and the transverse normal strain ϵ_z is as shown.

When the stress transformations, Eqs. (1.3-1) are substituted into the equilibrium equations, Eq. (1.1-4) we obtain

$$\frac{\partial \sigma'_x}{\partial x} + \frac{\partial \tau'_{xy}}{\partial y} + \frac{\partial \tau'_{xz}}{\partial z} - \frac{E\alpha}{1-\nu}\frac{\partial T}{\partial x} = 0$$

$$\frac{\partial \sigma'_y}{\partial y} + \frac{\partial \tau'_{yz}}{\partial z} + \frac{\partial \tau'_{xy}}{\partial x} - \frac{E\alpha}{1-\nu}\frac{\partial T}{\partial y} = 0 \qquad (1.3-3)$$

$$\frac{\partial \sigma'_z}{\partial z} + \frac{\partial \tau'_{xz}}{\partial x} + \frac{\partial \tau'_{yz}}{\partial y} - \frac{E\alpha}{1-\nu}\frac{\partial T}{\partial z} = \frac{\partial \tau'_{xz}}{\partial x} + \frac{\partial \tau'_{yz}}{\partial y} = 0$$

and when Eqs. (1.3-1) are set into the boundary equations, Eqs. (1.1-5) these become

ONE-DIMENSIONAL EQUIVALENT LOADS

$$\bar{X} + \frac{E\alpha T}{1-\nu}\ell = \sigma'_{xs}\ell + \tau'_{xys}m + \tau'_{xzs}n$$

$$\bar{Y} + \frac{E\alpha T}{1-\nu}m = \sigma'_{ys}m + \tau'_{yzs}n + \tau'_{xys}\ell \qquad (1.3\text{-}4)$$

$$\bar{Z} + \frac{E\alpha T}{1-\nu}n = \sigma'_{zs}n + \tau'_{xzs}\ell + \tau'_{yzs}m$$

An examination of the foregoing sets of equations shows that the use of the two-dimensional displacement stress formulation requires the application of equivalent thermal body forces of magnitude

$$X' = -\frac{E\alpha}{1-\nu}\frac{\partial T}{\partial x} \qquad Y' = -\frac{E\alpha}{1-\nu}\frac{\partial T}{\partial y} \qquad Z' = -\frac{E\alpha}{1-\nu}\frac{\partial T}{\partial z} \qquad (1.3\text{-}5)$$

and surface tractions

$$\bar{X}' = \frac{E\alpha T}{1-\nu}\ell \qquad \bar{Y}' = \frac{E\alpha T}{1-\nu}m \qquad \bar{Z}' = \frac{E\alpha T}{1-\nu}n \qquad (1.3\text{-}6)$$

These surface tractions may be expressed as a surface traction, normal to all surfaces, of magnitude

$$\bar{N}' = \frac{E\alpha T}{1-\nu} \qquad (1.3\text{-}7)$$

After determination of the displacement stresses the true stresses are obtained by the addition of the internal pressure, $\sigma'' = -\frac{E\alpha T}{1-\nu}$, in accordance with Eqs. (1.3-1).

1.4 ONE – DIMENSIONAL EQUIVALENT LOADS

A one-dimensional thermal stress problem is defined as one in which $\sigma_y = \sigma_z = 0$. There are no other restrictions. In the one-dimensional formulation, the relationships between the displacement stresses and true stresses are

ONE-DIMENSIONAL EQUIVALENT LOADS

$$\sigma_x = \sigma'_x - E\alpha T \qquad \sigma'_y = E\alpha T \qquad \sigma'_z = E\alpha T \qquad (1.4\text{-}1)$$

$$\tau_{xy} = \tau'_{xy} \qquad \tau_{xz} = \tau'_{xz} \qquad \tau_{yz} = \tau'_{yz}$$

When these are substituted into Eqs. (1.1-1), (1.1-4) and (1.1-5), we obtain the one-dimensional form of the stress-strain equations, equilibrium equations, and boundary equations, expressed in terms of displacement stresses.

The displacement stress-strain equations are

$$\epsilon_x = \frac{\sigma'_x}{E} \qquad \epsilon_y = \epsilon_z = -\frac{\nu}{E}\sigma'_x + (1+\nu)\alpha T \qquad \gamma_{ij} = \frac{\tau'_{ij}}{G} \qquad (1.4\text{-}2)$$

The displacement stress equilibrium equations are

$$\frac{\partial \sigma'_x}{\partial x} + \frac{\partial \tau'_{xy}}{\partial y} + \frac{\partial \tau'_{xz}}{\partial z} - E\alpha\frac{\partial T}{\partial x} = 0$$

$$\frac{\partial \sigma'_y}{\partial y} + \frac{\partial \tau'_{yz}}{\partial z} + \frac{\partial \tau'_{xy}}{\partial x} - E\alpha\frac{\partial T}{\partial y} = \frac{\partial \tau'_{yz}}{\partial z} + \frac{\partial \tau'_{xy}}{\partial x} = 0 \qquad (1.4\text{-}3)$$

$$\frac{\partial \sigma'_z}{\partial z} + \frac{\partial \tau'_{xz}}{\partial x} + \frac{\partial \tau'_{yz}}{\partial y} - E\alpha\frac{\partial T}{\partial z} = \frac{\partial \tau'_{xz}}{\partial z} + \frac{\partial \tau'_{yz}}{\partial y} = 0$$

The displacement stress boundary equations are

$$\overline{X} + E\alpha T = \sigma'_{xs}\ell + \tau'_{xys}m + \tau'_{xzs}n$$

$$\overline{Y} + E\alpha T = \sigma'_{ys}m + \tau'_{yzs}n + \tau'_{xys}\ell \qquad (1.4\text{-}4)$$

$$\overline{Z} + E\alpha T = \sigma'_{zs}n + \tau'_{xzs}\ell + \tau'_{yzs}m$$

It is observed from the foregoing sets of equations that a one-dimensional thermal stress analysis carried out in terms of displacement stresses, requires the application of body force of magnitude

$$X' = -E\alpha\frac{\partial T}{\partial x} \qquad Y' = -E\alpha\frac{\partial T}{\partial y} \qquad Z' = -E\alpha\frac{\partial T}{\partial z} \qquad (1.4\text{-}5)$$

and surface tractions of magnitude

$$\bar{N}' = E\alpha T \qquad (1.4\text{-}6)$$

As a result of the thermal loading, displacement stresses are generated. These are then transformed into the true stresses by the addition of an internal pressure, $\sigma'' = -E\alpha T$, in accordance with Eqs. (1.4-1).

1.5 EQUIVALENT LOAD SOLUTION PROCEDURE

A summary of the three-dimensional, two-dimensional, and one-dimensional displacement stress formulation is given in Table 1.5-1. The terms used in the equivalent load method of analysis are listed in the first column. <u>The free strain,</u> $\epsilon_f = \alpha T$, is the unit expansion of an unrestrained element at a uniform temperature. When a structural element, such as a bar, is heated uniformly to a temperature, T_o, a free strain, αT_o, is developed. This is true of all unrestrained uniformly heated structures, wherein each element expands individually and the structure expands as a whole. It is clear that no stresses are generated in uniformly heated structures that are free to expand.

<u>The equivalent internal pressure</u>, σ'', which is equal to $-\dfrac{E\alpha T}{1-2\nu}$ in a three-dimensional system, represents the pressure required to bring a freely expanded element back to its original size. In the equivalent load analysis, this quantity is added to the computed displacement stress σ', to obtain the net thermal stress. In a two-dimensional problem the equivalent internal pressure has the form, $-\dfrac{E\alpha T}{1-\nu}$, and in a one-dimensional system it is $-E\alpha T$. The equivalent internal pressure is not utilized in the analysis of a problem. It is added to the computed displacement stresses after these have been determined.

EQUIVALENT LOAD SOLUTION PROCEDURE

Table 1.5-1 Thermal Loading Summary

Equivalent Load Terms		One Dimensional	Two-Dimensional	Three-Dimensional
1. Free Strain	$\epsilon_f =$	αT	αT	αT
2. Internal Pressure	$\sigma'' =$	$-E\alpha T$	$-\dfrac{E\alpha T}{1-\nu}$	$-\dfrac{E\alpha T}{1-2\nu}$
3. Body Force	$X' =$	$-E\alpha \dfrac{\partial T}{\partial x}$	$-\dfrac{E\alpha}{1-\nu} \dfrac{\partial T}{\partial x}$	$-\dfrac{E\alpha}{1-2\nu} \dfrac{\partial T}{\partial x}$
	$Y' =$	$-E\alpha \dfrac{\partial T}{\partial y}$	$-\dfrac{E\alpha}{1-\nu} \dfrac{\partial T}{\partial y}$	$-\dfrac{E\alpha}{1-2\nu} \dfrac{\partial T}{\partial y}$
	$Z' =$	$-E\alpha \dfrac{\partial T}{\partial z}$	$-\dfrac{E\alpha}{1-\nu} \dfrac{\partial T}{\partial z}$	$-\dfrac{E\alpha}{1-2\nu} \dfrac{\partial T}{\partial z}$
4. Surface Traction, ($\overline{X}'$, $\overline{Y}'$, $\overline{Z}'$)	$\overline{N}' =$	$E\alpha T$	$\dfrac{E\alpha T}{1-\nu}$	$\dfrac{E\alpha T}{1-2\nu}$
5. Displacement Stress	$\sigma_x' =$	$\sigma_x + E\alpha T$	$\sigma_x + \dfrac{E\alpha T}{1-\nu}$	$\sigma_x + \dfrac{E\alpha T}{1-2\nu}$
	$\sigma_y' =$	$E\alpha T$	$\sigma_y + \dfrac{E\alpha T}{1-\nu}$	$\sigma_y + \dfrac{E\alpha T}{1-2\nu}$
	$\sigma_z' =$	$E\alpha T$	$\dfrac{E\alpha T}{1-\nu}$	$\sigma_z + \dfrac{E\alpha T}{1-2\nu}$
6. Displacement Strain	$\epsilon_x =$	$\dfrac{\sigma_x}{E} + \alpha T$	$\dfrac{1}{E}(\sigma_x - \nu \sigma_y) + \alpha T$	$\dfrac{1}{E}(\sigma_x - \nu\sigma_y - \nu\sigma_z) + \alpha T$
	$\epsilon_y =$	$-\dfrac{\nu \sigma_x}{E} + \alpha T$	$\dfrac{1}{E}(\sigma_y - \nu \sigma_x) + \alpha T$	$\dfrac{1}{E}(\sigma_y - \nu\sigma_z - \nu\sigma_x) + \alpha T$
	$\epsilon_z =$	$-\dfrac{\nu \sigma_x}{E} + \alpha T$	$-\dfrac{\nu}{E}(\sigma_x + \sigma_y) + \alpha T$	$\dfrac{1}{E}(\sigma_z - \nu\sigma_x - \nu\sigma_y) + \alpha T$
7. Thermal Stress	$\sigma_x =$	$\sigma_x' - E\alpha T$	$\sigma_x' - \dfrac{E\alpha T}{1-\nu}$	$\sigma_x' - \dfrac{E\alpha T}{1-2\nu}$
	$\sigma_y =$	0	$\sigma_y' - \dfrac{E\alpha T}{1-\nu}$	$\sigma_y' - \dfrac{E\alpha T}{1-2\nu}$
	$\sigma_z =$	0	0	$\sigma_z' - \dfrac{E\alpha T}{1-2\nu}$
8. Thermal Strain	$\epsilon_{xt} =$	$\dfrac{\sigma_x}{E}$	$\dfrac{1}{E}(\sigma_x - \nu\sigma_y)$	$\dfrac{1}{E}(\sigma_x - \nu\sigma_y - \nu\sigma_z)$
	$\epsilon_{yt} =$	$-\dfrac{\nu \sigma_x}{E}$	$\dfrac{1}{E}(\sigma_y - \nu\sigma_x)$	$\dfrac{1}{E}(\sigma_y - \nu\sigma_z - \nu\sigma_x)$
	$\epsilon_{zt} =$	$-\dfrac{\nu \sigma_x}{E}$	$-\dfrac{\nu}{E}(\sigma_x + \sigma_y)$	$\dfrac{1}{E}(\sigma_z - \nu\sigma_x - \nu\sigma_y)$

EQUIVALENT LOAD SOLUTION PROCEDURE

In a three-dimensional problem the <u>equivalent body forces</u>

$$X' = -\frac{E\alpha}{1-2\nu}\frac{\partial T}{\partial x}, \quad Y' = -\frac{E\alpha}{1-2\nu}\frac{\partial T}{\partial y}, \quad Z' = -\frac{E\alpha}{1-2\nu}\frac{\partial T}{\partial z},$$

and the equivalent <u>surface tractions</u>, $\overline{N}' = \frac{E\alpha T}{1-2\nu}$, represent the loading on the structure. These forces should be viewed as if they were real loads acting on a body. The thermal body forces and surface tractions represent a set of loads under which a structure is in static equilibrium. They produce internal stresses, σ', which are called <u>displacement stresses</u>, since a body having isothermal stresses, of magnitude σ', would develop the actual thermal displacements or deformations. The body loads and surface tractions generate <u>displacement strains</u> which are the true strains or measurable strains in the body. In a two-dimensional stress problem the body forces have the form, $-\frac{E\alpha}{1-\nu}\frac{\partial T}{\partial x}, -\frac{E\alpha}{1-\nu}\frac{\partial T}{\partial y}, -\frac{E\alpha}{1-\nu}\frac{\partial T}{\partial z}$ and the surface tractions are $\overline{N}' = \frac{E\alpha T}{1-\nu}$. In a one-dimensional problem the body forces are $-E\alpha\frac{\partial T}{\partial x}, -E\alpha\frac{\partial T}{\partial y}, -E\alpha\frac{\partial T}{\partial z}$, and the surface tractions are $\overline{N}' = E\alpha T$. The analytical part of the equivalent load solution involves the determination of the displacement stresses and strains that arise from the thermal loading. When there are present real body loads and (or) surface tractions, they are added to the thermal body forces and surface tractions, and the displacement stresses and strains are computed from the combined real loads and thermal loads.

<u>The thermal stresses</u> are obtained by adding the internal pressures, σ'', listed on line 2 of Table 1.5-1, to the computed displacement stresses. The <u>thermal strains</u> ϵ_t, are related to the thermal stresses through isothermal stress-strain relationships as shown in Eqs. (1.1-2). For example, in a one-dimensional stress problem the thermal stress is σ_x and the corresponding thermal strain is σ_x/E. Defining a thermal strain quantity is useful in some fatigue analysis computations. When the strains are due to thermal cycling,

fatigue failure is produced by the thermal strain and not the true strain. Consider, for example, the bar shown in Fig. 1.5-1 which is held between rigid walls and is subjected to uniform thermal cycling. It is clear that the displacement strain or true strain will be zero since there is no axial displacement of any point. There is present however a thermal stress, and also a thermal strain which could cause fatigue failure.

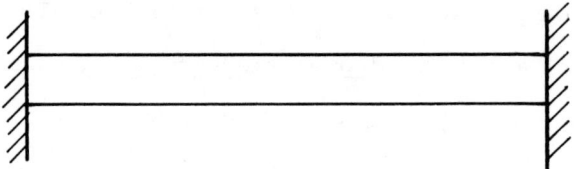

Fig. 1.5-1

1.6 MODIFIED ELASTIC CONSTANTS

To facilitate the treatment of some common types of two-dimensional and three-dimensional stress problems, elastic constants E', ν', a', and E'', termed modified elastic constants, can be employed. These have the effect of reducing three-dimensional problems to two-dimensional problems, and of reducing two-dimensional problems to one-dimensional problems. The most commonly used transformation is that which is used to reformulate a three-dimensional plane strain problem so that it can be analyzed as if it were a two-dimensional plane stress problem.

Plane Strain Reformulation

The three-dimensional thermoelastic stress-strain relationships are:

$$\epsilon_x = \frac{1}{E}(\sigma_x - \nu\sigma_y - \nu\sigma_z) + \alpha T$$

$$\epsilon_y = \frac{1}{E}(\sigma_y - \nu\sigma_z - \nu\sigma_x) + \alpha T$$

$$\epsilon_z = \frac{1}{E}(\sigma_z - \nu\sigma_x - \nu\sigma_y) + \alpha T$$

(1.6-1)

$$\gamma_{xy} = \frac{2(1+\nu)}{E}\tau_{xy}$$

$$\gamma_{xz} = \frac{2(1+\nu)}{E}\tau_{xz}$$

$$\gamma_{yz} = \frac{2(1+\nu)}{E}\tau_{yz}$$

In <u>plane strain</u> deformation, ϵ_z is constant and $\gamma_{xz} = \gamma_{yz} = 0$. When expansion in the z direction is prevented, ϵ_z is zero, and we have <u>zero plane strain</u>, and when there is no net load in the z direction, $\bar{\sigma}_z = 0$, and we have <u>free plane strain</u>. Zero plane strain and free plane strain are special cases of plane strain deformation.

From the third of Eqs. (1.6-1), the stress σ_z, is

$$\sigma_z = E\epsilon_z + \nu(\sigma_x + \sigma_y) - E\alpha T$$

(1.6-2)

Substitution into the first and second of Eqs. (1.6-1) yields

$$\epsilon_x + \nu\epsilon_z = \frac{1-\nu^2}{E}\sigma_x - \frac{\nu(1+\nu)}{E}\sigma_y + (1+\nu)\alpha T$$

$$\epsilon_y + \nu\epsilon_z = \frac{1-\nu^2}{E}\sigma_y - \frac{\nu(1+\nu)}{E}\sigma_x + (1+\nu)\alpha T$$

(1.6-3)

With the use of modified elastic constants, defined as

$$E' = \frac{E}{1-\nu^2}$$

$$\nu' = \frac{\nu}{1-\nu}$$

(1.6-4)

$$\alpha' = \alpha(1+\nu)$$

Eqs. (1.6-3) becomes

$$\epsilon'_x = \epsilon_x + \nu\epsilon_z = \frac{1}{E'}(\sigma_x - \nu'\sigma_y) + \alpha' T$$

$$\epsilon'_y = \epsilon_y + \nu\epsilon_z = \frac{1}{E'}(\sigma_y - \nu'\sigma_x) + \alpha' T$$

(1.6-5)

These are now in the form of the two-dimensional plane stress equations, and the techniques for the solution of plane stress problems can therefore be applied to the solution of plane strain problems. It should be noted that the shear modulus $G = \dfrac{E}{2(1 + \nu)}$ remains the same when the modified constants E' and ν' are substituted for E and ν. That is, $G = G'$, and $\tau_{xy} = G'\gamma_{xy} = G\gamma_{xy}$.

To compute σ_x, and σ_y, and τ_{xy}, the stresses which are present in the cross section planes of cylindrical or prismatic bodies that undergo plane strain deformation, one proceeds by substituting E', ν', and α' for E, ν, and α, in the comparable plane stress equations. This applies as well to solutions which utilize stress functions such as the Airy stress function φ, discussed in Sections 3.1 and 3.2. A general discussion of plane stress and plane strain is given in these sections.

To obtain the cross section strains ϵ_x and ϵ_y, the use of Eq. (1.6-5) is required. When the problem is that of free plane strain, with $\overline{\sigma}_z = 0$, it is shown in Section 3.2 that $\epsilon_z = \alpha \overline{T}$, so that the cross section strains, ϵ_x and ϵ_y in Eqs. (1.6-5) become

$$\epsilon_x = \frac{1}{E'} (\sigma_x - \nu'\sigma_y) + \alpha'T - \nu\alpha\overline{T} \tag{1.6-6}$$

$$\epsilon_y = \frac{1}{E'} (\sigma_y - \nu'\sigma_x) + \bar{\alpha}'T - \nu\alpha\overline{T} \tag{1.6-7}$$

In the reformulation which changes a plane strain problem into a plane stress problem, the equivalent body loads, surface tractions, and internal pressures will appear as

$$X' = -\frac{E'\alpha'}{1-\nu'} \frac{\partial T}{\partial x}, \qquad Y' = -\frac{E'\alpha'}{1-\nu'} \frac{\partial T}{\partial y}$$

$$\overline{N}' = \frac{E'\alpha'T}{1-\nu'} \tag{1.6-8}$$

and the stresses will be given as

$$\sigma_x = \sigma'_x - \frac{E'\alpha' T}{1-\nu'} \qquad \sigma_y = \sigma'_y - \frac{E'\alpha' T}{1-\nu'} \qquad \tau_{xy} = \tau'_{xy} \qquad (1.6\text{-}9)$$

When the solutions of plane strain problems are carried out by using equivalent thermal loading in the form given by Eqs. (1.6-8), no axial thermal tractions are applied. The plane strain problem is solved as if it were a plane stress problem. After σ_x and σ_y are found, σ_z is obtained via Eq. (1.6-2).

1.7 PURE EXTENSION AND PURE BENDING

Thermal stress problems are often concerned with plates and shells in which the temperature distribution is a function of the thickness z, coordinate only. Such bodies, when they are free of external restraints, undergo deformations which are termed pure extension and (or) pure bending. Pure extension is a uniform extensional deformation of a bar, plate, or shell without shear distortion. Pure bending is a uniform bending without shear distortion. Pure bending of a plate is referred to as spherical bending since a uniform curvature is produced.

The stress in this type of deformation is independent of the orientation of the plane coordinates, and invariant with x and y. Thus

$$\sigma(z) = \sigma_x(z) = \sigma_y(z) \qquad (1.7\text{-}1)$$

The stress normal to the plate and all shear stresses are zero. That is,

$$\sigma_z = 0 \qquad \tau_{xy} = \tau_{xz} = \tau_{yz} = 0 \qquad (1.7\text{-}2)$$

The stress-strain equations then become

$$\epsilon = \epsilon_x = \epsilon_y = \frac{1-\nu}{E}\sigma(z) + \alpha T(z) = \frac{\sigma(z)}{E''} + \alpha T(z) \qquad (1.7\text{-}3)$$

with E'' a modified modulus of elasticity (or E_i'' for a layered plate), equal to

$$E'' = \frac{E}{1-\nu} \qquad E_i'' = \frac{E_i}{1-\nu_i} \qquad (1.7-4)$$

Eq. (1.7-3) is the one-dimensional formulation of the stress-strain relationship applicable to pure extension and pure bending of plates. It is clear that the equations of equilibrium will be identical for any set of axes chosen in the plane of the plate, and that the compatibility conditions will automatically be satisfied since ϵ is constant, or a linear function of z.

In the case of pure extension we can integrate Eq. (1.7-3) over the thickness, h, of the plate or shell, and, recognizing that ϵ is constant, we obtain

$$\epsilon = \frac{1-\nu}{E}\bar{\sigma} + a\bar{T} \qquad (1.7-5)$$

Also as $\bar{\sigma} = 0$ in the absence of external loading, Eq. (1.7-5) yields

$$\epsilon = a\bar{T} \qquad (1.7-6)$$

so that for extensional deformation of a homogeneous plate or shell, Eqs. (1.7-3) and (1.7-6) yield

$$\sigma(z) = E'' a (\bar{T} - T(z)) \qquad (1.7-7)$$

In the case of a layered plate or shell under pure extension we have

$$\epsilon = \frac{\bar{\sigma}_i}{E_i''} + a_i T_i = \frac{\bar{\sigma}_1}{E_1''} + a_1 \bar{T}_1 = \frac{\bar{\sigma}_2}{E_2''} + a_2 T_2 = \ldots \text{etc.} \qquad (1.7-8)$$

and as there are no external loads

$$\bar{\sigma}_1 h_1 + \bar{\sigma}_2 h_2 + \ldots = 0 \qquad (1.7-9)$$

Eqs. (1.7-8) and (1.7-9) are sufficient to determine the stress distribution in a layered plate or shell undergoing pure extensional deformation. If we multiply each of Eqs. (1.7-8) by their respective $E_i A_i$ and sum up, we find that the extensional strain in a layered plate or shell is

$$\epsilon = \frac{\sum_{i=1}^{n} E_i'' \alpha_i \overline{T}_i h_i}{\sum_{i=1}^{n} E_i'' h_i} = \overline{\alpha T}' \qquad (1.7\text{-}10)$$

1.8 ONE – DIMENSIONAL SOLUTIONS

As all physical structures are three-dimensional, the thermal loading, in the form of body forces and surface tractions, would normally be three-dimensional in character. In practice, stress analysis problems are simplified and wherever possible stresses of small magnitude are neglected so that the problem can be treated by means of a two-dimensional or a one-dimensional analysis. The approximations that are necessary are fairly well established for certain types of common structural elements such as bars, beams and plates, and these simplifications can be utilized as well in the application of the equivalent load method of thermal stress analysis. In beam bending, for example, tranverse loads produce the axial stresses that are of interest, and the normal stresses in the transverse direction, are small and are disregarded. However, as shown in Sect. 1.10, the transverse thermal loading on a bar or beam is self-equilibrating and does not produce axial stresses. When the assumption is made that a bar or beam problem is one-dimensional in character, then the transverse loading can be disregarded and the thermal body forces and surface tractions taken as

$$X' = -E\alpha \frac{\partial T}{\partial x} \quad ; \qquad \overline{X}' = \overline{N}' = E\alpha T \qquad (1.8\text{-}1)$$

Similarly, if we assume in a plate that the stresses in the direction perpendicular to the surface are essentially zero, then the problem is two-dimensional, and the complete thermal loading is

$$X' = -\frac{E\alpha}{1-\nu}\frac{\partial T}{\partial x} \qquad Y' = -\frac{E\alpha}{1-\nu}\frac{\partial T}{\partial y} \qquad \overline{N}' = \frac{E\alpha T}{1-\nu} \qquad (1.8\text{-}2)$$

The stress-strain relationships that are used must, of course, be consistent with the assumed presence or absence of normal stresses in the specified coordinate directions.

Example (a)   Uniformly Heated Bar

The bar shown in Fig. (1.8-1) is unrestrained, and heated to a temperature T_o.

Fig. 1.8-1

Since the bar is uniformly heated there are no thermal body loads. These are present only when there are temperature gradients. In a three-dimensional analysis, normal surface tractions would be applied on all surfaces, but we can confine our attention to the axial direction and assume $\sigma_y = \sigma_z = 0$. Only the thermal loading in the axial direction need be considered. The surface tractions are, as shown in Fig. 1.8-1,

$$\overline{X}' = E\alpha T_o \qquad (1)$$

These tractions produce a uniform axial displacement stress

$$\sigma'_x = E\alpha T_o \qquad (2)$$

The displacement strain, or true strain is

$$\epsilon_x = \frac{\sigma'_x}{E} = \frac{E\alpha T_o}{E} = \alpha T_o. \qquad (3)$$

ONE-DIMENSIONAL SOLUTIONS

The thermal stress is

$$\sigma_x = \sigma'_x - E\alpha T_o = E\alpha T_o - E\alpha T_o = 0 \qquad (4)$$

and the thermal strain is

$$\epsilon_{xt} = \frac{\sigma_x}{E} = 0 \qquad (5)$$

We find the obvious result that the bar is stress-free and that it expands axially an amount αT_o per unit length.

Example (b)   Uniformly Heated Axially Restrained Bar

The bar shown in Fig. 1.8-2 has a cross section area A, and is heated to a temperature T_o. There are no axial body forces, since these are present only when there are temperature gradients. The thermal surface tractions, $E\alpha T_o$, are applied at the extremities, at the surfaces which are in contact with rigid walls. The surface traction load, $E\alpha T_o A$, is taken entirely by the rigid walls; no part of the load is taken by the elastic bar. As there is no axial displacement load in the bar the displacement stress σ'_x is zero.

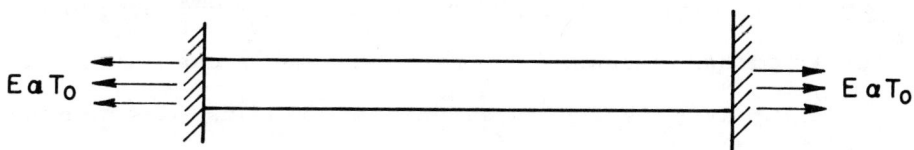

Fig. 1.8-2

The thermal stress is

$$\sigma_x = \sigma'_x - E\alpha T_o = 0 - E\alpha T_o = -E\alpha T_o \qquad (1)$$

and the thermal strain and true strain are respectively

$$\epsilon_{th} = \frac{\sigma_x}{E} = -\alpha T_o \qquad \epsilon = \frac{\sigma'_x}{E} = 0 \qquad (2)$$

Thus one obtains the fairly obvious result that there is no real strain in the bar; only a compressive stress, $-E\alpha T_o$, and a corresponding thermal strain $-\alpha T_o$.

Example (c) Unrestrained Bar with Axial Temperature Variation

In the two preceding examples the elements were at a uniform temperature. Consider now a rod in which the temperature changes gradually from one end to the other. The requirement of slow temperature variation is imposed, since as shall be seen when studying the problem of plate strips with rapidly changing temperatures along the length, only slow axial variations of temperature produce negligible transverse stresses. The magnitude of the transverse stresses that are produced by an axial temperature variation is a measure of the degree to which transverse dimensional compatibility is not satisfied. The magnitude of the neglected stress can be shown to be of the order $E\alpha(\Delta T)_h / 2$, where $(\Delta T)_h$ is the <u>axial</u> temperature change over an axial length equal to the beam depth, h.

Figure 1.8-3 shows the rod with the equivalent thermal body forces and surface tractions. Assume that the rod has a unit cross section area. The displacement stress is found by summing up the loads that act on the length of bar extending from x = 0 to any point x in the rod. Thus

$$\sigma'_x = E\alpha T_1 - \int_0^x (-E\alpha \frac{dT}{dx} \, dx) = E\alpha T \tag{1}$$

with T = T(x). Note that the body forces are algebraically $-E\alpha \, dT/dx$, assumed acting in the positive coordinate direction, to the right. If we sum up the loads acting to the right of the point x, we obtain

$$\sigma'_x = E\alpha T_2 + \int_x^L (-E\alpha \frac{dT}{dx} \, dx) = E\alpha T \tag{2}$$

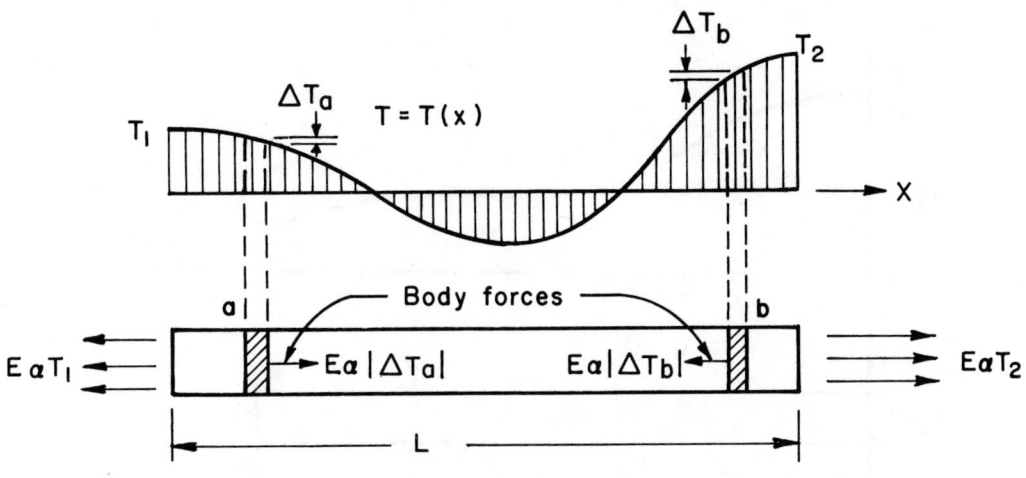

Fig. 1.8-3

Clearly, the same result, $\sigma_x' = E\alpha T$, is obtained by integrating from either end. The strain is

$$\epsilon = \frac{\sigma_x'}{E} = \alpha T \tag{3}$$

signifying that each element is free to expand longitudinally in accordance with the local temperature. The rod is stress-free since

$$\sigma_x = \sigma_x' - E\alpha T = 0 \tag{4}$$

Example (d)   Restrained Bar with Axial Temperature Variation

The bar shown in Fig. 1.8-4 has a temperature distribution $T = T(x)$ which changes slowly in the axial direction.

In this problem the solution is most easily obtained with the use of the classical equations. A one-dimensional state of stress is assumed with

$$\sigma_y = \sigma_z = 0 \qquad \sigma_x \neq 0 \tag{1}$$

The strain, ϵ_x is

ONE-DIMENSIONAL SOLUTIONS

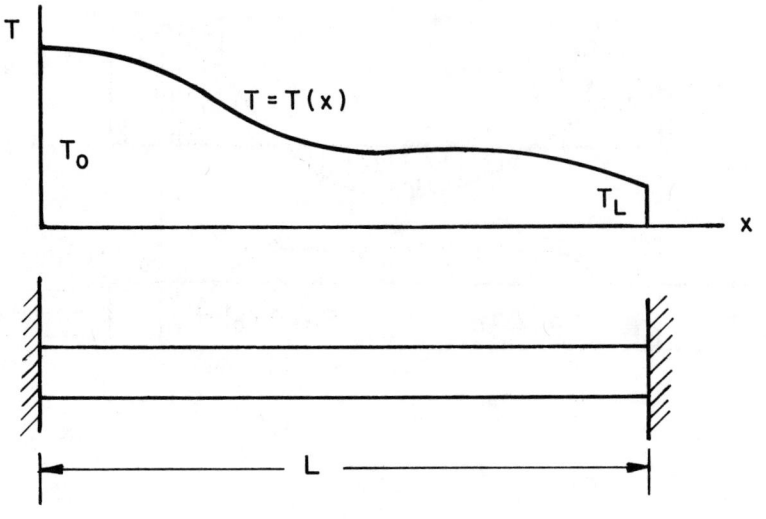

Fig. 1.8-4

$$\epsilon_x = \frac{\sigma_x}{E} + \alpha T \tag{2}$$

As there is no increase in the length of the bar

$$\int_0^L \epsilon_x dx = 0 \tag{3}$$

Substitution of Eq. (2) into Eq. (3) yields

$$\int_0^L \left(\frac{\sigma_x}{E} + \alpha T\right) dx = \frac{\sigma_x L}{E} + \alpha \overline{T} L = 0 \tag{4}$$

from which

$$\sigma_x = -E\alpha \overline{T} \tag{5}$$

Setting the above into Eq. (2) gives the strain as

$$\epsilon_x = -\alpha \overline{T} + \alpha T = -\alpha(\overline{T} - T) \tag{6}$$

The stress is thus proportional to the mean temperature, and the strain is equal to the difference between the local free strain, αT, and the mean free strain $\alpha\bar{T}$. Note that while all parts of the bar are under a compressive stress, a portion of the bar will have a tensile strain.

Example (e) Unrestrained Bar with Linear Transverse Temperature

The temperature distribution in the bar shown in Fig. 1.8-5 is $T = T_o y/h$. The depth of the bar is h and its length is L. Assume that it has a unit cross section area.

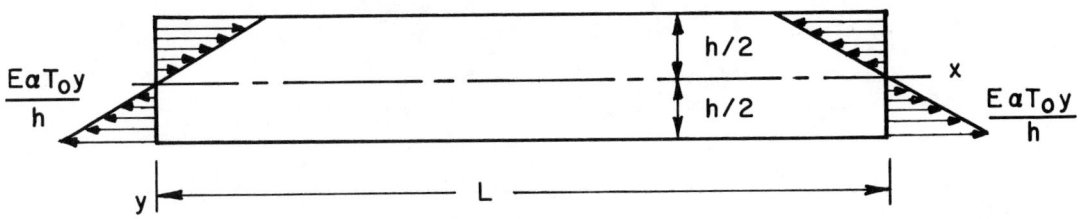

Fig. 1.8-5

In the transverse, y coordinate, direction there are self-equilibrating body forces and surface tractions, which, as will be demonstrated in Sect. 1.10, do not produce any stresses. In the axial direction there are no body forces, as the temperature is not a function of x. There are however, surface tractions at each end, linearly distributed as shown in Fig. 1.8-5. Their magnitude is

$$\bar{X}'_o = \bar{X}'_L = \frac{E\alpha T_o y}{h} \tag{1}$$

These surface tractions produce end moments

$$\bar{M}' = \int_{-h/2}^{h/2} \frac{E\alpha T_o y}{h} y dA = \frac{E\alpha T_o I}{h} \tag{2}$$

where I is the rectangular moment of inertia about the z axis. The displacement stress due to the end moments $\bar{M}'$ is

30 ONE-DIMENSIONAL SOLUTIONS

$$\sigma'_x = \frac{\overline{M}' y}{I} = \frac{E\alpha T_o y}{h} \tag{3}$$

The strain is

$$\epsilon_x = \frac{\sigma'_x}{E} = \frac{\alpha T_o y}{h} \tag{4}$$

and the curvature $1/R$ is

$$\frac{1}{R} = \frac{\epsilon_x}{y} = \frac{\alpha T_o}{h} . \tag{5}$$

The thermal stress, $\sigma_x = \sigma'_x - E\alpha T$, is

$$\sigma_x = \frac{E\alpha T_o y}{h} - \frac{E\alpha T_o y}{h} = 0 \tag{6}$$

An unrestrained bar with a transverse linear temperature distribution is thus seen to be stress-free. It is shown in Section 3.3 that any structural element that is unrestrained will be stress-free if the temperature distribution is linear.

Example (f)   Restrained Bar with Transverse Temperature Distribution

The bar shown in Fig. 1.8-6 has its ends rigidly fixed and is subjected to an arbitrary transverse temperature variation $T = T(y)$.

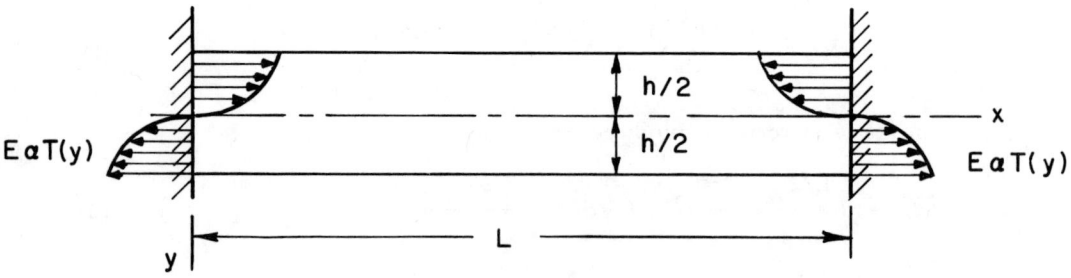

Fig. 1.8-6

As in the preceding example, the surface tractions $E\alpha T(y)$ are applied at the two ends. In the present example these surface tractions have no effect, since the ends are rigidly held. There are therefore no effective thermal body forces or surface tractions acting on the bar. Consequently there are no displacement stresses. That is

$$\sigma'_x = 0 \tag{1}$$

The thermal stress is

$$\sigma_x = \sigma'_x - E\alpha T(y) = -E\alpha T(y) \tag{2}$$

and the strain is

$$\epsilon_x = \frac{\sigma'_x}{E} = 0 \tag{3}$$

As the strains are zero there is no deformation, and we can conclude that any rigidly held bar with a transverse temperature distribution will not deform.

Example (g)   Rigidly Enclosed Bar with Axial Temperature Variation

The bar shown in Fig. 1.8-7 is in contact with rigid walls at all surfaces. The interfacial surfaces are lubricated so that cross section planes can be displaced axially. The bar is subjected to an axial temperature variation $T(x)$.

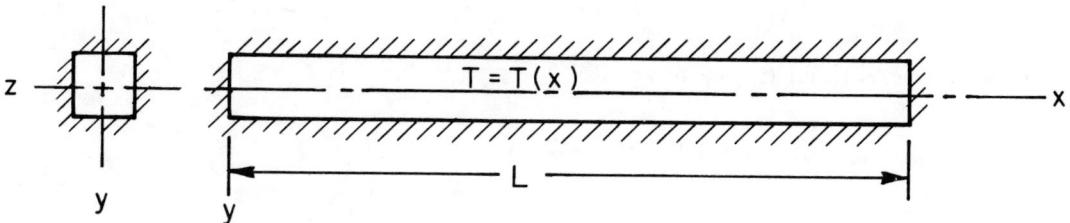

Fig. 1.8-7

ONE-DIMENSIONAL SOLUTIONS

Since axial displacement is prevented we have

$$\Delta L = \int \epsilon_x \, dx = 0 \tag{1}$$

There will be no displacements perpendicular to the axis of the bar; therefore

$$\epsilon_y = \frac{1}{E}(\sigma_y - \nu\sigma_z - \nu\sigma_x) + \alpha T = 0 \tag{2}$$

$$\epsilon_z = \frac{1}{E}(\sigma_z - \nu\sigma_y - \nu\sigma_x) + \alpha T = 0 \tag{3}$$

Also, the axial stress σ_x will be constant and the transverse and lateral stresses σ_y and σ_z will be functions of x only. Eqs. (1), (2) and (3), are integrated over the length of the bar. Thus

$$\frac{E}{L}\int_0^L \epsilon_x \, dx = \sigma_x - \nu\bar{\sigma}_y - \nu\bar{\sigma}_z + E\alpha\bar{T} = 0 \tag{4}$$

$$\frac{E}{L}\int_0^L \epsilon_y \, dx = \bar{\sigma}_y - \nu\bar{\sigma}_z - \nu\sigma_x + E\alpha\bar{T} = 0 \tag{5}$$

$$\frac{E}{L}\int_0^L \epsilon_z \, dx = \bar{\sigma}_z - \nu\bar{\sigma}_y - \nu\sigma_x + E\alpha\bar{T} = 0 \tag{6}$$

with $\bar{\sigma}_y$ and $\bar{\sigma}_z$ representing σ_y and σ_z averaged over the length of the bar.

From Eqs. (4), (5), and (6) we have

$$\sigma_x = \bar{\sigma}_y = \bar{\sigma}_z = -\frac{E\alpha\bar{T}}{1-2\nu} \tag{7}$$

and from Eqs. (2) and (5) we have

$$(1-\nu)\sigma_y + E\alpha T = (1-\nu)\bar{\sigma}_y + E\alpha\bar{T} \tag{8}$$

or

ONE-DIMENSIONAL SOLUTIONS   33

$$\sigma_y = \bar{\sigma}_y + \frac{E\alpha}{1-\nu}(\bar{T} - T) \tag{9}$$

Substitution of the value of $\bar{\sigma}_y$, from Eq. (7) yields

$$\sigma_y = -\frac{E\alpha}{1-\nu}\left(\frac{\nu \bar{T}}{1-2\nu} + T\right) \tag{9}$$

Similarly

$$\sigma_z = -\frac{E\alpha}{1-\nu}\left(\frac{\nu \bar{T}}{1-2\nu} + T\right) \tag{10}$$

It is observed that when the bar is uniformly heated, so that $\bar{T} = T$, the foregoing equations give $\sigma_x = \sigma_y = \sigma_z = -E\alpha T/(1-2\nu)$.

Example (h)   Pinned Two-Bar Assembly

The assembly is shown in Fig. 1.8-8. The bars are of different materials and are at temperatures T_1 and T_2.

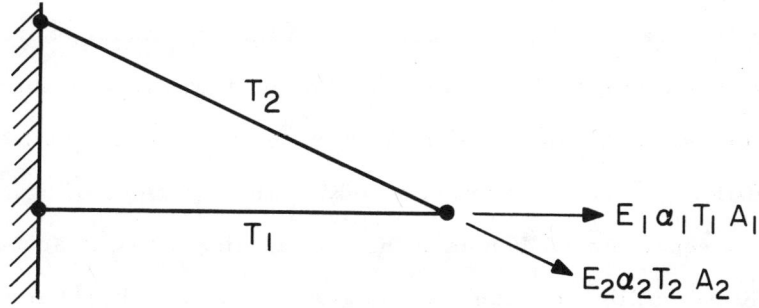

Fig. 1.8-8

The integrated surface traction loads are $E_1 \alpha_1 T_1 A_1$ and $E_2 \alpha_2 T_2 A_2$ as shown, and act in line with the respective bars. The displacement stresses in the bars are

$$\sigma'_1 = E_1 \alpha_1 T_1$$
$$\sigma'_2 = E_2 \alpha_2 T_2 \qquad (1)$$

The stresses are

$$\sigma_1 = \sigma_1' - E_1 \alpha_1 T_1 = 0$$
$$\sigma_2 = \sigma_2 - E_2 \alpha_2 T_2 = 0 \qquad (2)$$

The stresses in the bars are thus seen to be zero. It will be demonstrated in Section 3.3 that the stresses in statically determinate trusses in which the members are of dissimilar materials and at different temperatures, are always zero.

1.9 EQUILIBRIUM OF AXIAL THERMAL LOADS

The body forces and the surface tractions of equivalent thermal loading can be shown to be in static equilibrium. It is clear that in general, static equilibrium must be established when equivalent thermal body forces and surface tractions are applied, since the requirements of static equilibrium are locally fulfilled for each element of a body by the equations of equilibrium and the boundary equations. When each internal element and surface element is individually in equilibrium, the body as a whole must be in static equilibrium. This can be demonstrated from a macroscopic point of view, by examining the rod discussed in Section 1.8, Example (c).

The significance of equivalent thermal loads is clarified by examining the manner in which the static equilibrium of body forces and surface tractions is established in an unrestrained bar with an axially varying temperature.

EQUILIBRIUM OF AXIAL THERMAL LOADS 35

The unrestrained rod of unit cross section area is subjected to an axial temperature variation as shown in Fig. 1.9-1.

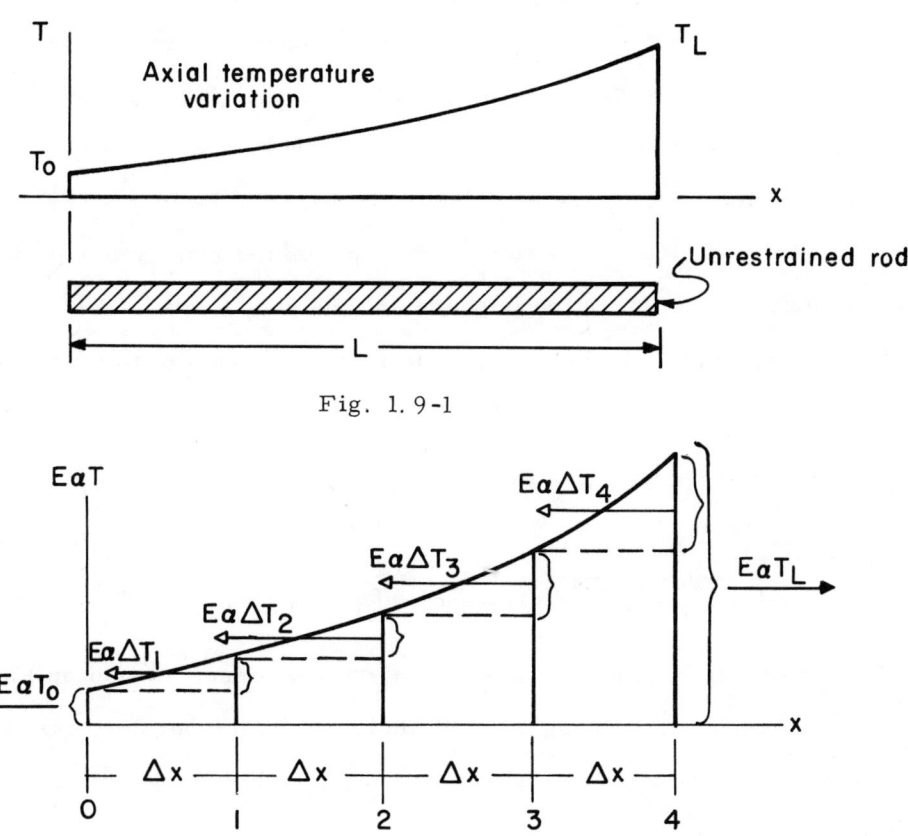

Fig. 1.9-1

Fig. 1.9-2

The surface tractions are $E\alpha T_o$ and $E\alpha T_L$, and the body force on each element of length Δx is $E\alpha \left(\frac{dT}{dx}\right)\Delta x$ or $E\alpha \Delta T$. Figure 1.9-2 shows the rod loaded with the surface tractions and the body forces.

It is seen that $E\alpha T_o$ represents a stress jump at the left side of the rod, and that it acts towards the left. Since the temperature gradient is positive at all points, the incremental stress jumps that are due to the body loads are negative, and also act toward the left. The magnitudes of the surface traction on the left side of the bar and the load increments due to the body forces, are proportional to the heights of the bracketed vertical lines (with $E\alpha$ the proportionality factor). On the right side of the bar, acting toward the right, is the surface traction $E\alpha T_L$. The magnitude of this surface traction is proportional to the height of the bracketed vertical line on the right. Since the vertical length of the latter is equal to the sum of the vertical lengths representing the surface traction jump on the left side of the bar and the four stress jumps due to the body loads, the surface traction on the right side equilibrates the surface traction on the left side and the body loads.

This can be shown algebraically as follows: The sum of the forces acting along the axis of the rod is

$$\sum F' = - E\alpha T_o - E\alpha \Delta T_1 - E\alpha \Delta T_2 - E\alpha \Delta T_3 - E\alpha \Delta T_4 + E\alpha T_L \quad (1.9\text{-}1)$$

or

$$\sum F' = E\alpha(-T_o - \Delta T_1 - \Delta T_2 - \Delta T_3 - \Delta T_4 + T_L) \quad (1.9\text{-}2)$$

The quantity in parenthesis is obviously zero, since, as seen in Fig. 1.9-1, the net temperature change, which includes the temperature rise at $x = 0$, the temperature increments along the rod, and temperature drop at $x = L$, is zero. We therefore have

$$\sum F' = -E\alpha T_o - E\alpha \Delta T_1 - E\alpha \Delta T_2 - E\alpha \Delta T_3 - E\alpha \Delta T_4 + E\alpha T_L = 0 \quad (1.9\text{-}3)$$

which establishes the static equilibrium of the bar under the action of the axial body forces and surface tractions.

1.10 TRANSVERSE THERMAL LOADS

In the application of the equivalent load method, in the examples of

TRANSVERSE THERMAL LOADS

Section 1.8, transverse body forces and surface tractions were tacitly disregarded; only the thermal loads in the axial direction were taken into account. Consider a temperature distribution in a bar which is a function of both the axial and the transverse coordinates, with the restriction that the axial temperature variation takes place slowly. The temperature distribution in a short length of bar can then be assumed to be a function of the transverse coordinate only, and approximated locally as $T = T(y)$. This is shown in Fig. 1.10-1 below.

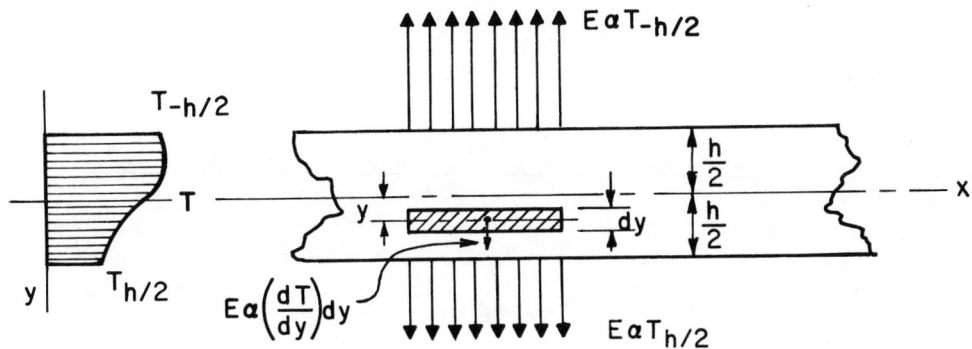

Fig. 1.10-1

The displacement stress σ'_y, summing up the loading from the bottom of the bar, is

$$\sigma'_y = E\alpha T_{h/2} + \int_y^{\frac{h}{2}} \left(-E\alpha \frac{dT}{dy}\right) dy \tag{1.10-1}$$

which gives

$$\sigma'_y = E\alpha T_{h/2} - (E\alpha T_{h/2} - E\alpha T) = E\alpha T \tag{1.10-2}$$

The transverse thermal stress σ_y at any point is then

$$\sigma_y = \sigma'_y = E\alpha T = 0 \qquad (1.10\text{-}3)$$

The transverse stress due to a local transverse temperature variation in a bar (or beam) is thus zero. A similar analysis can be performed to show that transverse stresses are zero when the problem is a two-dimensional one, as in a flat plate. The surface loads are then taken as $\dfrac{E\alpha T_{h/2}}{1-\nu}$ and $\dfrac{E\alpha T_{-h/2}}{1-\nu}$, and the body loads taken as $-\dfrac{E\alpha}{1-\nu}\left(\dfrac{\partial T}{\partial z}\right)dz$, with z the normal plate coordinate.

PROBLEMS

1-1 Using the Lamé notations

$$\lambda = \frac{E\nu}{(1+\nu)(1-2\nu)} \qquad \mu = G = \frac{E}{2(1+\nu)} \qquad 3\lambda + 2\mu = \frac{E}{1-2\nu}$$

and the real dilatation, e, the thermal dilatation, e_t, and the principal stress sum, θ, defined as

$$e = \epsilon_x + \epsilon_y + \epsilon_z = \frac{1-2\nu}{E}(\sigma'_x + \sigma'_y + \sigma'_z) = \frac{1-2\nu}{E}\theta + 3\alpha T$$

$$e_t = \epsilon_{xt} + \epsilon_{yt} + \epsilon_{zt} = \frac{1-2\nu}{E}(\sigma_x + \sigma_y + \sigma_z)$$

$$\theta = \sigma_x + \sigma_y + \sigma_z$$

show that the thermal stresses are expressed in terms of strains and temperature as follows:

$$\sigma_x = \lambda e_t + 2\mu\epsilon_{xt} = \lambda e + 2\mu\epsilon_x - \frac{E\alpha T}{1-2\nu} \qquad \tau_{xy} = \mu\gamma_{xy}$$

$$\sigma_y = \lambda e_t + 2\mu\epsilon_{yt} = \lambda e + 2\mu\epsilon_y - \frac{E\alpha T}{1-2\nu} \qquad \tau_{yz} = \mu\gamma_{yz}$$

$$\sigma_z = \lambda e_t + 2\mu\epsilon_{zt} = \lambda e + 2\mu\epsilon_z - \frac{E\alpha T}{1-2\nu} \qquad \tau_{zx} = \mu\gamma_{zx}$$

1-2 The strain compatibility relationships are given as

$$\frac{\partial^2 \epsilon_x}{\partial y^2} + \frac{\partial^2 \epsilon_y}{\partial x^2} = \frac{\partial^2 \gamma_{xy}}{\partial x \partial y} \qquad \frac{2 \partial^2 \epsilon_x}{\partial y \partial z} = \frac{\partial}{\partial x}\left(-\frac{\partial \gamma_{yz}}{\partial x} + \frac{\partial \gamma_{xz}}{\partial y} + \frac{\partial \gamma_{xy}}{\partial z}\right)$$

$$\frac{\partial^2 \epsilon_y}{\partial z^2} + \frac{\partial^2 \epsilon_z}{\partial y^2} = \frac{\partial^2 \gamma_{yz}}{\partial y \partial z} \qquad \frac{2 \partial^2 \epsilon_y}{\partial x \partial z} = \frac{\partial}{\partial y}\left(\frac{\partial \gamma_{yz}}{\partial x} - \frac{\partial \gamma_{xz}}{\partial y} + \frac{\partial \gamma_{xy}}{\partial z}\right)$$

$$\frac{\partial^2 \epsilon_z}{\partial x^2} + \frac{\partial^2 \epsilon_x}{\partial z^2} = \frac{\partial^2 \gamma_{xz}}{\partial x \partial z} \qquad \frac{2 \partial^2 \epsilon_z}{\partial x \partial y} = \frac{\partial}{\partial z}\left(\frac{\partial \gamma_{yz}}{\partial x} + \frac{\partial \gamma_{xz}}{\partial y} - \frac{\partial \gamma_{xy}}{\partial z}\right)$$

With the use of the stress-strain relations and the equations of equilibrium show that the compatibility equations can be expressed in terms of stresses as follows:

$$(1 + \nu) \nabla^2 \sigma_x + \frac{\partial^2 \theta}{\partial x^2} + E\alpha \left(\frac{1 + \nu}{1 - \nu} \nabla^2 T + \frac{\partial^2 T}{\partial x^2}\right) = 0$$

$$(1 + \nu) \nabla^2 \sigma_y + \frac{\partial^2 \theta}{\partial y^2} + E\alpha \left(\frac{1 + \nu}{1 - \nu} \nabla^2 T + \frac{\partial^2 T}{\partial y^2}\right) = 0$$

$$(1 + \nu) \nabla^2 \sigma_z + \frac{\partial^2 \theta}{\partial z^2} + E\alpha \left(\frac{1 + \nu}{1 - \nu} \nabla^2 T + \frac{\partial^2 T}{\partial z^2}\right) = 0$$

$$(1 + \nu) \nabla^2 \tau_{xy} + \frac{\partial^2 \theta}{\partial x \partial y} + E\alpha \frac{\partial^2 T}{\partial x \partial y} = 0$$

$$(1 + \nu) \nabla^2 \tau_{yz} + \frac{\partial^2 \theta}{\partial y \partial z} + E\alpha \frac{\partial^2 T}{\partial y \partial z} = 0$$

$$(1 + \nu) \nabla^2 \tau_{xz} + \frac{\partial^2 \theta}{\partial x \partial z} + E\alpha \frac{\partial^2 T}{\partial x \partial z} = 0$$

1-3 Show that the equations of equilibrium can be expressed in terms of strains, in rectangular coordinates as

$$(\lambda + \mu) \frac{\partial e}{\partial x} + \mu \nabla^2 u - (3\lambda + 2\mu) \alpha \frac{\partial T}{\partial x} = 0$$

$$(\lambda + \mu) \frac{\partial e}{\partial y} + \mu \nabla^2 v - (3\lambda + 2\mu) \alpha \frac{\partial T}{\partial y} = 0$$

$$(\lambda + \mu) \frac{\partial e}{\partial y} + \mu \nabla^2 w - (3\lambda + 2\mu) \alpha \frac{\partial T}{\partial z} = 0$$

and in cylindrical coordinates as

$$(\lambda + 2\mu) \frac{\partial e}{\partial r} - 2\mu \left(\frac{\partial \omega_z}{r \partial \varphi} - \frac{\partial \omega_\varphi}{\partial z} \right) - (3\lambda + 2\mu) \alpha \frac{\partial T}{\partial r} = 0$$

$$(\lambda + 2\mu) \frac{\partial e}{r \partial \varphi} - 2\mu \left(\frac{\partial \omega_r}{\partial z} - \frac{\partial \omega_z}{\partial r} \right) - (3\lambda + 2\mu) \alpha \frac{\partial T}{r \partial \varphi} = 0$$

$$(\lambda + 2\mu) \frac{\partial e}{\partial z} - 2\mu \left(\frac{\partial (r \omega_\varphi)}{\partial r} - \frac{\partial \omega_r}{\partial \varphi} \right) - (3\lambda + 2\mu) \alpha \frac{\partial T}{\partial z} = 0$$

1-4 If the temperature distribution in an unrestrained body is such that all of the stresses are zero, show that the compatibility equations of Problem 1-2 take the form

$$\frac{\partial^2 T}{\partial x^2} + \frac{\partial^2 T}{\partial y^2} = 0 \qquad \frac{\partial^2 T}{\partial y \, \partial z} = 0$$

$$\frac{\partial^2 T}{\partial z^2} + \frac{\partial^2 T}{\partial y^2} = 0 \qquad \frac{\partial^2 T}{\partial x \, \partial z} = 0$$

$$\frac{\partial^2 T}{\partial x^2} + \frac{\partial^2 T}{\partial z^2} = 0 \qquad \frac{\partial^2 T}{\partial x \, \partial y} = 0$$

What conclusions can be drawn regarding the stresses in an unrestrained body with a temperature distribution

$$T = T_0 + T_1 x + T_2 y + T_3 z$$

PROBLEMS

1-5 What are the displacement stress equilibrium equations, in rectangular coordinates, for the case of:

a) Plane stress

b) Plane strain.

1-6 What are the displacement stress equilibrium equations, in cylindrical coordinates, for the case of

a) Plane stress

b) Plane strain.

1-7 When a stress function φ is used, the compatibility equation for thermal plane stress takes the form

$$\frac{\partial^4 \varphi}{\partial y^4} + 2 \frac{\partial^4 \varphi}{\partial x^2 \partial y^2} + \frac{\partial^4 \varphi}{\partial x^4} + E\alpha \left(\frac{\partial^2 T}{\partial x^2} + \frac{\partial^2 T}{\partial y^2} \right) = 0$$

Deduce the form of this equation for the case of plane strain, using the concept of modified elastic constants.

1-8 In the plane stress analysis of a circular flat plate with a radial temperature variation, the radial displacement u is found to be

$$u = \alpha \left[(1-\nu) \frac{r}{b^2} \int_0^b Tr\, dr + \frac{1+\nu}{r} \int_0^r Tr\, dr \right]$$

Show that the radial displacement in a circular cylinder, with a radial temperature variation, which is restrained from axial expansion ($\epsilon_z = 0$) is

$$u = \frac{1+\nu}{1-\nu} \alpha \left[(1-2\nu) \frac{r}{b^2} \int_0^b Tr\, dr + \frac{1}{r} \int_0^r Tr\, dr \right]$$

1-9 Perform a two-dimensional analysis of a uniformly heated unrestrained bar. Apply surface tractions as shown in Fig. 1-9 below. Show that the results are the same as those obtained in the one-dimensional analysis of Example (a), Section 1.8.

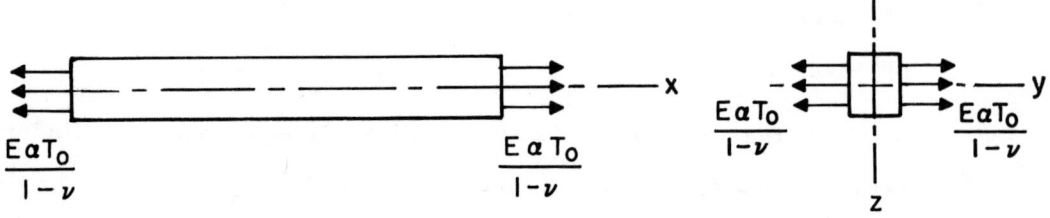

Fig. 1-9

1-10 Perform a three-dimensional analysis on a uniformly heated unrestrained bar, applying surface tractions as shown in Fig. 1-10. Show that the results are the same as those obtained in Problem 1-9.

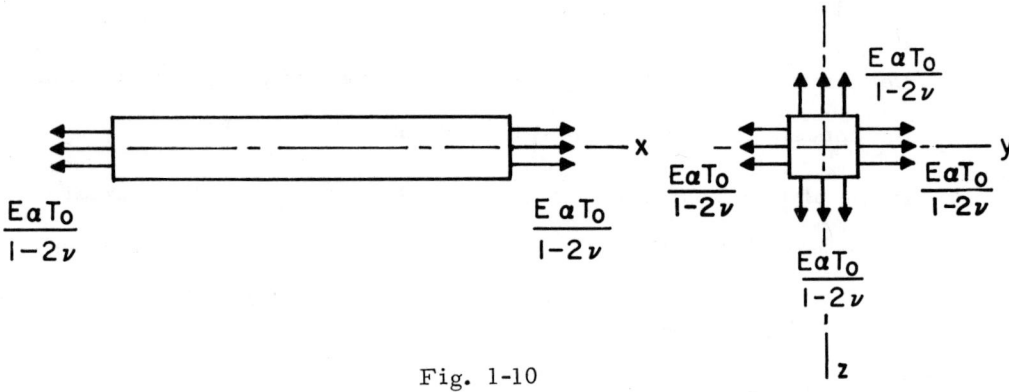

Fig. 1-10

1-11 Perform a three-dimensional classical analysis on an axially restrained bar subjected to a slowly varying axial temperature. Compute the strains in the y and z coordinate directions.

1-12 Using equivalent load analysis find the stresses and strains in the three-bar assembly when the central bar is heated to a temperature T_o. The equivalent thermal load $E\alpha T_o A_1$ acts as shown.

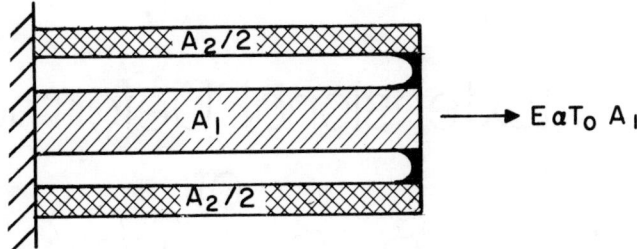

Fig. 1-12

1-13   A bar is restrained between two rigid walls. The left half of the bar is heated to a temperature T_o. Find the stresses and strains in each part of the bar.

1-14   One end of the bar shown in Fig. 1-14 is attached to a rigid wall. The bar has a transverse temperature distribution $T = T_o y/h$. Find the deflection of the end of the bar using equivalent load analysis.

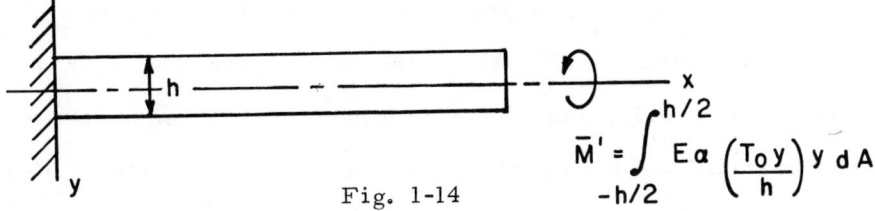

Fig. 1-14 $\bar{M}' = \int_{-h/2}^{h/2} E\alpha \left(\frac{T_o y}{h}\right) y\, dA$

1-15   Using two-dimensional equivalent loads, find the stress distribution in a thin flat plate subjected to a temperature variation $T = T_o z^3/h^3$. The plate is a part of a box assembly in which the four walls to which it is attached can be assumed to be rigid.

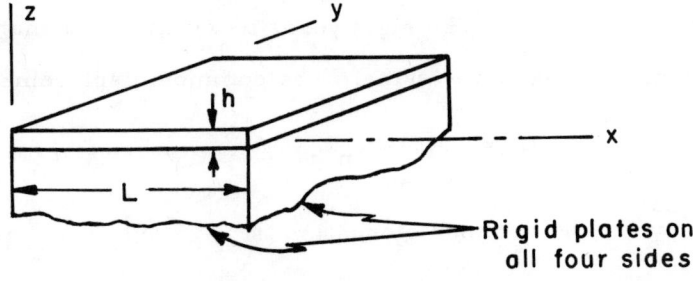

Fig. 1-15

2 | Extensional Deformations of Bars and Rings

2.1 PLANE DISPLACEMENT OF A BAR ASSEMBLY

When a structure is composed of one-dimensional elements in a parallel array, and the deformation is restricted to plane strain deformation or plane displacement, the structure is designated a parallel element assembly. Concentric tubes attached to each other at the extremities is an example of a discrete parallel assembly, and a bar with a symmetrical temperature distribution is an example of a homogeneous parallel element assembly, the elements being the parallel fibres. One requirement for plane displacement or plane strain, is a symmetrical temperature distribution. In the case of a non-homogeneous assembly of elements, a further requirement is that there be a symmetrical distribution of material and material properties; otherwise the structure would be subject to bending. A parallel element assembly is shown in Fig. 2.1-1 below. The rigid yokes to which the elements are attached constrains them to elongate the same. The common displacement is

$$\Delta L = \epsilon_1 L_1 = \epsilon_2 L_2 = \ldots = \epsilon_n L_n \tag{2.1-1}$$

and as $\epsilon_m = \dfrac{\sigma'_m}{E_m}$ the common elongation is

PLANE DISPLACEMENT OF A BAR ASSEMBLY

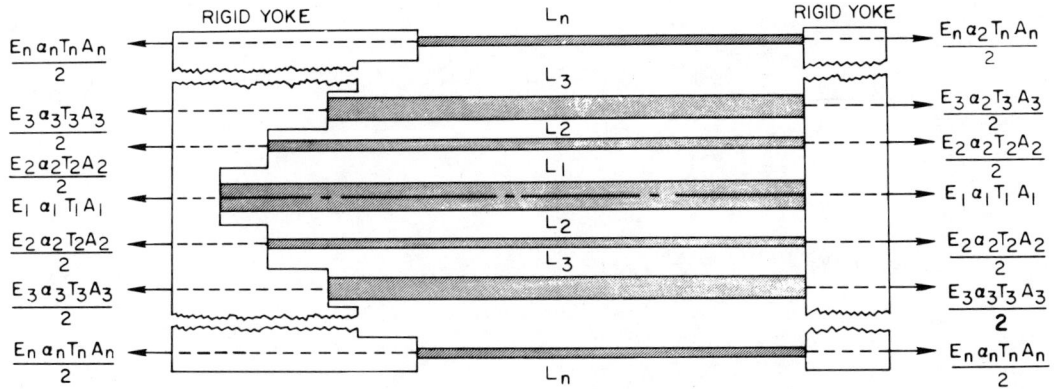

Fig. 2.1-1

$$\Delta L = \frac{\sigma_1' L_1}{E_1} = \frac{\sigma_2' L_2}{E_2} = \ldots \frac{\sigma_n' L_n}{E_n} \tag{2.1-2}$$

The displacement stresses in each of the bars is expressed in terms of the displacement stress in bar 1 as

$$\sigma_m' = \frac{E_m L_1}{E_1 L_m} \sigma_1' \tag{2.1-3}$$

The surface tractions acting over the bar areas yield

$$\overline{P}' = E_1 \alpha_1 T_1 A_1 + E_2 \alpha_2 T_2 A_2 + \ldots + E_n \alpha_n T_n A_n \tag{2.1-4}$$

which must be equal to the sum of the displacement loads in the bars, or

$$\overline{P}' = \sigma_1' A_1 + \sigma_2' A_2 + \ldots + \sigma_n' A_n \tag{2.1-5}$$

Substitution of Eq. (2.1-3) into Eq. (2.1-5) yields

$$\overline{P}' = \frac{\sigma_1' L_1}{E_1} \left(\frac{E_1 A_1}{L_1} + \frac{E_2 A_2}{L_2} + \ldots + \frac{E_n A_n}{L_n} \right) \tag{2.1-6}$$

From the foregoing equation and Eq. (2.1-4) the displacement stress in Bar 1 is obtained as

$$\sigma'_1 = \frac{E_1}{L_1} \frac{\sum_{1}^{n} E_j \alpha_j T_j A_j}{\sum_{1}^{n} E_j A_j / L_j} \tag{2.1-7}$$

The strain in Bar 1 is $\frac{\sigma'_1}{E_1}$, or

$$\epsilon_1 = \frac{1}{L_1} \frac{\sum_{1}^{n} E_j \alpha_j T_j A_j}{\sum_{1}^{n} E_j A_j / L_j} \tag{2.1-8}$$

and the stress in Bar 1 is

$$\sigma_1 = \sigma'_1 - E_1 \alpha_1 T_1 = \frac{E_1}{L_1} \cdot \frac{\sum_{1}^{n} E_j \alpha_j T_j A_j}{\sum_{1}^{n} E_j A_j / L_j} - E_1 \alpha_1 T_1 \tag{2.1-9}$$

The displacement strain and thermal stress in any bar, m, obtained in the same manner, is

$$\epsilon_m = \frac{1}{L_m} \cdot \frac{\sum_{1}^{n} E_j \alpha_j T_j A_j}{\sum_{1}^{n} E_j A_j / L_j} \tag{2.1-10}$$

and

$$\sigma_m = E_m \left[\frac{\sum_{1}^{n} E_j \alpha_j T_j A_j}{L_m \sum_{1}^{n} E_j A_j / L_j} - \alpha_m T_m \right] \tag{2.1-11}$$

2.2 PLANE STRAIN IN BAR ASSEMBLY

In many problems the length of the elements are equal, and the strain in each element is the same; that is, the deformation becomes that of plane strain. Eq. (2.1-10) then becomes

$$\epsilon = \frac{\sum_{1}^{n} E_j \alpha_j T_j A_j}{\sum_{1}^{n} E_j A_j} = \overline{\alpha T} \qquad (2.2\text{-}1)$$

where $\overline{\alpha T}$ represents the free expansions, $\alpha_j T_j$, averaged over the rigidities $E_j A_j$. The stress for the case of equal length elements (plane strain), is

$$\sigma_m = E_m \left[\frac{\sum_{1}^{n} E_j \alpha_j T_j A_j}{\sum_{1}^{n} E_j A_j} - \alpha_m T_m \right] = E_m (\overline{\alpha T} - \alpha_m T_m) \qquad (2.2\text{-}2)$$

In the case of two attached parallel elements, Eqs. (2.2-1) and (2.2-2) become

$$\epsilon = \frac{E_1 \alpha_1 T_1 A_1 + E_2 \alpha_2 T_2 A_2}{E_1 A_1 + E_2 A_2} \qquad (2.2\text{-}3)$$

and

$$\sigma_1 = \frac{E_1 E_2 A_2 (\alpha_2 T_2 - \alpha_1 T_1)}{E_1 A_1 + E_2 A_2} \qquad (2.2\text{-}4)$$

$$\sigma_2 = \frac{E_1 E_2 A_1 (\alpha_1 T_1 - \alpha_2 T_2)}{E_1 A_1 + E_2 A_2} \qquad (2.2\text{-}5)$$

Eqs. (2.2-4) and (2.2-5) are the commonly used forms to express the axial stresses in an assembly consisting of two elements only. When there are three or more parallel elements, the form given by Eq. (2.2-2) is used.

2.3 SHEAR STRESS IN PARALLEL ELEMENT ATTACHMENTS

The stresses given by Eqs. (2.2-4) and (2.2-5) would be generated, for example, in the concentric tube assembly, welded together at the extremities as shown in Fig. 2.3-1. The thermal surface loads, per unit circumferential length are $E_1 \alpha_1 T_1 h_1$ and $E_2 \alpha_2 T_2 h_2$.

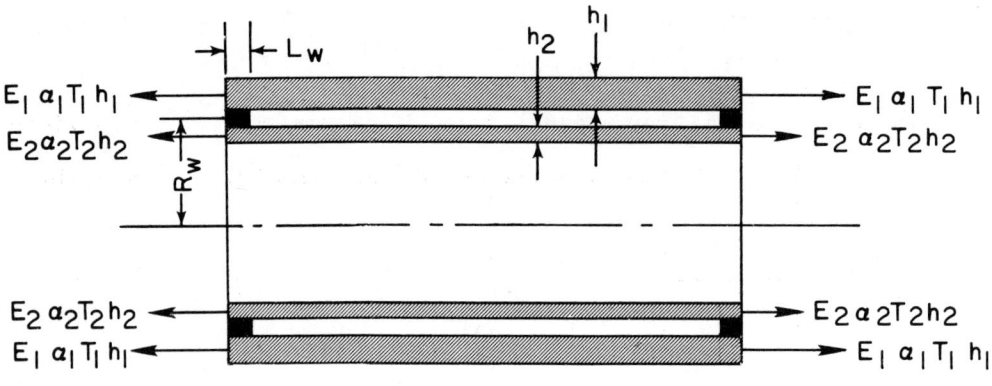

Fig. 2.3-1

The force in each of the tubes is the same, one being tensile and the other compressive. This force, F, transmitted through the end welds, is equal to

$$F = |\sigma_1 A_1| = |\sigma_2 A_2| = \tau A_s \qquad (2.3-1)$$

where A_s is the shear area of the weld, equal to $2\pi R_w L_w$. The shear stress in the weld is then

$$\tau = \frac{\sigma_2 A_2}{A_s} = \frac{\sigma_2 \cdot 2\pi R_2 h_2}{2\pi R_w L_w} \approx \left|\frac{\sigma_2 h_2}{L_w}\right| \approx \left|\frac{\sigma_1 h_1}{L_w}\right| \qquad (2.3-2)$$
$$\hspace{5em} L_w < h_2 < h_1 \quad L_w < h_1 < h_2$$

The restrictions indicate that the expressions are valid only if the length of the weld is smaller than the thickness of the thinner tube. The shear stress in the attachment will then be greater than either of the normal stresses. However, if L_w is greater than one of the tube thicknesses Eq. (2.3-2) is not valid.

This restriction is required because the shear stress is confined to a short length near the end of the assembly, and an increase in the length of the weld, so that it is greater than the thickness of the thinnest tube, does not reduce the stress. This problem is discussed in Section 2.10.

2.4 BIAXIAL STRESSES IN BARS AND RINGS

In the analysis of thermal stresses in bars and rings in the sections which follow, the bars and rings are treated as one-dimensional elements. That is, it is assumed that normal stresses are present in one coordinate direction only, the normal stresses along the other two coordinates being assumed equal to zero. Such an assumption is valid in the case of parallel elements which are attached at their extremities. However, in the case of homogeneous bars and rings with transverse temperature variations, and in the case of bimetallic bars and rings, the assumption of a uniaxial normal stress is an approximation. Because of the internal lateral and transverse expansion restraints that are present, the stress can be greater than that computed by one-dimensional analysis. The amount of increase in the computed one-dimensional stress in rectangular bars, that is due to the presence of lateral and transverse stresses, can be estimated from known solutions of pertinent plane stress and plane strain problems. The stress modifications that are recommended here are based to a large extent on the solutions presented in Sections 3.4 and 3.5.

In the case of a homogeneous bar with a transverse temperature variation, the one-dimensional analysis will be quite accurate when there is no abrupt change in the symmetrical transverse temperature distribution, and when the lateral dimension, w, is small in comparison with the transverse dimension, h, of the bar, along which the temperature varies. Fig. 2.4-1

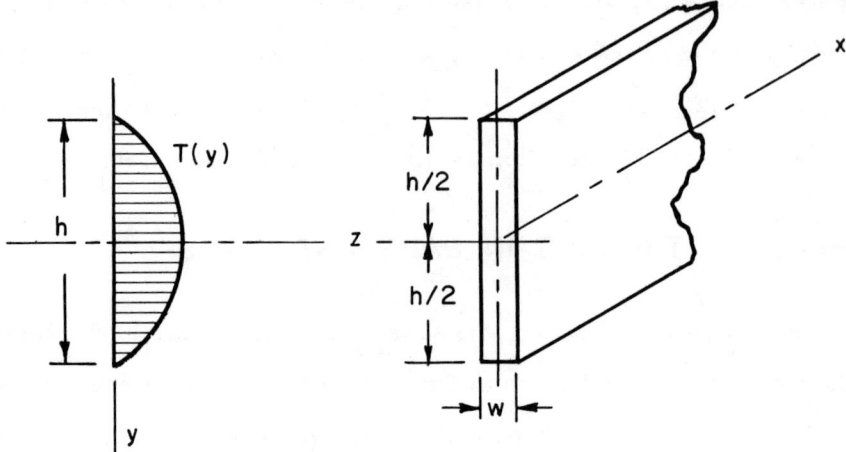

Fig. 2.4-1

shows such an element. It is clear that when w << h, lateral expansion can take place freely, and the normal stresses in the y and z coordinate directions will be small. On the other hand if the lateral dimension, w, becomes large, as in Fig. 2.4-2, the bar begins to assume the configuration of a

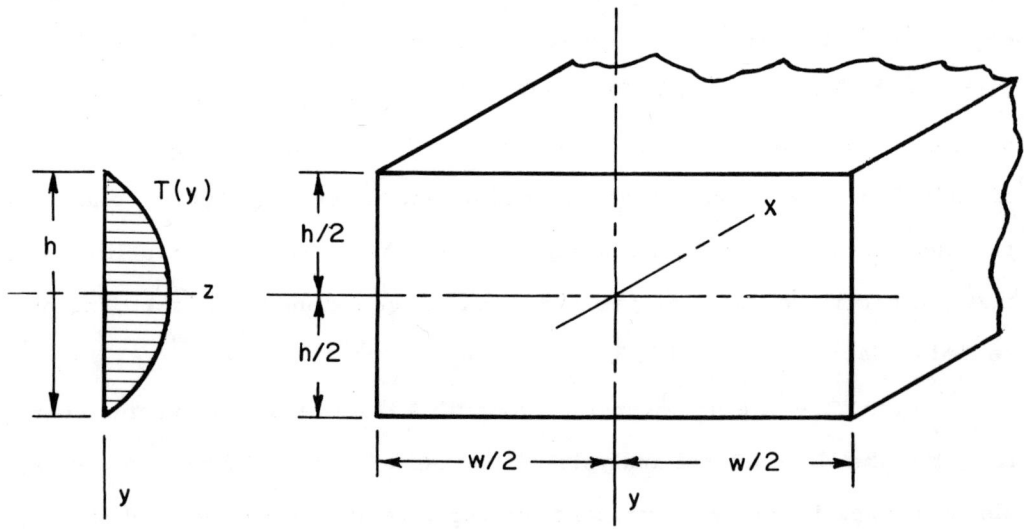

Fig. 2.4-2

plate having a symmetrical transverse temperature variation. In the plate with a transverse temperature variation we have, as a result of the uniform omnidirectional stress in the plane of each plate limina, a two-dimensional state of stress wherein $\sigma_z = \sigma_x$. From the stress-strain relationships we readily obtain the result that the biaxial stress in a plate is larger than the uniaxial stress in a laterally thin bar, by a factor $1/(1 - \nu)$. The accurate determination of the stress in a bar such as that shown in Fig. 2.4-3, whose width is

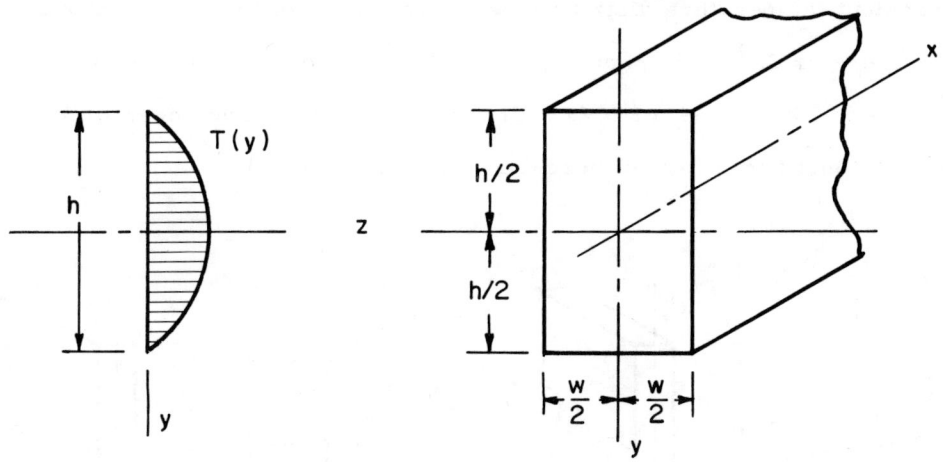

Fig. 2.4-3

neither very small nor very large, can be a complex problem. A study of the solutions presented in Sections 3.4 and 3.5 show that a conservative estimate of the modified axial stress in an unrestrained rectangular bar, taking into account the stresses in the plane of the cross section, can be obtained by assuming a modified modulus of elasticity, E^*, defined as follows:

$$E^* = \frac{E}{1 - \nu \frac{w}{h_t}} \qquad \frac{w}{h_t} \le 1 \qquad (2.4-1)$$

$$E^*_{max} = \frac{E}{1-\nu} \qquad \frac{w}{h_t} > 1 \qquad (2.4\text{-}2)$$

The modified modulus of elasticity, E^*, replaces E in the expressions for stress computed by a one-dimensional analysis. The transverse distance, h_t, represents the length over which a temperature reversal takes place. Steady-state temperature distributions due to internal heat generation will generally appear in the form shown in Fig. 2.4-3, and as the temperature reversal takes place over the depth of the bar, we have $h_t = h$. In the case of transient temperature distributions which may take the form shown in Figs. 2.4-4 or 2.4-5, h_t represents the distance for a full temperature reversal as in Fig. 2.4-4; or $h_t/2$ will represent the distance over which there is a monotonic temperature increase as in Fig. 2.4-5.

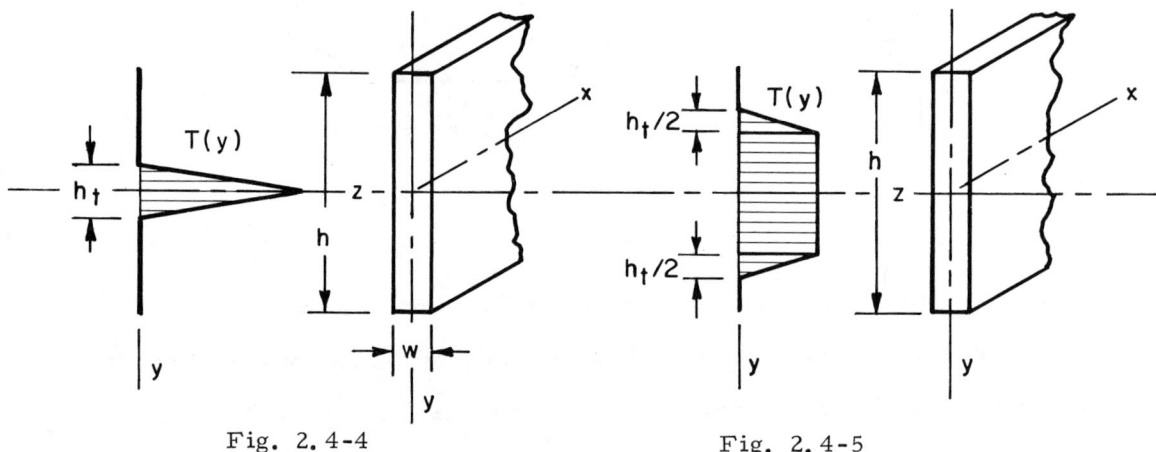

Fig. 2.4-4   Fig. 2.4-5

When there is a temperature variation simultaneously in the lateral and transverse directions, then from pragmatic considerations, the one-dimensional stress should be increased in accordance with Eqs. (2.4-1) and (2.4-2) using **either** the transverse or lateral temperature variation depending upon which gives the maximum stress increase.

In the case of homogeneous rings, treated in Sections 2.12 through 2.15, Eqs. (2.4-1) and (2.4-2) also apply. In rings, h represents the dimension along which the temperature variation takes place. Thus a radially thin ring (a short cylinder) with an axial symmetrical temperature variation requires only a minor one-dimensional stress modification, while the stress in a radially thin ring with a radial temperature variation will be increased by a factor $1/(1 - \nu)$. On the other hand an axially thin ring (circular plate with a very large central hole) with a radial temperature variation will require little stress modification while the stress in an axially thin ring with an axial symmetrical temperature variation will be modified by a factor $1/(1 - \nu)$.

The qualification, stated earlier, that the biaxial thermal stress modification is applicable to <u>unrestrained</u> bars and rings is necessary since this modification is due to thermal effects alone. When boundary restraints are present, the stresses due to the restraints are isothermal stresses to which the biaxial modification does not apply. The use of stress modification in a thermal stress analysis of a restrained bar or beam would therefore require a two-step procedure in which the modified thermal stresses are first computed for the unrestrained body and the unmodified stresses due to the boundary restraints then added.

In steady-state transverse and lateral heat conduction, in a simply connected bar without internal heat generation, the temperature distribution is a plane harmonic function of the cross section coordinates. It is demonstrated, in Section 3.3, that under these conditions no lateral or transverse stresses are developed; only axial stresses are present. In this type of problem no modification of the axial stress is required since lateral and transverse stresses are not present.

In most of the bar problems treated in the present chapter, symmetrical temperature distributions which arise from internal heat generation are assumed. In these types of problems the temperature distribution is not a plane harmonic function and

$$\frac{\partial^2 T}{\partial y^2} + \frac{\partial^2 T}{\partial z^2} \neq 0 \qquad (2.4-3)$$

Lateral and transverse stresses will then be developed.

2.5 BARS WITH TRANSVERSE AND AXIAL TEMPERATURE VARIATION

It is possible to apply the analysis of Sections 2.1 and 2.2, developed for the computation of thermal stresses and strains in discrete parallel elements, to a homogeneous bar with a symmetrical transverse temperature variation such as shown in Fig. 2.5-1.

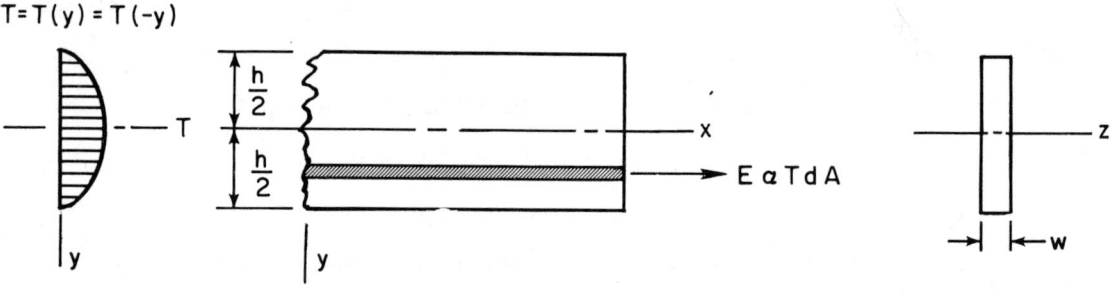

Fig. 2.5-1

The restriction of a symmetrical transverse temperature variation is imposed in order to exclude bars which undergo bending. The analysis which follows applies also for the case of non-symmetrical transverse temperature variations, provided flexural deformations are prevented by the end restraints.

Unrestrained Bar with Transverse Temperature Variation

The plate strip or narrow bar in Fig. 2.5-1 can be assumed to consist of a set of parallel fibres of equal length, each having the same material properties. Eqs. (2.2-1) and (2.2-2) give the strains and stresses in parallel elements as

$$\epsilon = \frac{\sum_{1}^{n} E_j \alpha_j T_j A_j}{\sum_{1}^{n} E_j A_j} \tag{2.5-1}$$

and

$$\sigma_m = E_m \frac{\sum_{1}^{n} E_j \alpha_j T_j A_j}{\sum_{1}^{n} E_j A_j} - \alpha_m T_m \tag{2.5-2}$$

Consider the homogeneous bar as consisting of individual parallel elements of cross section dA, each having the same modulus of elasticity E and coefficient of expansion α. The temperatures of the parallel elements are now given as $T = T(y) = T(-y)$. The summation signs in Eqs. (2.5-1) and (2.5-2) become integral signs, and A_j becomes dA. The strain in the homogeneous bar, Eq. (2.5-1), can then be written as

$$\epsilon = \frac{E\alpha \int_{-h/2}^{h/2} T \, dA}{E \int_{-h/2}^{h/2} dA} = \alpha \overline{T} \tag{2.5-3}$$

and the thermal stress becomes

$$\sigma = E(\epsilon - \alpha T) = E\alpha (\overline{T} - T) \tag{2.5-4}$$

Thus the strain is uniform and equal to $\alpha \overline{T}$, and the stress is the product of

56 TRANSVERSE AND AXIAL TEMPERATURE VARIATION

the modulus of elasticity and the difference between the mean cross section strain and the local free strain.

Unrestrained Bar with Transverse and Axial Temperature Variation

When the symmetrical transverse temperature variation undergoes a slow variation in the x direction, the displacement stress will be due to both the surface tractions and the integrated body forces. Fig. 2.5-2 shows the temperature distribution, $T = T(x, y) = T(x, -y)$, and the associated thermal body forces and surface tractions.

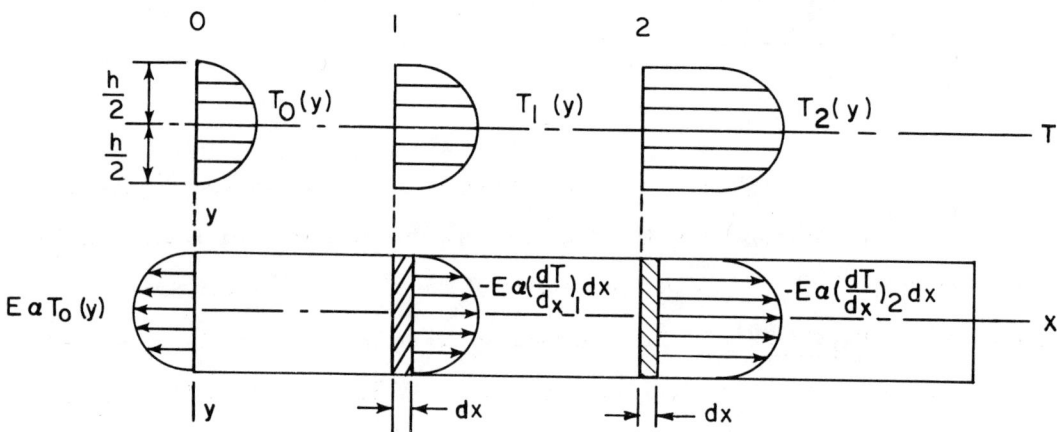

Fig. 2.5-2

At any cross section the displacement stress is the sum of the surface tractions and the integrated body loads between the cross section location and the end of the bar. Thus

$$\sigma'_x = \frac{E\alpha}{A} \int_{-\frac{h}{2}}^{\frac{h}{2}} T_0(y)\, dA - \frac{E\alpha}{A} \int_{-\frac{h}{2}}^{\frac{h}{2}} \int_0^x -\frac{\partial T}{\partial x}\, dx\, dA \qquad (2.5\text{-}5)$$

or

$$\sigma'_x = E\alpha \overline{T}_0 + E\alpha \int_0^x \frac{\partial \overline{T}(x)}{\partial x}\, dx = E\alpha \overline{T}(x) \qquad (2.5\text{-}6)$$

where $\overline{T}(x)$ is the mean cross section temperature. The net stress is therefore

$$\sigma_x = \sigma'_x - E\alpha T = E\alpha\left[\overline{T}(x) - T(x,y)\right] \tag{2.5-7}$$

or more accurately

$$\sigma_x = E^*\left[\overline{T}(x) - T(x,y)\right] \tag{2.5-8}$$

when biaxiality of stress is taken into account. E^* is given by Eqs. (2.4-1) and (2.4-2).

Transverse and Axial Temperature Variation in Restrained Bar

The bar shown in Fig. 2.5-3 is held between rigid walls and is subjected to a temperature variation $T = T(x,y)$. As before it is assumed that the axial variation is gradual.

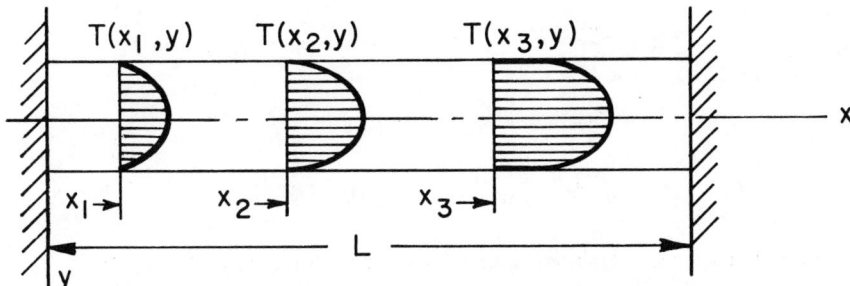

Fig. 2.5-3

The axial strain is

$$\epsilon = \frac{\sigma}{E} + \alpha T \tag{2.5-9}$$

with σ a function of x and y. As the strain is uniform over a section, and the mean stress, $\bar{\sigma}$, does not vary with x, the integration of Eq. (2.5-9) over the cross section yields

$$\epsilon = \frac{\bar{\sigma}}{E} + \alpha \bar{T}(x) \tag{2.5-10}$$

where $\bar{T}(x)$ represents the mean cross section temperature at a specified location, x.

There is no change in the length of the bar. Therefore

$$\int_0^L \epsilon \, dx = \int_0^L (\frac{\bar{\sigma}}{E} + \alpha \bar{T}(x)) \, dx = 0 \tag{2.5-11}$$

This yields

$$\bar{\sigma} = -E\alpha\bar{T} \tag{2.5-12}$$

where $\bar{T}$ is the temperature averaged over the whole bar.

From Eqs. (2.5-9) and (2.5-10) we have

$$\sigma - \bar{\sigma} = -E\alpha (T - \bar{T}(x)) \tag{2.5-13}$$

and as $\bar{\sigma} = -E\alpha\bar{T}$, the distributed stress σ(x, y) becomes

$$\sigma = -E\alpha (\bar{T} - \bar{T}(x) + T(x, y)) \tag{2.5-14}$$

Note that when T is a function of x only we have $T(x, y) = \bar{T}(x)$, and σ = $-E\alpha\bar{T}$ as in Example (d) of Sect. 1.8. When there is a transverse temperature variation only, $\bar{T}(x) = \bar{T}$ and we obtain σ = $-E\alpha T(y)$ as in Example (f) of Sect. 1.8.

2.6 BARS WITH THREE-DIMENSIONAL TEMPERATURE VARIATION

The temperature in a rectangular bar is taken as a slowly varying function of the coordinate, x, and as a symmetrical function of the cross section coordinates y and z. The temperature of the end face at x = 0, will be $T_o = T_o(y, z)$ and at any point x it will be $T_x = T_x(y, z)$ as shown in Fig. 2.6-1. The temperature distribution is taken symmetrical about the y and z

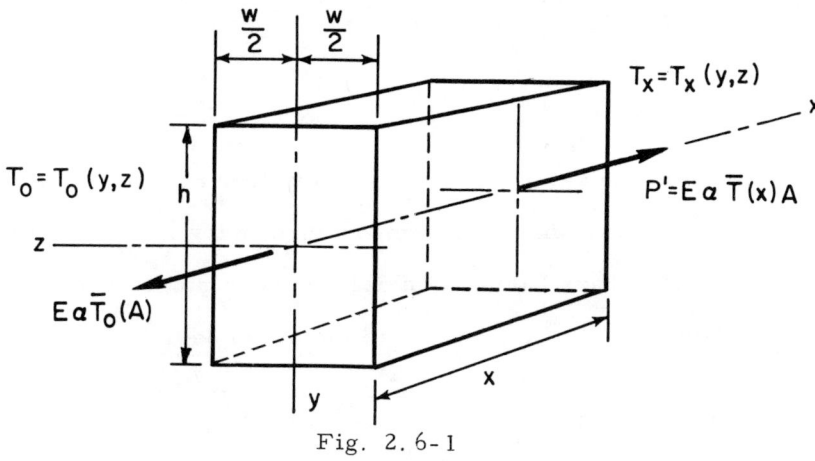

Fig. 2.6-1

axes in order to preclude bending, that is, $T = T(x, y, z) = T(x, -y, -z)$. However, if restraints are present which prevent bending, this restriction is not required. The axial displacement load due to the surface tractions is additive to the axially distributed thermal body forces. At the end face the surface tractions produce an axial thermal load $E\alpha \bar{T}_o A$, and at a location, x, the displacement load P' is

$$P' = E\alpha \int_{-h/2}^{h/2} \int_{-w/2}^{w/2} T_o \, dy \, dz - E\alpha \int_{o}^{x} \int_{-h/2}^{h/2} \int_{-w/2}^{w/2} -\frac{\partial T}{\partial x} dx \, dy \, dz \quad (2.6\text{-}1)$$

which gives

$$P' = E\alpha \bar{T}(x) A \quad , \quad \sigma' = E\alpha \bar{T}(x) \quad (2.6\text{-}2)$$

with $\bar{T}(x)$ the mean section temperature at a given x location. The net axial stress is then

$$\sigma = \sigma' - E\alpha T = E\alpha[\bar{T}(x) - T] \qquad (2.6\text{-}3)$$

T being a function of x, y, and z.

If the function $T(y, z)$, at a specified x location, is a plane harmonic function satisfying the equation

$$\frac{\partial^2 T}{\partial z^2} + \frac{\partial^2 T}{\partial y^2} = 0 \qquad (2.6\text{-}4)$$

then Eq. (2.6-3) gives the axial stress accurately (see Section 3.3). This will not be the case when the symmetrical temperature is the result of internal heat generation. In circular bars having an axially symmetrical temperature distribution (see Section 4.5) the axial stress is found to be

$$\sigma = \frac{E\alpha}{1-\nu}(\bar{T} - T) \qquad (2.6\text{-}5)$$

It can be inferred from the circular rod analysis that a rectangular bar with a symmetrical temperature distribution and with a uniform surface temperature, will develop normal axial stresses of more or less the same magnitude as those given by Eq. (2.6-5). In general, an adjustment of the axial normal stress due to the presence of triaxial stresses can be made by means of the expressions in Eqs. (2.4-1) and (2.4-2). Either the lateral or transverse temperature distributions should be utilized in accordance with which gives the greatest stress increase.

It is clear that plane strain deformations will be produced by the symmetrical temperature distribution that is specified in the present problem and that an accurate and complete solution will be obtained only when the problem is analyzed as a plane strain problem. Plane strain solutions of circular rods are given in Sect. 4.5, and plane strain solutions of rectangular bars are given in Sect. 3.5.

2.7 LAYERED BARS

The layered bar shown in Fig. 2.7-1 has a symmetrical temperature and material distribution. The temperatures of the constituents are $T_1 = T_1(x, y) = T_1(x, -y)$ and $T_2 = T_2(x, y) = T_2(x, -y)$. The net displacement

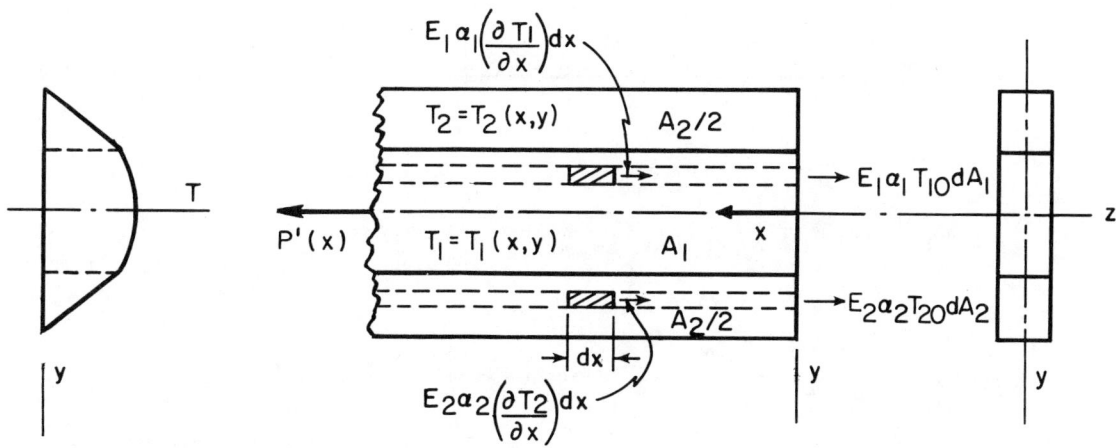

Fig. 2.7-1

load P' at any cross section is the sum of the surface traction and integrated body forces. It is

$$P' = \int_{A_1} E_1 \alpha_1 T_{10} dA_1 + \int_{A_2} E_2 \alpha_2 T_{20} dA_2 + \int_o^x \int_{A_1} E_1 \alpha_1 \frac{\partial T_1}{\partial x} dx dA_1 + \int_o^x \int_{A_2} E_2 \alpha_2 \frac{\partial T_2}{\partial x} dx dA_2$$

(2.7-1)

or

$$P'(x) = E_1 \alpha_1 \overline{T}_1(x) A_1 + E_2 \alpha_2 \overline{T}_2(x) A_2 \qquad (2.7-2)$$

The load P' distributes itself in components 1 and 2 in accordance with the rigidities $E_1 A_1$ and $E_2 A_2$. Thus the displacement loads P'_1 and P'_2 in the components are

$$P'_1 = \frac{P' E_1 A_1}{E_1 A_1 + E_2 A_2} = E_1 A_1 \frac{E_1 \alpha_1 \overline{T}_1(x) A_1 + E_2 \alpha_2 \overline{T}_2(x) A_2}{E_1 A_1 + E_2 A_2} = E_1 A_1 \overline{\alpha T}(x)$$

$$P'_2 = \frac{P' E_2 A_2}{E_1 A_1 + E_2 A_2} = E_2 A_2 \frac{E_1 \alpha_1 \overline{T}_1(x) A_1 + E_2 \alpha_2 \overline{T}_2(x) A_2}{E_1 A_1 + E_2 A_2} = E_2 A_2 \overline{\alpha T}(x)$$

(2.7-3)

where

$$\epsilon = \overline{\alpha T}(x) = \frac{E_1 \alpha_1 \overline{T}_1(x) A_1 + E_2 \alpha_2 \overline{T}_2(x) A_2}{E_1 A_1 + E_2 A_2} \tag{2.7-4}$$

The stresses are

$$\sigma_1 = \sigma'_1 - E_1 \alpha_1 T_1 = \frac{P'_1}{A_1} - E_1 \alpha_1 T_1 = E_1 \left[\overline{\alpha T}(x) - \alpha_1 T_1(x, y) \right]$$

(2.7-5)

$$\sigma_2 = \sigma'_2 - E_2 \alpha_2 T_2 = \frac{P'_2}{A_2} - E_2 \alpha_2 T_2 = E_2 \left[\overline{\alpha T}(x) - \alpha_2 T_2(x, y) \right]$$

As the interfacial stresses in composite elements are always biaxial, the normal stresses at the interface are more accurately given as

$$\sigma_{1i} = \frac{E_1}{1 - \nu_1} \left[\overline{\alpha T}(x) - \alpha_1 T_1(x, y_i) \right]$$

(2.7-6)

$$\sigma_{2i} = \frac{E_2}{1 - \nu_2} \left[\overline{\alpha T}(x) - \alpha_2 T_2(x, y_i) \right]$$

If the ends of a layered bar are held rigidly so that they cannot rotate, and the temperature is a function of the transverse coordinate only, the restriction of symmetrical material and temperature distribution is not required. The foregoing solution then applies for any transverse material distribution

2.8 SHEAR STRESSES IN HOMOGENEOUS BARS

In a bar in which there is no axial temperature variation, there are no shear stresses present except at extreme ends where the fall-off of the normal stress to zero produces local shear stresses. However, when there is a combined transverse and axial temperature variation, shear stresses will be generated throughout the bar.

We compute the shear stresses in the bar, of which a length dx is shown in Fig. 2.8-1. The temperature and material distribution are symmetrical

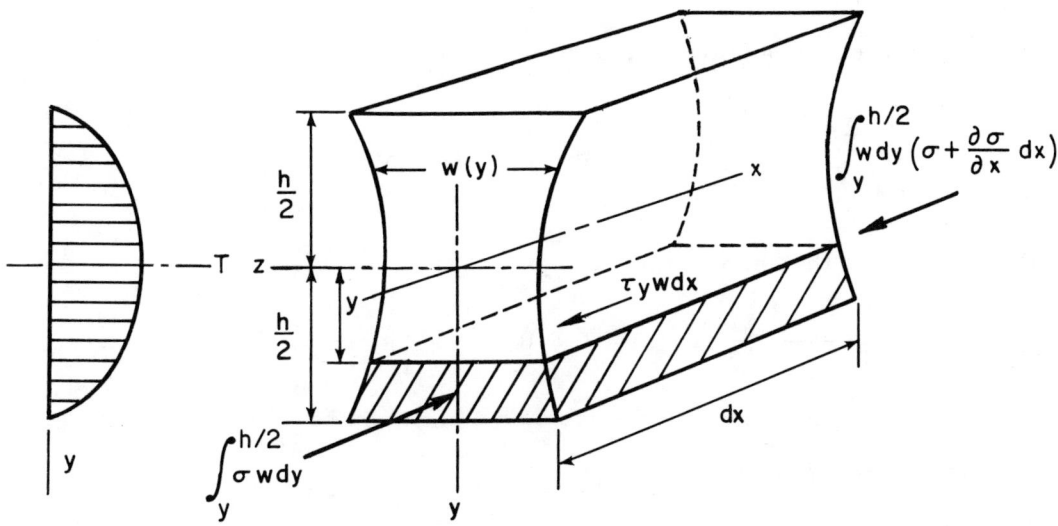

Fig. 2.8-1

about the mid-plane in order that no bending be induced. The balance of (real) loads acting on the shaded portion of the element yields

$$\tau w\, dx = -\int_y^{\frac{h}{2}} \frac{\partial \sigma_x}{\partial x}\, dx\, w\, dy = -\int_y^{\frac{h}{2}} \frac{\partial \sigma_x}{\partial x}\, dx\, dA \qquad (2.8-1)$$

It is clear that $\tau = \tau(y)$. From Eq. (2.8-1),

$$\tau = -\frac{1}{w} \int_y^{h/2} \frac{\partial \sigma_x}{\partial x} dA \qquad (2.8-2)$$

Substitution of σ_x from Eq. (2.5-7) gives

$$\tau = -\frac{E\alpha}{w} \frac{\partial}{\partial x} \left[\int_y^{h/2} \overline{T}(x) dA - \int_y^{h/2} T(x, y) dA \right] \qquad (2.8-3)$$

We designate A_y as the part of the cross section area lying between the plane y = constant and the surface $y = \frac{h}{2}$; and $\overline{T}_y(x)$ is the mean temperature of this area, i.e.,

$$A_y = \int_y^{h/2} dA \qquad \overline{T}_y(x) = \frac{1}{A_y} \int_y^{h/2} T(x, y) dA \qquad (2.8-4)$$

Over the full cross section, we have

$$A = \int_{-h/2}^{h/2} dA \qquad \overline{T}(x) = \frac{1}{A} \int_{-h/2}^{h/2} T(x, y) dA \qquad (2.8-4)(a)$$

Eq. (2.8-3) can then be written as

$$\tau = -\frac{E\alpha A_y}{w} \frac{\partial}{\partial x} \left[\overline{T}(x) - \overline{T}_y(x) \right] \qquad (2.8-5)$$

When the temperature distribution $T(x, y)$ can be expressed as the product of two functions,

$$T(x, y) = f(x) T(y) \qquad (2.8-6)$$

the shear stress given by Eq. (2.8-5) can be written as

$$\tau = -\frac{E\alpha A_y}{w} f'(x) \left[\overline{T} - \overline{T}_y \right] \qquad (2.8-7)$$

and the normal stress as

SHEAR STRESSES IN HOMOGENEOUS BARS

$$\sigma = E\alpha f(x)\left[\overline{T} - T(y)\right] \qquad (2.8\text{-}8)$$

The temperature parameters in Eqs. (2.8-7) and (2.8-8) are

$$\overline{T}_y = \frac{1}{A_y} \int_y^{\frac{h}{2}} T(y)\, dA \qquad \overline{T} = \frac{1}{A} \int_{-\frac{h}{2}}^{\frac{h}{2}} T(y)\, dA$$

The shear stress τ in Eqs. (2.8-5) and (2.8-7) are zero at the upper and lower surfaces $y = \pm \frac{h}{2}$ since A_y is zero at these points, and $\overline{T}(x) - \overline{T}_y(x)$ or $\overline{T} - \overline{T}_y$ are equal to zero at the center of the bar because of the symmetry of temperature and material about the z axis. When temperature distributions are due to internal heat generation or heat conduction, the peak normal stresses generally occur at the surface or the center of the bar. The fact that the shear stresses are zero at these points eliminates the need, in bars of more or less uniform width, to combine the normal and shear stresses. On the other hand in members of non-uniform width such as I beams or wide flange beams (not subject to bending) combined peak stress values can be obtained at the junction of web and flange. This is because a large value for A_y is obtained near the surfaces, $y = \pm \frac{h}{2}$, where the normal stress tends to be large. A similar situation is encountered in isothermal and thermal bending of structural beams having thin webs and heavy flanges.

The comparative magnitude of the shear stresses and normal stresses may be obtained by examining a rectangular beam of width w and depth h, in which the axial variation of temperature, $f(x)$, in Eq. (2.8-6), is given as

$$f(x) = \frac{x^n}{L^n} \qquad (2.8\text{-}9)$$

Eq. (2.8-7) then yields

$$\tau = E\alpha \left(\frac{h}{2} - y\right) \frac{nx^{n-1}}{L^n} \left[\overline{T} - \overline{T}_y\right] \qquad (2.8\text{-}10)$$

SHEAR STRESSES IN HOMOGENEOUS BARS

The ratio of shear stress to normal stress is

$$\frac{\tau}{\sigma} = \frac{E\alpha n(\frac{h}{2} - y)\frac{x^{n-1}}{L^n}(\overline{T} - T_y)}{E\alpha \frac{x^n}{L^n}(\overline{T} - T(y))} \qquad (2.8\text{-}11)$$

Near the end of the bar, $x = L$, where the normal stress is a maximum, the order of magnitude of the stress ratio is

$$\frac{\tau}{\sigma} \approx \frac{nh}{L} \qquad (2.8\text{-}12)$$

This indicates that thermal shear stresses are important in bars with large depth to length ratios and also when the rate of change of temperature in the axial direction is relatively large.

Example (a) Shear Stress in a Rectangular Bar

Consider a homogeneous rectangular bar of width w, depth, h, and length L, having a temperature distribution

$$T = 4T_o \frac{y^2}{h^2} \cdot \frac{x^2}{L^2} \qquad (1)$$

The mean temperature distribution is

$$\overline{T}(x) = f(x)\overline{T} = \left(\frac{x^2}{L^2}\right)\left(\frac{T_o}{3}\right) \qquad (2)$$

As $T(y) = 4T_o y^2/h^2$ and $f(x) = x^2/L^2$, the normal stress, from Eq. (2.8-8) is

$$\sigma = \frac{E\alpha x^2}{L^2}(\overline{T} - T) = \frac{E\alpha T_o x^2}{L^2}\left(\frac{1}{3} - \frac{4y^2}{h^2}\right) \qquad (3)$$

The shear stress, from Eq. (2.8-7) is

$$\tau = -\frac{E\alpha T_o}{3} \frac{2x}{L^2}\left[\left(\frac{h}{2} - y\right) - \left(\frac{h}{2} - \frac{4y^3}{h^2}\right)\right] \qquad (4)$$

or

$$\tau = -\frac{2E\alpha T_o}{3} \frac{xy}{L^2} \left(\frac{4y^2}{h^2} - 1\right) \tag{5}$$

The point at which the shear stress is a maximum is obtained by setting the derivative of τ with respect to y equal to zero. This location is found to be at $y = \pm \frac{h\sqrt{3}}{6}$. When set into Eq. (5) we obtain the maximum shear stress, at $x = L$, as

$$\tau = \frac{2\sqrt{3}}{27} \frac{h}{L} E\alpha T_o \tag{6}$$

The maximum normal stress is

$$\sigma = \frac{1}{3} E\alpha T_o \tag{7}$$

and the ratio of shear stress to normal stress is

$$\frac{\tau}{\sigma} = \frac{2\sqrt{3}}{9} \frac{h}{L} \tag{8}$$

2.9 INTERFACIAL SHEAR STRESS IN LAYERED BARS

We compute here the shear stress in a layered bar having a symmetrical transverse distribution of material and temperature. An element of length dx is shown in Fig. 2.9-1.

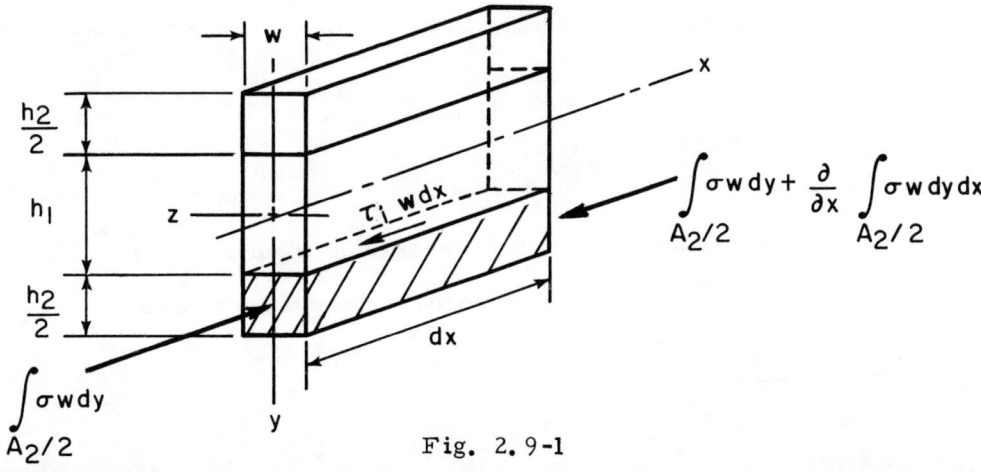

Fig. 2.9-1

The normal stresses, in accordance with Eq. (2.7-5) are

$$\sigma_1 = E_1\left[\overline{\alpha T}(x) - \alpha_1 T_1(x,y)\right]$$

$$\sigma_2 = E_2\left[\overline{\alpha T}(x) - \alpha_2 T_2(x,y)\right] \qquad (2.9\text{-}1)$$

with $\overline{\alpha T}(x)$ given by Eq. (2.7-4). The equilibrium of loads on the length dx of the outer layer, having a cross section area, $A_2/2$, is expressed as

$$\tau_i\, w\, dx = -\frac{\partial}{\partial x}\int_{A_2/2}\sigma_2 w\, dy\, dx \qquad (2.9\text{-}2)$$

where τ_i is the interfacial shear stress. As w is constant, we have

$$\tau_i = -\frac{\partial}{\partial x}\int_{A_2/2}\sigma_2\, dy \qquad (2.9\text{-}3)$$

Substitution of σ_2 from Eqs. (2.9-1) gives

$$\tau_i = -\frac{E_2 h_2}{2}\frac{\partial}{\partial x}\left[\overline{\alpha T}(x) - \alpha_2 \overline{T}_2(x)\right] \qquad (2.9\text{-}4)$$

Example (a)   Interfacial Shear Stress in Layered Bar

If one takes a temperature distribution of the form $T = T(y)\, x^n/L^n$ Eq. (2.9-4) becomes

$$\tau_i = -\frac{E_2 h_2}{2}\cdot\frac{n x^{n-1}}{L^n}\left[\overline{\alpha T} - \alpha_2 \overline{T}_2\right] \qquad (1)$$

with the barred quantities no longer functions of x. The normal stress, from Eq. (1), is

$$\sigma_2 = \frac{E_2 x^n}{L^n}\left[\overline{\alpha T} - \alpha_2 T_2\right] \qquad (2)$$

If we take the temperature T_2 to be a function of x only, then $T_2 = \overline{T}_2$ and the ratio of interfacial shear stress to normal stress at $x = L$ is

$$\frac{\tau_i}{\sigma_2} = \frac{n h_2}{2L} \qquad (3)$$

The relative shear stress is thus proportional to the ratio of the outer layer thickness to the length of element, and is also dependent upon the rate of temperature variation in the axial direction, i.e., the exponent n.

2.10 END SHEAR STRESSES

The sketch in Fig. 2.10-1, represents a bimetallic sandwich beam.

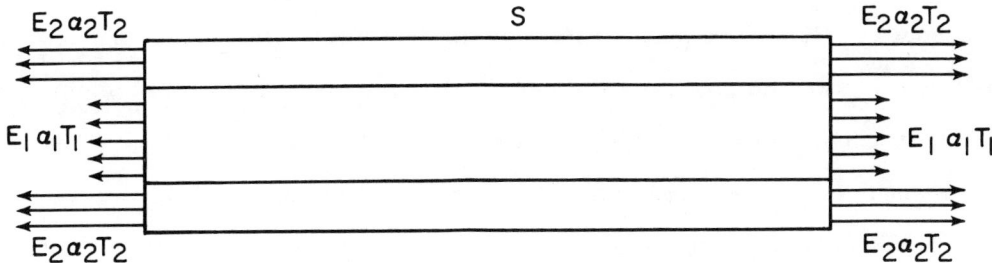

Fig. 2.10-1

The transverse temperature and material distributions are symmetrical, and do not vary axially. The constituents are joined along their entire length.

The normal axial stresses in such an assembly are computed in Section 2.7. In the one-dimensional computation of the normal stresses in attached elements, the analysis remains the same whether the elements are attached along the entire length, as in Section 2.7, or attached only at their extremities as in Section 2.2.

An estimate can be made of the magnitude of the shearing stresses at the interfacial surface, near the ends of the assembly shown in Fig. 2.10-1. In rectangular coordinates, the two-dimensional equation of equilibrium is

$$\frac{\partial \sigma_x}{\partial x} + \frac{\partial \tau_{xy}}{\partial y} = 0 \qquad (2.10-1)$$

As the axial stress does not change along the length of each of the constituents, the term $\frac{\partial \sigma_x}{\partial x}$ is zero. From Eq. (2.10-1) it is clear that $\frac{\partial \tau_{xy}}{\partial y}$ must therefore also be zero. At the surface, S, there is no shear stress, and as $\frac{\partial \tau_{xy}}{\partial y} = 0$, i.e., there is no transverse variation in the shear stress, then the shear stress must remain zero as we proceed inward from the outer surface. The shear stress is therefore also zero at the interface.

At the extreme ends of the assembly the axial stress falls to zero, since the ends are unloaded. This change in the axial stress near the end of the bar produces a localized shear stress at the interface. In Section 2.3 the shear stress was computed for the case of attachment areas or weldments smaller in cross section area than each of the assembly elements. When the attachment exists over a considerable interfacial length, or along the entire length of the assembly, the shear stress falls off rapidly with distance from the ends, since there is no shear stress present at internal points away from the ends.

It has been found in end-problem analysis that the axial stress increases from zero to the computed normal stress (as given by Eqs. (2.7-5)) at a distance from the end more or less equal to the thickness or depth of the assembly. From St. Venant's principle we would also arrive at this conclusion. The shear loads will therefore be of the same order of magnitude as the normal axial loads and the shear stress at the extremity of the assembly will correspondingly be of the same order of magnitude as the normal stress.

When joining together two one-dimensional elements, a reduction in the shear stress will not be obtained by making the interfacial attachment areas at the ends large. An interface attachment length of about two times the thickness of the assembly will serve to reduce the interfacial shear stress to its lowest possible value. On the other hand, if the length of attachment is less than the thickness of the assembly, shear stresses which are larger than the normal axial stress may be developed. The magnitude of the shear stress is then given by Eq. (2.3-2).

Example (a)   Riveted Sandwich Strip

The components of the sandwich assembly shown in Fig. 2.10-2 are heated uniformly to temperatures T_1 and T_2. We wish to determine the thermal stresses in each of the strips and the shear stresses in the rivets.

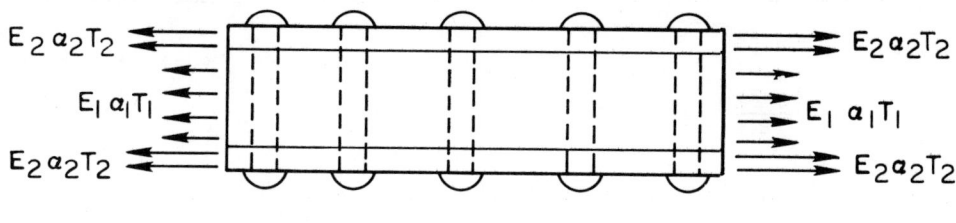

Fig. 2.10-2

This is a one-dimensional problem in which only axial surface tractions are present and these produce the end tractions shown in Fig. 2.10-2. The thermal stresses, in accordance with Eqs. (2.2-4) and (2.2-5) are

$$\sigma_1 = \frac{E_1 E_2 A_2 (\alpha_2 T_2 - \alpha_1 T_1)}{E_1 A_1 + E_2 A_2} \tag{1}$$

$$\sigma_2 = \frac{E_1 E_2 A_1 (\alpha_1 T_1 - \alpha_2 T_2)}{E_1 A_1 + E_2 A_2} \tag{2}$$

These axial stresses are present everywhere along the length of the strips, excepting the ends where the axial stress falls off to zero. Consider the shear load on an interior rivet. Since the normal load in each of the constituents does not change axially from one side of a rivet to the other, it is clear that the shear load on an interior rivet must be zero. This is shown in Fig. 2.10-3(a). At the end of the assembly the normal load falls from $\dfrac{\sigma_2 A_2}{2}$ to zero. A load

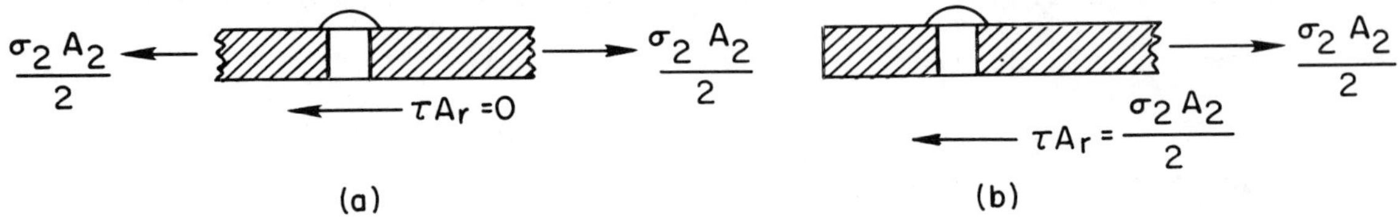

END SHEAR STRESSES

Fig. 2.10-3

balance on a short length of cover plate, near the end of the assembly, is shown in Fig. 2.10-3(b). It is clear that the change in normal load is carried by the end rivet. The shear load in the end rivets, at each of the interfacial planes is therefore

$$\tau A_r = \frac{\sigma_2 A_2}{2} \tag{3}$$

Substitution of Eq. (2) into the above gives the shear stress as

$$\tau = \frac{\sigma_2 A_2}{2 A_r} = \frac{E_1 E_2 A_1 A_2 (\alpha_1 T_1 - \alpha_2 T_2)}{2 A_r (E_1 A_1 + E_2 A_2)} \tag{4}$$

The assumption that only the end rivet takes the entire shear load is proper only if the rivet pitch is equal to or greater than the thickness of the assembly, since the full normal stress is developed over an axial distance more or less equal to the thickness of the strip. If more than one rivet is present within this distance, the load is distributed among these. Even in the case of a single rivet being located within this distance from the end, the total load will be taken by the rivet only if the resulting bearing stress on the plate and shear stress in the rivet does not exceed the yield stress. The practical design of riveted connections often permits the yield stress to be exceeded. The load then redistributes itself partially to rivets farther from the ends. In the case of sustained (isothermal) loads, the assumption is made that the load redistributes itself equally to all of the rivets in the connection. In the thermal stress problem the yielding of the rivets is accompanied by a reduction in the axial load, which gives a self-relieving character to the assembly.

2.11 CIRCULAR RING WITH AXIAL TEMPERATURE VARIATION

A ring will undergo extensional deformation when the temperature distribution is symmetrical about the central plane of the ring. If the symmetry were not present the ring would rotate. The ring can be viewed as composed of axially stacked lamina of the same material, each at a different temperature. Fig. 2.11-1(a) shows the temperature distribution and ring cross section. Fig. 2.11-1(b) shows the radially acting thermal surface tractions and tangential

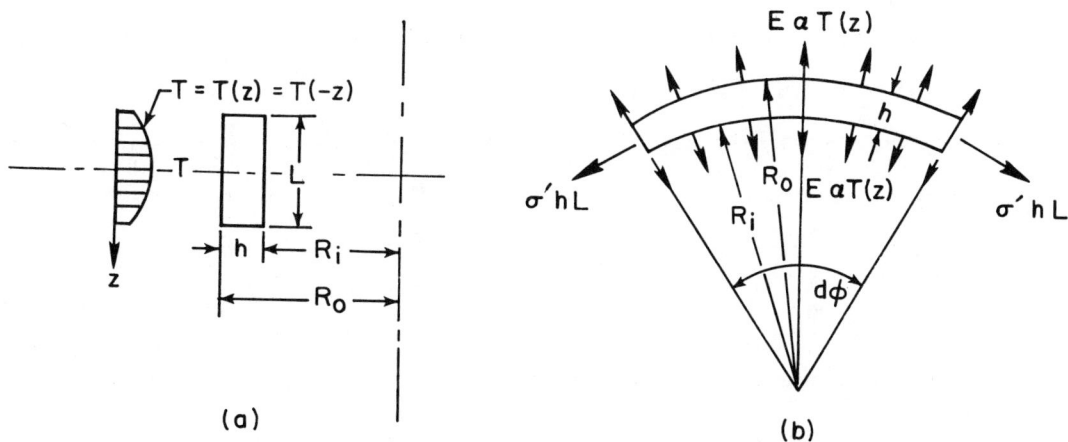

Fig. 2.11-1

displacement stresses. In the axial, z, direction the thermal body loads and surface tractions are locally self-equilibrating do not produce any axial stresses. (See Section 1.10). We assume that σ_r and σ_z are zero, and use the one-dimensional form of the thermal loading. The radial equilibrium of the ring segment requires that

$$\sigma' h L d\varphi = \int_{-\frac{L}{2}}^{\frac{L}{2}} E\alpha T\, dz\, R_o d\varphi - \int_{-\frac{L}{2}}^{\frac{L}{2}} E\alpha T\, dz\, R_i d\varphi \qquad (2.11-1)$$

which yields

$$\sigma' h L = E\alpha \overline{T} L R_o - E\alpha \overline{T} L R_i = E\alpha \overline{T} L h \qquad (2.11-2)$$

with $\overline{T}$ the mean temperature. The displacement stress and strain are, from Eq. (2.11-2)

$$\sigma' = E\alpha \overline{T} \qquad (2.11-3)$$

and

$$\epsilon = \frac{\sigma'}{E} = \alpha \overline{T} \qquad (2.11-4)$$

The thermal stress distribution in the ring follows. It is

$$\sigma = \sigma' - E\alpha T = E\alpha(\overline{T} - T) \qquad (2.11-5)$$

The effect of biaxiality of stress, in circular rings, is similar to its effect in bars. The increase in the computed uniaxial stress due to biaxiality can be included by the use of a modified modulus of elasticity E^*. For the case of axial temperature variation in radially thin rings $E^* = E$.

The modified tangential stress is expressed as

$$\sigma = E^* \alpha (\overline{T} - T) \qquad (2.11-6)$$

For rings of intermediate thickness the modified modulus of elasticity is

$$E^* = \frac{E}{1 - \frac{h}{L}\nu} \qquad \frac{h}{L} \leq 1 \qquad (2.11-7)$$

and for radially thick rings the modified modulus of elasticity is

$$E^* = \frac{E}{1 - \nu} \qquad \frac{h}{L} \geq 1 \qquad (2.11-8)$$

Shear Stress

The mean shear stresses due to the axially varying temperature are readily obtained from the equilibrium of the ring segment of thickness $\frac{L}{2} - z$,

CIRCULAR RING WITH AXIAL TEMPERATURE VARIATION

shown in Fig. 2.11-2.

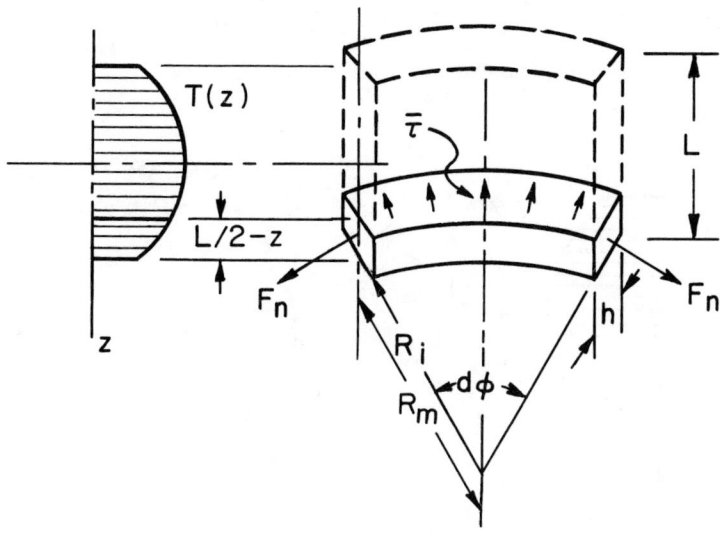

Fig. 2.11-2

The normal forces F_n at the ends of the element are

$$F_n = \int_z^{L/2} \sigma_t \, h \, dz \tag{2.11-9}$$

The tangential stress, from Eq. (2.11-5) is $E\alpha(\overline{T} - T)$. This is set into Eq. (2.11-9) and we obtain

$$F_n = E\alpha (\overline{T} - \overline{T}_z) \left(\frac{L}{2} - z\right) h \tag{2.11-10}$$

where $\overline{T}_z$ represents the mean temperature of the portion of the ring lying between z and $L/2$.

The mean shear stress $\overline{\tau}$ is in equilibrium with the radial component of the loads, F_n. Equilibrium requires that

$$\overline{\tau} = \frac{F_n}{hR_m} = \frac{E\alpha(\overline{T} - \overline{T}_z)\left(\frac{L}{2} - |z|\right)}{R_m} \tag{2.11-11}$$

It is interesting to note that at $z = 0$ and $z = L/2$ the shear stress goes to zero. The derivative of $\bar{\tau}$ at $z = 0$ is also seen to be zero. The shear stress distribution is therefore as shown in Fig. 2.11-3. This is the same pattern of shear stress distribution as was found in bars.

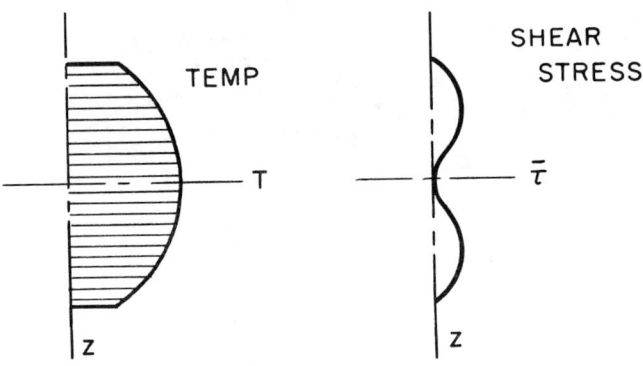

Fig. 2.11-3

It should be noted that the shear stress is an order of magnitude lower than the normal stress. That is

$$\tau \approx \frac{L}{R_m} \sigma \tag{2.11-12}$$

2.12 AXIALLY LAYERED CIRCULAR RING

A symmetrical distribution of material and temperature is stipulated in order that we obtain tangential extensional deformation only. Fig. 2.12(a), (b) and (c) show the surface tractions and internal displacement loads acting on a ring segment. The axial thermal body loads and surface tractions are self-equilibrating, as in the case of the thermal transverse loads on bars (Sect. 1.10). These do not generate axial stresses. The radial stresses are also taken to be zero. The temperature of the central ring is $T_1(z)$ and the temperature of the upper and lower rings is $T_2(z)$.

AXIALLY LAYERED CIRCULAR RING

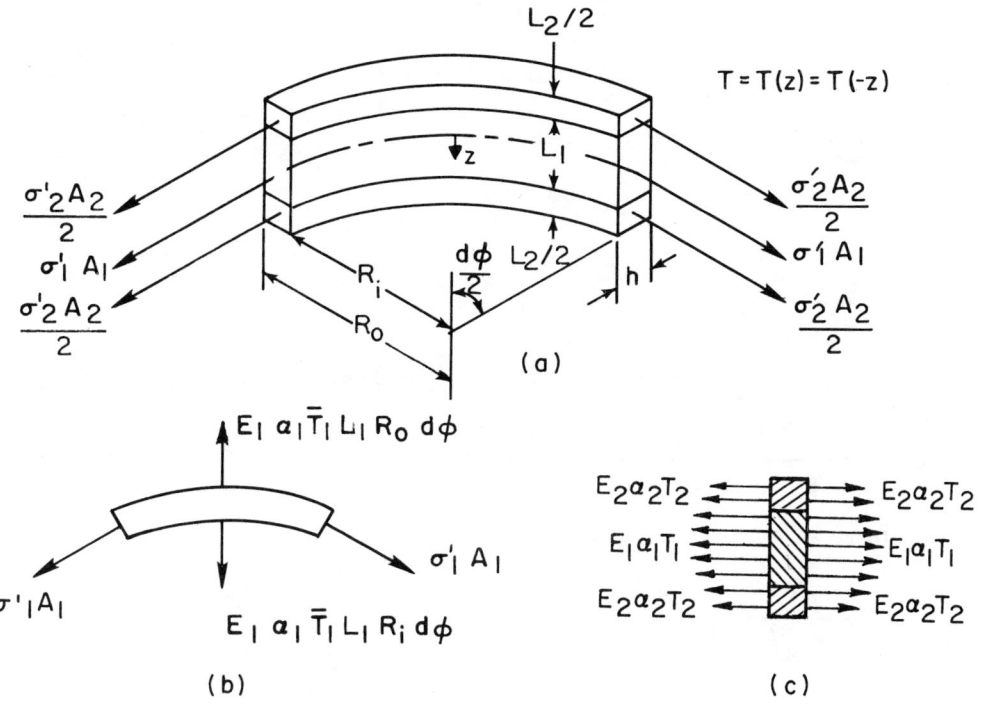

Fig. 2.12-1

Thermal surface loads $E_1 \alpha_1 \overline{T}_1 L_1 R_i \, d\varphi$ and $E_1 \alpha_1 \overline{T}_1 L_1 R_o \, d\varphi$ act on the inner and outer surfaces of the central ring, in accordance with the one-dimensional form of thermal loading. This is shown in Figs. 2.12-1(b). Thermal surface loads $E_2 \alpha_2 \overline{T}_2 \frac{L_2}{2} R_i \, d\varphi$ and $E_2 \alpha_2 \overline{T}_2 \frac{L_2}{2} R_o \, d\varphi$ act on the inner and outer surfaces of the upper and lower rings. Radial equilibrium along the center line of the ring segment requires that

$$(\sigma_1' A_1 + \sigma_2' A_2) \, d\varphi = E_1 \alpha_1 \overline{T}_1 L_1 R_o d\varphi + E_2 \alpha_2 \overline{T}_2 L_2 R_o d\varphi$$

$$- E_1 \alpha_1 \overline{T}_1 L_1 R_i d\varphi - E_2 \alpha_2 \overline{T}_2 L_2 R_o d\varphi$$

or

$$\sigma'_1 A_1 + \sigma'_2 A_2 = E_1 \alpha_1 \overline{T}_1 A_1 + E_2 \alpha_2 \overline{T}_2 A_2 \qquad (2.12\text{-}1)$$

Since the strains in each of the rings are the same, we have

$$\epsilon_t = \frac{\sigma'_1}{E_1} = \frac{\sigma'_2}{E_2} \qquad (2.12\text{-}2)$$

and

$$\sigma'_2 = \frac{\sigma'_1 E_2}{E_1}$$

Substitution into Eq. (2.12-1) gives the displacement stress, σ'_1, as

$$\sigma'_1 = E_1 \frac{E_1 \alpha_1 \overline{T}_1 A_1 + E_2 \alpha_2 \overline{T}_2 A_2}{E_1 A_1 + E_2 A_2} = E_1 \overline{\alpha T} \qquad (2.12\text{-}3)$$

The strain is σ'_1 / E_1 or

$$\epsilon = \frac{E_1 \alpha_1 \overline{T}_1 A_1 + E_2 \alpha_2 \overline{T}_2 A_2}{E_1 A_1 + E_2 A_2} = \overline{\alpha T} \qquad (2.12\text{-}4)$$

and the stresses, $\sigma_1 = \sigma'_1 - E_1 \alpha_1 T_1$ and $\sigma_2 = \sigma'_2 - E_2 \alpha_2 T_2$, are

$$\sigma_1 = E_1 (\overline{\alpha T} - \alpha_1 T_1) \qquad \sigma_2 = E_2 (\overline{\alpha T} - \alpha_2 T_2) \qquad (2.12\text{-}5)$$

The mean stresses, obtained by integrating the above over the cross section areas, are

$$\overline{\sigma}_1 = E_1 (\overline{\alpha T} - \alpha_1 \overline{T}_1) \qquad \overline{\sigma}_2 = E_2 (\overline{\alpha T} - \alpha_2 \overline{T}_2) \qquad (2.12\text{-}6)$$

The expressions for the stresses and strains in axially stacked rings which expand in unison are thus seen to be identical to the expressions obtained for layered bars. As in the case of layered bars the stress σ_m in any one of an assembly of axially stacked rings which are not subject to out-of-plane bending, is given by the expression

$$\sigma_m = E_m (\overline{\alpha T} - \alpha_m T_m) \qquad (2.12\text{-}7)$$

with $\overline{\alpha T}$ the mean free strain as given by Eq. (2.2-1).

The interfacial stress in a brazed or welded bimetallic ring assembly is alway biaxial at the interfacial plane. The normal stresses at the interfacial surface are therefore more properly given as

$$\sigma_i = \frac{E_i}{1 - \nu_i} (\overline{\alpha T} - \alpha_i T_i) \tag{2.12-8}$$

Shear Stress

The shear stress in the bolts or weldments that hold the rings together is obtained from the equilibrium of the ring segment shown in Fig. 2.12-2.

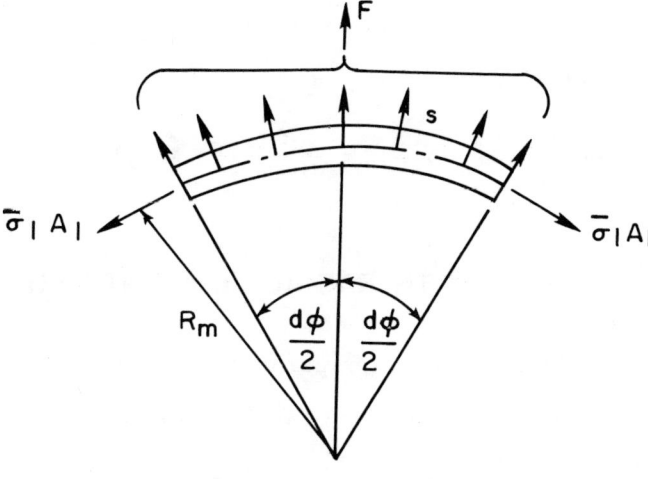

Fig. 2.12-2

The normal load acting on a cross section of the ring segment is $\overline{\sigma}_1 A_1$, and the resisting shear load per unit circumferential length is s. The resultant of the shear loads, acting radially at the center of the segment, $r = R_m$, is

$$F = s R_m d\varphi \tag{2.12-9}$$

and static equilibrium of the ring segment requires that

$$s R_m d\varphi = \bar{\sigma}_1 A_1 d\varphi \qquad (2.12\text{-}10)$$

This gives the shear load, s, per unit circumferential length as

$$s = \frac{\bar{\sigma}_1 A_1}{R_m} \qquad (2.12\text{-}11)$$

When the area of attachment is the full plane area of the ring, $2\pi R_m h$, we obtain the shear stress, from Eqs. (2.12-6) and (2.12-11) as

$$\bar{\tau} = \frac{s}{2h} = \left|\frac{\bar{\sigma}_1 L_1}{2R_m}\right| = \left|\frac{\bar{\sigma}_2 L_2}{2R_m}\right| = \frac{E_1 L_1}{2R_m}(\overline{\alpha T} - \alpha_1 \overline{T}_1) = \frac{E_2 L_2}{2R_m}(\overline{\alpha T} - \alpha_2 \overline{T}_2) \qquad (2.12\text{-}12)$$

On the other hand if the double shear area is "a" per unit circumferential length, the shear stress in the fasteners is

$$\tau = \frac{\bar{\sigma}_1 A_1}{a R_m} = \frac{A_1 E_1}{a R_m}(\overline{\alpha T} - \alpha_1 \overline{T}_1) = \frac{A_2 E_2}{a R_m}(\overline{\alpha T} - \alpha_2 \overline{T}_2) \qquad (2.12\text{-}13)$$

2.13 CIRCULAR RING WITH RADIAL TEMPERATURE VARIATION

A circular ring with a radial temperature variation can be viewed as a set of concentric rings composed of the same material, with each ring at a different temperature.

Fig. 2.13-1(a) shows the one-dimensional thermal body forces and surface tractions acting on a segment of such a ring, and Fig. 2.13-1(b) shows the ring cross section and radial temperature variation.

CIRCULAR RING WITH RADIAL TEMPERATURE VARIATION

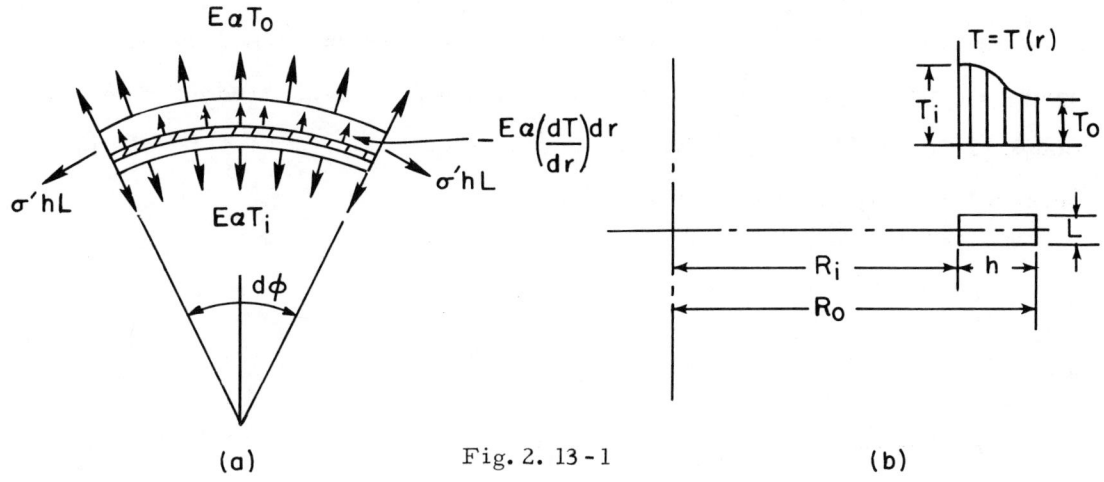

Fig. 2.13-1

A balance of loads, taken at the center of the ring segment, gives

$$\sigma' h L d\varphi = E\alpha T_o L R_o d\varphi - E\alpha T_i L R_i d\varphi - \int_{R_i}^{R_o} E\alpha \frac{dT}{dr} dr L r d\varphi \qquad (2.13-1)$$

The last term is integrated by parts and we obtain

$$\sigma' h = E\alpha(T_o R_o - T_i R_i) - E\alpha \left[T r \right]_{R_i}^{R_o} + E\alpha \int_{R_i}^{R_o} T dr$$

or

$$\sigma' = \frac{E\alpha}{h} \int_{R_i}^{R_o} T dr \approx E\alpha \overline{T} \qquad (2.13-2)$$

The strain is

$$\epsilon = \frac{\sigma'}{E} = \alpha \overline{T} \qquad (2.13-3)$$

and the thermal stress is

$$\sigma = \sigma' - E\alpha T = E\alpha(\overline{T} - T) \qquad (2.13-4)$$

Eq. (2.13-4) is accurate for rings in which the ratio of ring thickness h, to mean radius, is small. When the ring thickness becomes large it becomes a circular plate with a central hole. That problem is treated in Section 4.2. In the present problem the modification of Eq. (2.13-4) to take into account biaxiality of stress, takes the form

$$\sigma = E^* \alpha (\overline{T} - T) \qquad (2.13\text{-}5)$$

with

$$E^* = \frac{E}{1 - \nu} \qquad \frac{L}{h} \geq 1 \qquad (2.13\text{-}6)$$

and

$$E^* = \frac{E}{1 - \frac{L}{h}\nu} \qquad \frac{L}{h} \leq 1 \qquad (2.13\text{-}7)$$

Radial Stress

A part of the ring segment of Fig. 2.13-1(a), extending from R_i to r, is shown in Fig. 2.13-2.

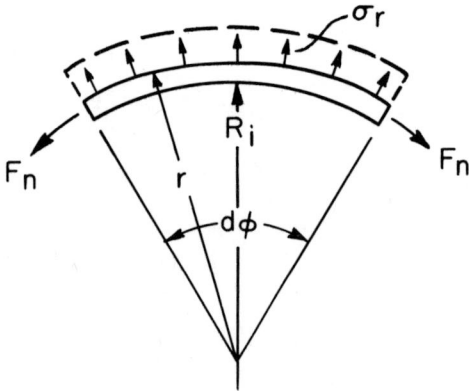

Fig. 2.13-2

The tangential loads F_n are

$$F_n = \int_{R_i}^{r} \sigma_t L \, dr \qquad (2.13\text{-}8)$$

Substitution of $\sigma = \sigma_t = E\alpha(\overline{T} - T)$, from Eq. (2.13-4) yields

$$F_n = E\alpha (\overline{T} - \overline{T}_r) (r - R_i) L \qquad (2.13\text{-}9)$$

with $\overline{T}_r$ representing the mean temperature of the part of the ring between R_i and r. Equilibrium requires that

$$F_n \, d\varphi = \sigma_r L r \, d\varphi \qquad (2.13\text{-}10)$$

from which we obtain

$$\sigma_r = \frac{F_n}{Lr} = E\alpha (\overline{T} - \overline{T}_r)(1 - \frac{R_i}{r}) \qquad (2.13\text{-}11)$$

This can also be written as

$$\sigma_r = E\alpha(\overline{T} - \overline{T}_r) \frac{\Delta r}{r} \qquad (2.13\text{-}12)$$

with Δr the radial distance between R_i and r.

It is clear from the foregoing formula that the radial stress is an order of magnitude less than the tangential stress, since in order to qualify as a ring, the maximum value of $\Delta r/r$, which is h/r, must be no greater than about one tenth. A similar conclusion was reached in Sect. 2.12 with regard to the shear stress in a ring with an axially varying temperature. It is clear then that radial and shear stresses in rings with radial or axial temperature variations can be neglected in comparison to the tangential stresses.

2.14 CONCENTRIC RINGS WITH RADIAL TEMPERATURE VARIATION

When two rings are arranged concentrically, as in Fig. 2.14-1, with the thicknesses of the rings, h_1 and h_2 small in comparison to the radius, the approximation may be made that the ring assembly has a constant radius. The solution is then obtained in the same manner as in axially attached rings, in Section 2.12. The mean stresses in Rings 1 and 2 are as given by Eqs. (2.12-5) and (2.12-6) or

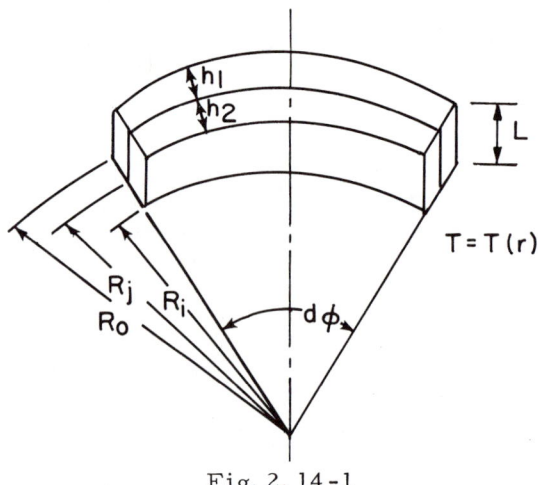

Fig. 2.14-1

$$\sigma_1 = E_1(\overline{\alpha T} - \alpha_1 T_1)$$
$$\sigma_2 = E_2(\overline{\alpha T} - \alpha_2 T_2)$$
(2.14-1)

with the tangential strain, $\epsilon_t = \overline{\alpha T}$, given as

$$\overline{\alpha T} = \frac{E_1 \alpha_1 \overline{T}_1 A_1 + E_2 \alpha_2 \overline{T}_2 A_2}{E_1 A_1 + E_2 A_2}$$
(2.14-2)

This approximation introduces an error in the computed stresses in Rings 1 and 2 of approximately $\dfrac{\sigma_1 h_1}{2R_j}$, and $\dfrac{\sigma_2 h_2}{2R_j}$ respectively. These corrections may be added to the stresses given by Eqs. (2.14-1) and (2.14-2), to obtain the maximum stresses in the rings.

When there are more than two concentric rings, and the radial thickness of the assembly is small in comparison to the mean radius, the stress σ_m in a ring is

$$\sigma_m = E_m (\overline{\alpha T} - \alpha_m T_m) \tag{2.14-3}$$

These results are accurate if the rings do not restrain each other from axial expansion. If there is a mutual axial expansion restraint, the stress at the interface is biaxial and is properly given as

$$\sigma_i = \frac{E_i}{1 - \nu_i} (\overline{\alpha T} - \alpha_i T_i) \tag{2.14-4}$$

Radial Stress

The radial stress at the interface can be obtained from the equation of equilibrium of the outermost segment of thickness, h_1. Fig. 2.14-2 shows the loads acting on this element.

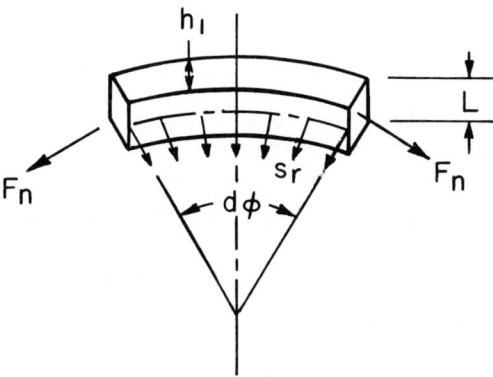

Fig. 2.14-2

The end forces F_n are the resultants of the tangential stresses. Thus

$$F_n = \int_{R_j}^{R_o} \sigma_1 L\, dr \qquad (2.14\text{-}5)$$

and with the use of Eq. (2.14-1) we have

$$F_n = \int_{R_j}^{R_o} E_1(\overline{\alpha T} - \alpha_1 T_1) L\, dr = E_1 A_1 (\overline{\alpha T} - \alpha_1 \overline{T}_1) \qquad (2.14\text{-}6)$$

The sum of the radial loads, acting on the element in Fig. 2.14-2 is zero. Thus

$$F_n d\varphi + s_r R_j d\varphi = 0 \qquad (2.14\text{-}7)$$

yielding the radial load per unit length as

$$s_r = -\frac{F_n}{R_j} = -\frac{E_1 A_1 (\overline{\alpha T} - \alpha_1 \overline{T}_1)}{R_j} \qquad (2.14\text{-}8)$$

When this radial load is distributed over the entire interface, we obtain an interfacial radial stress σ_{rj} as

$$\sigma_{rj} = -\frac{E_1 h_1 (\overline{\alpha T} - \alpha_1 \overline{T}_1)}{R_j} \qquad (2.14\text{-}9)$$

On the other hand, if the two rings are attached by bolts or rivets having a cross section area "a" per unit circumferential length, the radial stress in the fasteners is

$$\sigma_{ri} = -\frac{E_1 A_1 (\overline{\alpha T} - \alpha_1 \overline{T}_1)}{a R_j} \qquad (2.14\text{-}10)$$

RING WITH THREE-DIMENSIONAL TEMPERATURE

It is clear from Eq. (2.14-9) that when concentric rings are fully attached at their interface, the radial stress at the interface will be an order of magnitude lower than the tangential stress. However, when there is a partial or intermittent attachment at the radial joint, then we note from Eq. (2.14-10) that the stress will be large in the attachment when its area, a, is small.

2.15 RING WITH THREE-DIMENSIONAL TEMPERATURE VARIATION

Fig. 2.15-1(a) shows a ring segment loaded with radial thermal body forces and surface tractions, and Fig. 2.15-1(b) shows the ring cross section and temperature distribution at the free surfaces, $r = R_i$ and $r = R_o$. Radial body forces and surface tractions are present in accordance with the prescribed radial and symmetrical axial temperature distribution.

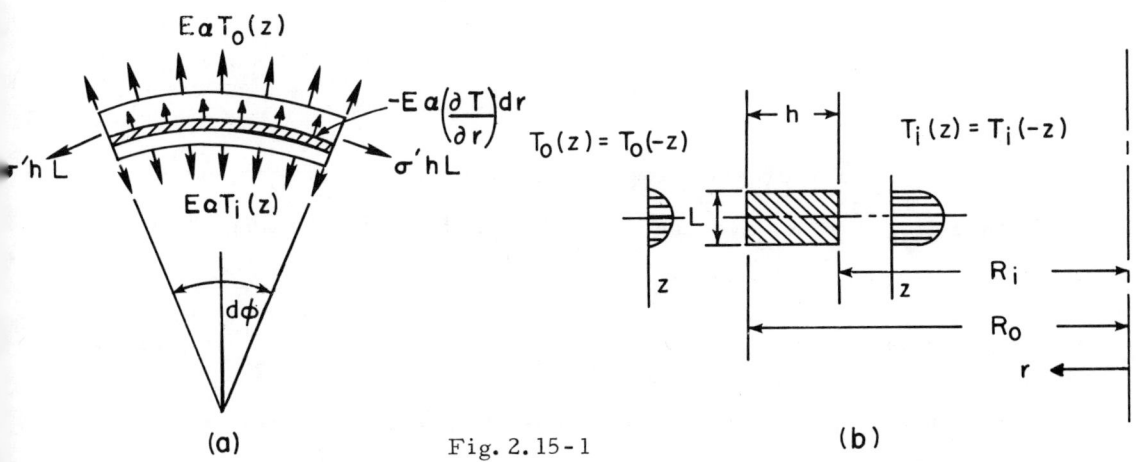

Fig. 2.15-1

In a radially thin ring the assumption can be made that the radial loads produce uniform hoop stresses. The radial equilibrium of the ring segment is expressed as

$$\sigma' hL d\varphi = E\alpha \left[R_o d\varphi \int_{-\frac{L}{2}}^{\frac{L}{2}} T_o(z) dz - R_i d\varphi \int_{-\frac{L}{2}}^{\frac{L}{2}} T_i(z) dz \right.$$

$$\left. - \int_{-\frac{L}{2}}^{\frac{L}{2}} \int_{R_i}^{R_o} \frac{\partial T}{\partial r} dr \, r \, d\varphi \, dz \right] \qquad (2.15\text{-}1)$$

Let $\overline{T}(r)$ represent the temperature averaged over the height of the ring, L, but varying with r. The foregoing equation is then written as

$$\sigma' hL = E\alpha R_o \overline{T}_o L - E\alpha R_i \overline{T}_i L - E\alpha L \int_{R_i}^{R_o} \frac{d\overline{T}(r)}{dr} r \, dr \qquad (2.15\text{-}2)$$

Integration of the last term by parts yields

$$\sigma' h = E\alpha \left\{ \overline{T}_o R_o - \overline{T}_i R_i - \left[\overline{T}(r) r \right]_{R_i}^{R_o} + \int_{R_i}^{R_o} \overline{T}(r) dr \right\} \qquad (2.15\text{-}3)$$

or

$$\sigma' = E\alpha \overline{\overline{T}} \qquad (2.15\text{-}4)$$

In Eq. (2.15-4), $\overline{\overline{T}}$ represents the average temperature over the cross section of the ring; that is, averaged over both r and z. The tangential strain is $\alpha \overline{\overline{T}}$ and the tangential stress is

$$\sigma = \sigma' - E\alpha T = E\alpha(\overline{\overline{T}} - T) \qquad (2.15\text{-}5)$$

It is seen that the expression for stress is of the same form as obtained for a ring with a radial temperature variation only, or a ring with a symmetrical axial temperature variation only.

As in the case of bars the effect of the presence of stresses σ_r and σ_z is to increase the computed tangential stress. The stress is more accurately given as

$$\sigma = E^* \alpha (\overline{T} - T) \tag{2.15-6}$$

with the modified modulus of elasticity E^* given Eqs. (2.4-1) and (2.4-2). The effect of the axial stress and of the radial stress should be considered separately, and the modification, Eq. (2.15-6), applied so as to produce the largest stress increment.

2.16 ELEMENTS IN SERIES

In thermal stress problems concerned with elements in parallel, it is often expedient to treat the problem by means of equivalent load analysis. On the other hand, when elements are in series, it is preferable to base a solution on the classical equations of elasticity. Elements of different materials and cross sections, and at different temperatures, are attached to each other in series, and the whole assembly lies in series with another isothermal elastic structure, shown schematically in Fig. 2.16-1, as an elastic

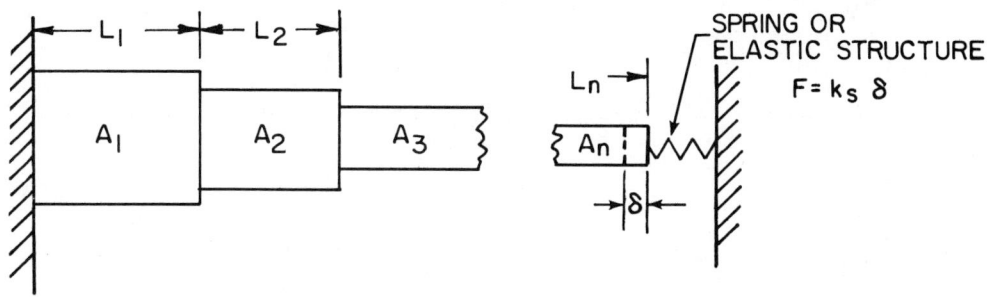

Fig. 2.16-1

spring. Due to temperature and cross section discontinuities at the junctions of the series elements, lateral stresses will be developed whose magnitude is of the order of $\frac{E_m E_n}{E_m + E_n} \Delta(\alpha T)$, where $\Delta(\alpha T)$ is the discontinuity in the free strain in the adjoining elements whose moduli of elasticity are E_m and E_n. It is quite possible that these lateral stresses will be larger than the axial stresses that we are presently concerned with. However, these are localized stresses and will have little effect on the axial stress computation when the length of each of the constituents is considerably greater than their respective transverse or lateral dimensions.

At the end of the assembly there is an elastic structure having a spring constant k_s. The axial force is constant across each cross section. This is expressed as

$$F = k_s \delta = \sigma_1 A_1 = \sigma_2 A_2 = \cdots = \sigma_n A_n \qquad (2.16\text{-}1)$$

In terms of σ_1, the axial stress in element 1, the stresses and the end deflection are

$$\sigma_2 = \frac{\sigma_1 A_1}{A_2}, \; \cdots \; \sigma_m = \frac{\sigma_1 A_1}{A_m} \qquad \delta = \frac{\sigma_1 A_1}{k_s} \qquad (2.16\text{-}2)$$

The strain in any bar is

$$\epsilon_m = \frac{\sigma_m}{E_m} + \alpha_m T_m \qquad (2.16\text{-}3)$$

and, as the integrated strain over the length of the assembly is equal to the displacement, δ, we have

$$\delta = \sum_1^n \epsilon_j L_j = \sum_1^n \left(\frac{\sigma_j}{E_j} + \alpha_j T_j \right) L_j \qquad (2.16\text{-}4)$$

ELEMENTS IN SERIES

Substitution of δ and σ_m from Eqs. (2.16-2) yields

$$\frac{\sigma_1 A_1}{k_s} = \sum_1^n \left(\frac{\sigma_1 A_1 L_j}{E_j A_j} + \alpha_j T_j L_j \right) \tag{2.16-5}$$

from which

$$\sigma_1 = \frac{\sum_1^n \alpha_j T_j L_j}{A_1 \left(\frac{1}{k_s} - \sum_1^n \frac{L_j}{E_j A_j}\right)} \qquad \sigma_m = \frac{\sum_1^n \alpha_j T_j L_j}{A_m \left(\frac{1}{k_s} - \sum_1^n \frac{L_j}{E_j A_j}\right)} \tag{2.16-6}$$

The deflection of the end of the assembly, from Eq. (2.16-2), is

$$\delta = \frac{\sigma_m A_m}{k_s} = \frac{\sum_1^n \alpha_j T_j L_j}{1 - k_s \sum_1^n \frac{L_j}{E_j} A_j} \tag{2.16-7}$$

The spring constant for a bar is $k_m = E_m A_m / L_m$, and for the assembly, which can be considered a set of springs in series, the overall spring constant is k_a, is expressed as

$$\frac{1}{k_a} = \sum_1^n \frac{1}{k_j} \tag{2.16-8}$$

since $F = k_a \delta = \delta / \left(\frac{1}{k_1} + \frac{1}{k_2} + \cdots + \frac{1}{k_n}\right)$. Eqs. (2.16-6) and (2.16-7) can be written as

$$\sigma_m = \frac{\delta_f}{A_m \left(\frac{1}{k_s} - \frac{1}{k_a}\right)} \qquad \delta = \frac{\delta_f}{1 - \frac{k_s}{k_a}} \tag{2.16-9}$$

where $\delta_f = \sum_1^n \alpha_j T_j L_j$ is the lengthening due to the free expansion.

When the assembly of rods is held between rigid walls, Eq. (2.16-6), written for any bar m, becomes

$$\sigma_m = -\frac{\sum_{j=1}^{n} \alpha_j T_j L_j}{\frac{L_m}{E_m} + A_m \sum_{\substack{j=1 \\ j \neq m}}^{n} \frac{L_j}{E_j A_j}} \quad (2.16-10)$$

Eq. (2.16-10) shows that the stress in an element will be high if its length L_m and its cross section area A_m are <u>both</u> small. Conversely, stress reduction in an element which is part of a series assembly can be obtained by either lengthening the element or increasing its cross section area. To avoid high stresses, an assembly of elements in series would require closely matching bar spring constants, $\frac{E_m A_m}{L_m}$.

An important difference between series element assemblies and parallel element assemblies, which are subjected to thermal stress, is the limitation of the stress that can be attained in parallel element assemblies to $E(\overline{\alpha T} - \alpha T)$, and the absence of any limit on the stress that can be developed in a series assembly. When series elements are subject to bending, short elements of small cross section are highly vulnerable. When yielding occurs in such an element it is called a plastic "hinge".

Example (a) Elements in Parallel and Series

An assembly consists of bar-type elements that are in a parallel-series arrangement as shown in Fig. 2.16-2.

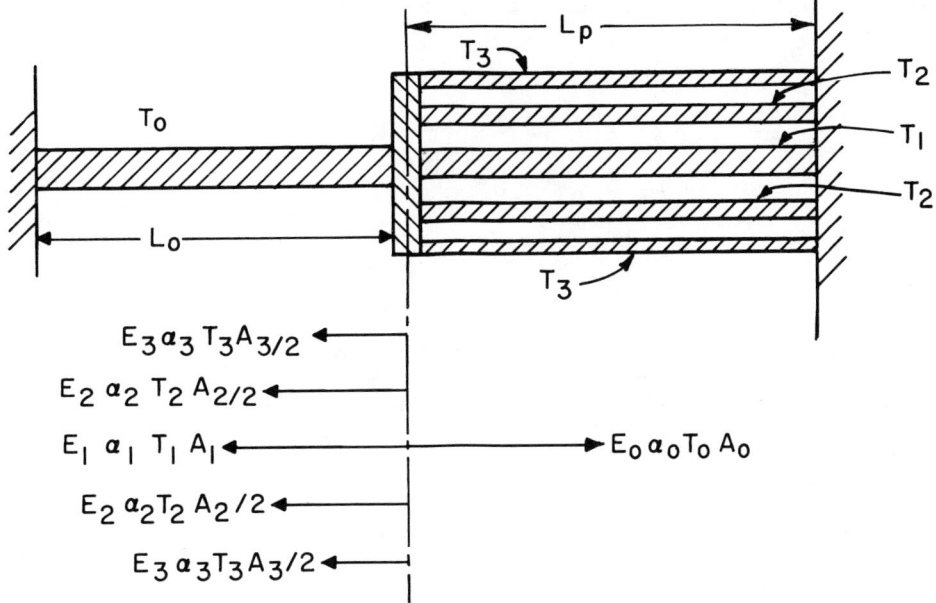

Fig. 2.16-2

The surface tractions which act on the rigid walls at the ends of the assembly have no effect, and are not shown. At the junction of the parallel and series sections, the surface tractions act as shown below the sketch. A load balance of the internal displacement stresses and surface traction loads gives

$$\sigma'_o A_o + E_1 \alpha_1 T_1 A_1 + E_2 \alpha_2 T_2 A_2 + E_3 \alpha_3 T_3 A_3$$
$$= \sigma'_1 A_1 + \sigma'_2 A_2 + \sigma'_3 A_3 + E_o \alpha_o T_o A_o \tag{1}$$

The length of the assembly remains unchanged so that

$$\Delta L = \epsilon_o L_o + \epsilon_p L_p = 0 \tag{2}$$

Also, the uniform strains, ϵ_o and ϵ_p, in the left hand and right hand sections are

$$\epsilon_o = \frac{\sigma'_o}{E_o}$$
$$\epsilon_p = \epsilon_1 = \epsilon_2 = \epsilon_3 \tag{3}$$

In terms of the displacement stresses σ'_1, the strain, ϵ_p, is

$$\epsilon_p = \frac{\sigma'_1}{E_1} = \frac{\sigma'_2}{E_2} = \frac{\sigma'_3}{E_3} \tag{4}$$

The displacement stresses σ'_2 and σ'_3 are written in terms of σ'_1 as

$$\sigma'_2 = \frac{\sigma'_1 E_2}{E_1} \qquad \sigma'_3 = \frac{\sigma'_1 E_3}{E_1} \tag{5}$$

Eq. (2) is expressed in terms of displacement stress as

$$\frac{\sigma'_o L_o}{E_o} + \frac{\sigma'_1 L_p}{E_1} = 0 \tag{6}$$

and substitution of σ'_2 and σ'_3, from Eq. (5) into Eq. (1) gives

$$\sigma'_o A_o = \frac{\sigma'_1}{E_1}(E_1 A_1 + E_2 A_2 + E_3 A_3)$$
$$+ E_o \alpha_o T_o A_o - (E_1 \alpha_1 T_1 A_1 + E_2 \alpha_2 T_2 A_2 + E_3 \alpha_3 T_3 A_3) \tag{7}$$

From Eqs. (6) and (7) we obtain

$$\sigma'_o = E_o \frac{E_o \alpha_o T_o A_o - (E_1 \alpha_1 T_1 A_1 + E_2 \alpha_2 T_2 A_2 + E_3 \alpha_3 T_3 A_3)}{E_o A_o + \frac{L_o}{L_p}(E_1 A_1 + E_2 A_2 + E_3 A_3)} \tag{8}$$

$$\sigma'_1 = E_1 \frac{E_1 \alpha_1 T_1 A_1 + E_2 \alpha_2 T_2 A_2 + E_3 \alpha_3 T_3 A_3 - E_o \alpha_o T_o A_o}{\frac{L_p}{L_o} E_o A_o + E_1 A_1 + E_2 A_2 + E_3 A_3} \tag{9}$$

The strains in each of the assembly sections are

$$\epsilon_o = \frac{E_o \alpha_o T_o A_o - (E_1 \alpha_1 T_1 A_1 + E_2 \alpha_2 T_2 A_2 + E_3 \alpha_3 T_3 A_3)}{E_o A_o + \frac{L_o}{L_p}(E_1 A_1 + E_2 A_2 + E_3 A_3)} \tag{10}$$

$$\epsilon_1 = \epsilon_p = \frac{E_1 \alpha_1 T_1 A_1 + E_2 \alpha_2 T_2 A_2 + E_3 \alpha_3 T_3 A_3 - E_o \alpha_o T_o A_o}{\frac{L_p}{L_o} E_o A_o + E_1 A_1 + E_2 A_2 + E_3 A_3} \tag{11}$$

and the stresses are

$$\sigma_o = E_o \left[\frac{E_o \alpha_o T_o A_o - (E_1 \alpha_1 T_1 A_1 + E_2 \alpha_2 T_2 A_2 + E_3 \alpha_3 T_3 A_3)}{E_o A_o + \frac{L_o}{L_p}(E_1 A_1 + E_2 A_2 + E_3 A_3)} - \alpha_o T_o \right] \tag{12}$$

$$\sigma_m = E_m \left[\frac{E_1 \alpha_1 T_1 A_1 + E_2 \alpha_2 T_2 A_2 + E_3 \alpha_3 T_3 A_3 - E_o \alpha_o T_o A_o}{\frac{L_p}{L_o} E_o A_o + E_1 A_1 + E_2 A_2 + E_3 A_3} - \alpha_m T_m \right] \tag{13}$$

where $m = 1, 2, 3$.

Note in these expressions that when $A_o = 0$ we obtain the proper expression for a parallel element assembly, and when $A_2 = A_3 = 0$ we obtain the proper expression for a series element assembly.

Example (b)   Sectionally Heated Bar

Fig. 2.16-3 shows a bar held between rigid walls, with each section $L_1, L_2, \ldots L_n$, having a different symmetrical temperature distribution

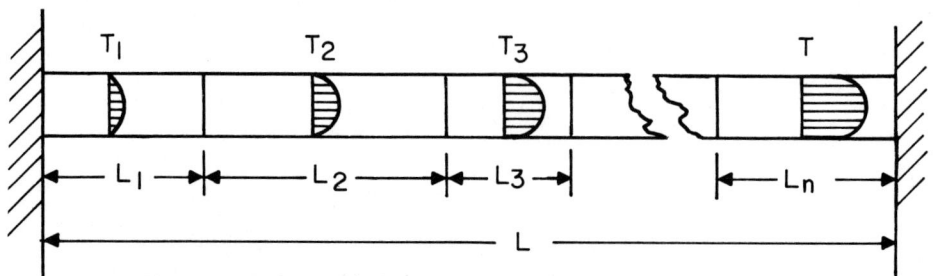

Fig. 2.16-3

$T_1(y) = T_1(-y)$, $T_2(y) = T_2(-y)$, ... $T_n(-y)$. The stress-strain equation for any section, i, is

$$\epsilon_i = \frac{\sigma_i}{E} + \alpha T_i \qquad (1)$$

As the axial strain is uniform in each section, integration of Eq. (1) yields

$$\epsilon_i = \frac{\overline{\sigma}_i}{E} + \alpha \overline{T}_i \qquad (2)$$

The length of the assembly does not change. Therefore

$$\sum_1^n \epsilon_j L_j = 0 \qquad (3)$$

Substitution of ϵ_i from Eq. (3) yields

$$\sum_1^n \left(\frac{\overline{\sigma}_j L_j}{E} + \alpha \overline{T}_j L_j \right) = 0 \qquad (4)$$

The axial load in each section is the same. This is expressed as

$$\int_A \sigma_i \, dA = \text{const}$$

or

$$\bar{\sigma}_1 = \bar{\sigma}_2 = \bar{\sigma}_3 = \ldots = \bar{\sigma}_n \tag{5}$$

Eq. (4) can then be written as

$$\frac{\bar{\sigma}_i}{E} \sum_1^n L_j = -\alpha \sum_1^n \bar{T}_j L_j \tag{6}$$

from which we obtain $\bar{\sigma}_i$ as

$$\bar{\sigma}_i = -E\alpha \frac{\sum_1^n \bar{T}_j L_j}{L} = -E\alpha \bar{T} \tag{7}$$

with $\bar{T}$ the mean temperature of the entire bar. From Eqs. (2) we obtain the strain in any section as

$$\epsilon_i = -\alpha [\bar{T} - \bar{T}_i] \tag{8}$$

and from Eq. (1) we obtain the distributed stress as

$$\sigma_i = -E\alpha [\bar{T} - \bar{T}_i + T_i] \tag{9}$$

For the case of a single bar with a transverse temperature distribution, held between rigid walls, $\bar{T} = \bar{T}_i$, and Eq. (9) becomes $\sigma = -E\alpha T(y)$. For the case of uniformly heated sections, $\bar{T}_i = T_i$, and Eq. (9) becomes $\sigma_i = -E\alpha \bar{T}$. These results are consistent with those obtained in the preceding sections.

Example (c)   Bars in Series with Transverse Temperature Variation

Fig. 2-16-4 shows two bars in series held between rigid walls with the left bar heated uniformly to a temperature T_o and the right bar having

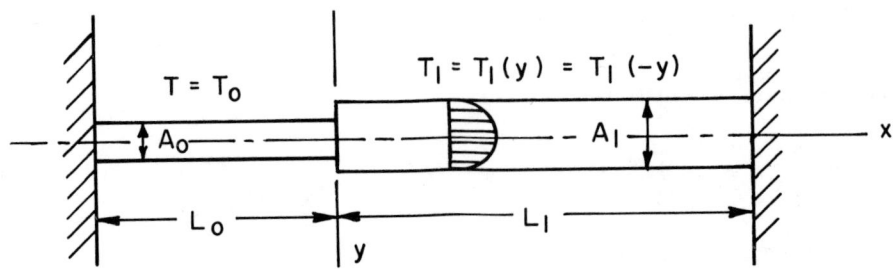

Fig. 2.16-4

transverse symmetrical temperature distribution $T = T(y) = T(-y)$. The stress-strain equations are

$$\epsilon_o = \frac{\sigma_o}{E_o} + \alpha_o T_o \tag{1}$$

$$\epsilon_1 = \frac{\sigma_1}{E_1} + \alpha_1 T_1 \tag{2}$$

The strains ϵ_o and ϵ_1 are constant, the stress σ_o and the temperature T_o are constant, and the stress σ_1 and temperature T_1 are functions of y. Integration of Eq. (2) over the cross section area gives

$$\epsilon_1 = \frac{\overline{\sigma_1}}{E_1} + \alpha_1 \overline{T_1} \tag{3}$$

The length of the composite bar assembly remains unchanged so that

$$\epsilon_o L_o + \epsilon_1 L_1 = 0 \tag{4}$$

Setting Eqs. (1) and (3) into the above we have

$$\frac{\sigma_o L_o}{E_o} + \alpha_o T_o L_o + \frac{\overline{\sigma_1} L_1}{E_1} + \alpha_1 \overline{T_1} L_1 = 0 \tag{5}$$

ELEMENTS IN SERIES

The axial load in each section of the bar must be the same. Therefore

$$\sigma_o A_o = \int_{A_1} \sigma_1 dA_1$$

or

$$\sigma_o A_o - \bar{\sigma}_1 A_1 = 0 \tag{6}$$

From Eqs. (5) and (6) we obtain σ_o and $\bar{\sigma}_1$ as

$$\sigma_o = - \frac{\alpha_o T_o L_o + \alpha_1 T_1 L_1}{\frac{L_o}{E_o} + \frac{A_o}{A_1} \frac{L_1}{E_1}} \tag{7}$$

$$\bar{\sigma}_1 = - \frac{\alpha_o T_o L_o + \alpha_1 T_1 L_1}{\frac{L_1}{E_1} + \frac{A_1}{A_o} \frac{L_o}{E_o}} \tag{8}$$

In the section of the bar with the varying temperature, the relationship between the distributed stress and the mean stress is

$$\sigma_1 = \bar{\sigma}_1 + E_1 \alpha_1 (\bar{T}_1 - T_1) \tag{9}$$

Substitution of $\bar{\sigma}_1$, from Eq. (8) gives

$$\sigma_1 = - \frac{\alpha_o T_o L_o + \alpha_1 T_1 L_1}{\frac{L_1}{E_1} + \frac{A_1}{A_o} \frac{L_o}{E_o}} + E_1 \alpha_1 (\bar{T}_1 - T_1) \tag{10}$$

The strains in each of the sections are

$$\epsilon_o = \alpha T_o - \frac{\alpha_o T_o L_o + \alpha_1 T_1 L_1}{L_o + \frac{A_o E_o}{A_1 E_1} L_1} \tag{11}$$

$$\epsilon_1 = \alpha_1 \overline{T}_1 - \frac{\alpha_o T_o L_o + \alpha_1 T_1 L_1}{L_1 + \frac{A_1 E_1}{A_o E_o} L_o} \tag{12}$$

2.17 BAR WITH NON-UNIFORM CROSS SECTION

The expressions derived in the preceding section, which apply to bars in series, can also be applied to homogeneous bars of non-uniform cross section. A restriction that is imposed on the present analysis is that the temperature vary slowly in the axial direction; otherwise significant stresses would be developed due to lateral expansion incompatibilities. As before, an elastic structure with a spring constant, k_s, is placed in series with the bar. The bar with varying cross section, shown in Fig. 2.17-1, can be viewed as consisting of short lengths of uniform cross section, such as L_1, L_2, etc.

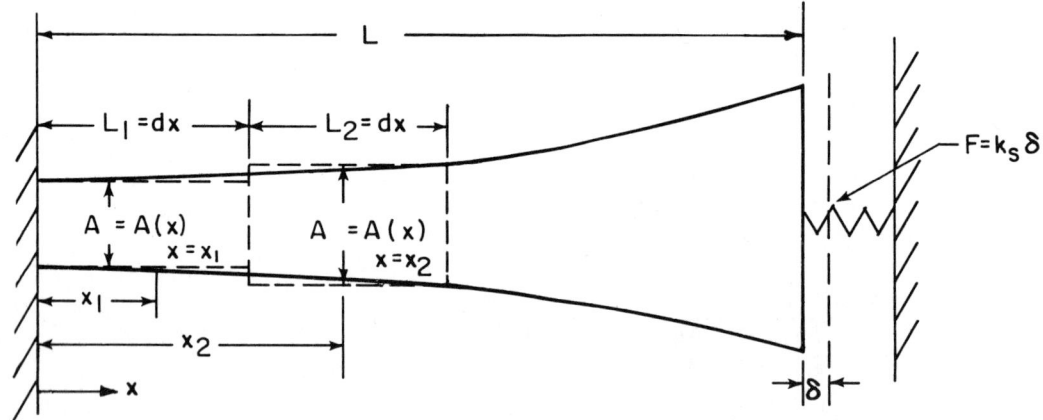

Fig. 2.17-1

In Eq. (2.16-6), the length, L_m, of any section becomes dx, and the cross section area A_m, becomes A(x). The bar is assumed homogeneous so that the modulus of elasticity and coefficient of expression are the same for each differential length of bar. The summation signs of the preceding section now become integral signs, and the expression for stress, Eq. (2.16-6), becomes

$$\sigma = \sigma_x = \frac{E\alpha \int_0^L T \, dx}{A\left(\frac{E}{k_s} - \int_0^L \frac{dx}{A}\right)} \qquad (2.17-1)$$

The elongation of the bar, δ, obtained from Eq. (2.16-7) is

$$\delta = \frac{\sigma A}{k_s} = \frac{\alpha \int_0^L T \, dx}{1 - \frac{k_s}{E} \int_0^L \frac{dx}{A}} \qquad (2.17-2)$$

Note that when the cross section is uniform, Eq. (2.17-1) gives the stress as

$$\sigma = -\frac{E\alpha \overline{T}}{1 - \frac{EA}{k_s L}} \qquad (2.17-3)$$

and when the bar is held between rigid walls ($k_s \to \infty$), the stress is simply $-E\alpha\overline{T}$. When the temperature is uniform, and equal to T_o throughout the bar Eq. (2.17-1) becomes

$$\sigma = -\frac{E\alpha T_o}{\frac{A}{L} \int_0^L \frac{dx}{A} - \frac{EA}{k_s L}} \qquad (2.17-4)$$

Example (a) Bar with Axial Parabolic Area and Temperature Variation

A bar has a cross section area variation and temperature distribution given as

$$A = A_o \frac{x^2}{L^2} \qquad \frac{L}{2} < x < L \tag{1}$$

$$T = T_o \frac{x^2}{L^2} \qquad \frac{L}{2} < x < L \tag{2}$$

It is held rigidly between two walls, at $x = \frac{L}{2}$ and $x = L$, as shown in Fig. 2.17-2.

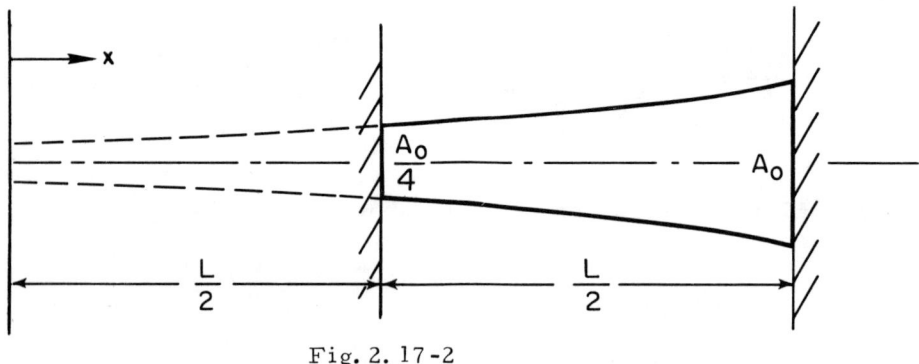

Fig. 2.17-2

From Eq. (2.17-1) with $k_s \to \infty$, the stress is

$$\sigma = - \frac{E\alpha \int_{L/2}^{L} \frac{T_o x^2}{L^2} dx}{A \int_{L/2}^{L} \frac{dx}{A_o x^2 / L^2}} = - \frac{7}{24} \frac{A_o}{A(x)} E\alpha T_o \tag{3}$$

and is a function of x. The axial force in the bar is constant and equal to

$$F = \sigma \cdot A(x) = - \frac{7}{24} E\alpha T_o A_o \tag{4}$$

Example (b) Bar with Axial Sinusoidal Area and Temperature Variation

The round bar shown in Fig. 2.17-3, held between rigid walls, has a cross section area variation and temperature distribution

$$A = A_o \sin \frac{\pi x}{L} \qquad \frac{L}{4} < x < \frac{3}{4} L \tag{1}$$

$$T = T_o \sin \frac{\pi x}{L} \qquad \frac{L}{4} < x < \frac{3}{4} L \tag{2}$$

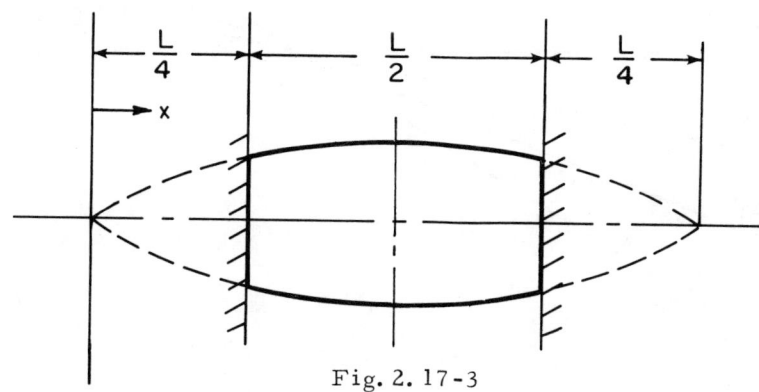

Fig. 2.17-3

From Eq. (2.17-1) the stress is

$$\sigma = - \frac{E\alpha \int_{L/4}^{3L/4} T_o \sin \frac{\pi x}{L} \, dx}{A \int_{L/4}^{3L/4} \frac{dx}{A_o \sin \frac{\pi x}{L}}} \tag{3}$$

or

$$\sigma = \frac{E\alpha T_o A_o}{A} \frac{\left[\cos \frac{\pi x}{L}\right]_{L/4}^{3L/4}}{\ln\left[\csc \frac{\pi x}{L} - \cot \frac{\pi x}{L}\right]_{L/4}^{3L/4}} = - \frac{\sqrt{2} \, E\alpha T_o}{(\sin \frac{\pi x}{L}) \ln(\frac{\sqrt{2}+1}{\sqrt{2}-1})} \tag{4}$$

The maximum stress occurs at $x = \frac{L}{4}$ and $x = \frac{3L}{4}$ and is

$$\sigma_{max} = -\frac{2E\alpha T_o}{\ln 5.83} = -1.35 \, E\alpha T_o \, . \tag{5}$$

2.18 STATICALLY INDETERMINATE BAR ASSEMBLY

The bars in the bar assembly shown in Fig. 2.18-1 are at temperatures T_1, T_2, T_2 and T_4. The analysis is carried out by means of

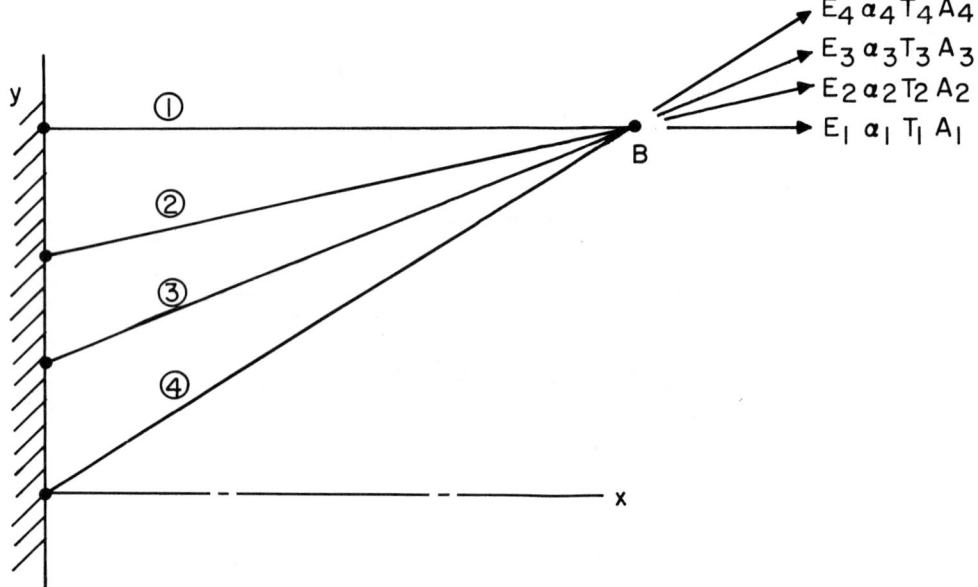

Fig. 2.18-1

equivalent thermal loads. The surface tractions are as shown.

Under the action of the surface traction loads whose vector sum is

$$\vec{F}' = \sum_1^4 E_j \alpha_j \vec{T}_j A_j \tag{2.18-1}$$

point B is displaced an amount u and v in the x and y directions.

Let m_i and n_i represent the direction cosines of bar i with respect to the x and y axes respectively. As a result of the displacements u and v it is readily determined that the increase in the length of a bar is

$$\Delta L_i = u m_i + v n_i \qquad (2.18\text{-}2)$$

The strain is then

$$\epsilon_i = \frac{\Delta L_i}{L_i} = \frac{u m_i + v n_i}{L_i} \qquad (2.18\text{-}3)$$

and the displacement stress is

$$\sigma'_i = E_i \epsilon_i = \frac{E_i}{L_i}(u m_i + v n_i) \qquad (2.18\text{-}4)$$

The displacement load in a bar is

$$F'_i = \sigma'_i A_i = \frac{E_i A_i}{L_i}(u m_i + v n_i) \qquad (2.18\text{-}5)$$

and the horizontal and vertical components of these loads are

$$F'_{ih} = F'_i m_i = \frac{E_i A_i m_i}{L_i}(u m_i + v n_i)$$

$$F'_{iv} = F'_i n_i = \frac{E_i A_i n_i}{L_i}(u m_i + v n_i) \qquad (2.18\text{-}6)$$

The sum of the horizontal and vertical displacement loads, in terms of displacements, are therefore

$$F'_h = \sum_1^4 \frac{E_j A_j m_j}{L_j}(u m_j + v n_j)$$

$$F'_v = \sum_1^4 \frac{E_j A_j n_j}{L_j}(u m_j + v n_j) \qquad (2.18\text{-}7)$$

The vertical and horizontal components of the surface traction loads acting at point B, are as seen from the sketch,

$$F'_h = \sum_1^4 E_j \alpha_j T_j A_j m_j$$

$$F'_v = \sum_1^4 E_j \alpha_j T_j A_j n_j$$

(2.18-8)

Equating these with the expressions of Eq. (2.18-7), and generalizing for the case of **s** bars, we have

$$u \sum_1^s \frac{E_j A_j m_j^2}{L_j} + v \sum_1^s \frac{E_j A_j m_j n_j}{L_j} = \sum_1^s E_j \alpha_j T_j A_j m_j$$

$$u \sum_1^s \frac{E_j A_j m_j n_j}{L_j} + v \sum_1^s \frac{E_j A_j n_j^2}{L_j} = \sum_1^s E_j \alpha_j T_j A_j n_j$$

(2.18-9)

From these equations we determine u and v as

$$u = \frac{\begin{vmatrix} \sum_1^s E_j \alpha_j T_j A_j m_j & \sum_1^s \frac{E_j A_j m_j n_j}{L_j} \\ \sum_1^s E_j \alpha_j T_j A_j n_j & \sum_1^s \frac{E_j A_j n_j^2}{L_j} \end{vmatrix}}{\begin{vmatrix} \sum_1^s \frac{E_j A_j m_j^2}{L_j} & \sum_1^s \frac{E_j A_j m_j n_j}{L_j} \\ \sum_1^s \frac{E_j A_j m_j n_j}{L_j} & \sum_1^s \frac{E_j A_j n_j^2}{L_j} \end{vmatrix}}$$

(2.18-10)

$$v = \frac{\begin{vmatrix} \sum_1^s \frac{E_j A_j m_j^2}{L_j} & \sum_1^s E_j \alpha_j T_j A_j m_j \\ \sum_1^s \frac{E_j A_j m_j n_j}{L_j} & \sum_1^s E_j \alpha_j T_j A_j n_j \end{vmatrix}}{\begin{vmatrix} \sum_1^s \frac{E_j A_j m_j^2}{L_j} & \sum_1^s \frac{E_j A_j m_j n_j}{L_j} \\ \sum_1^s \frac{E_j A_j m_j n_j}{L_j} & \sum_1^s \frac{E_j A_j n_j^2}{L_j} \end{vmatrix}} \qquad (2.18\text{-}11)$$

With the use of Eq. (2.18-4) the stress in any bar is expressed in terms of u and v as

$$\sigma_i = \sigma_i' - E_i \alpha_i T_i = E_i \left(\frac{u m_i + v n_i}{L_i} - \alpha_i T_i \right) \qquad (2.18\text{-}12)$$

The usefulness of the foregoing analysis is limited to redundant bar assemblies which are joined together at a common point. Thermal stress analysis of the more general type of redundant truss is discussed in Sects. 5.2, 5.3, and 5.4.

PROBLEMS

2-1 Using the classical approach derive the expressions for the stresses and strains in a parallel element assembly in which the members are of different lengths but undergo equal increase in length. Such an assembly is shown in Fig. 2.1-1. Assume symmetrical distribution of material, physical properties, and temperature.

2-2 A steel reinforced concrete bar has dimensions as shown in Fig. 2-2.

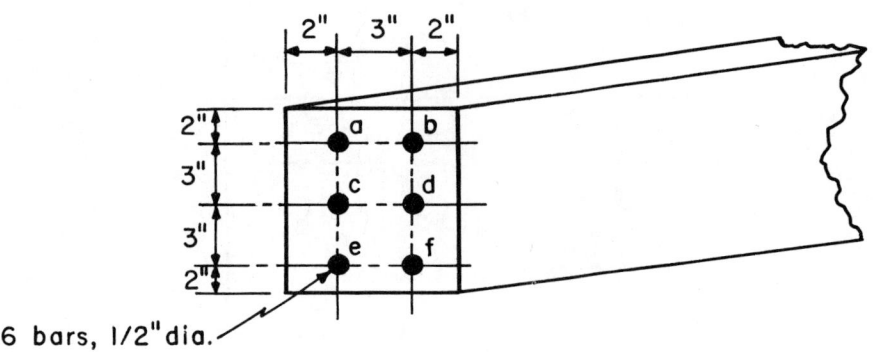

Fig. 2-2

An electric current, passed through bars c and d, heats them to a temperature T_o. If the tensile strength of concrete is 400 psi, find the temperature T_o at which the concrete will fail. Properties: $E_s = 30 \times 10^6$ psi, $E_c = 10^6$ psi, $\alpha_s = 6.5 \times 10^{-6}/°F$, $\alpha_c = 5.5 \times 10^{-6}/°F$.

2-3 A concentric tube assembly such as that shown in Fig. 2-3 has a stainless steel (ss) inner tube and carbon steel (cs) outer tube. The moduli of elasticity and coefficients of expansion are: $E_{ss} = 28 \times 10^6$ psi, $E_{cs} = 30 \times 10^6$ psi, $\alpha_{ss} = 10 \times 10^{-6}/°F$, $\alpha_{cs} = 6.5 \times 10^{-6}/°F$, and the cross section areas are $A_{ss} = 0.25$ in^2 and $A_{cs} = 0.40$ in^2. The

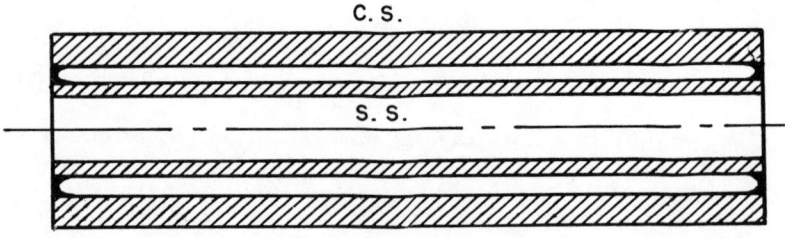

Fig. 2-3

assembly is heated uniformly. At what temperature will yielding in either of the tubes take place if $\sigma_{y,ss} = 40,000$ psi, $\sigma_{y,cs} = 20,000$ psi?

2-4　In the assembly of Problem 2-3, (Fig. 2.3-1), the cross section area of each of the stainless steel weldments, $A_w = 2\pi R_w L_w$, is equal to the cross section area of the stainless steel tube; i.e., $A_w = 0.25$ in^2. What is the shear stress in the weldment at the yielding temperature determined in Problem 2-3? What is the required shear area in order to reduce the stress in the weldment to a minimum?

2-5　In the unsymmetrical four bar assembly of Fig. 2-5, free axial expansion takes place but bending of the assembly is prevented by the end restraints. Find the expression for the shear stress in attachment of length L_w joining bars 2 and 3.

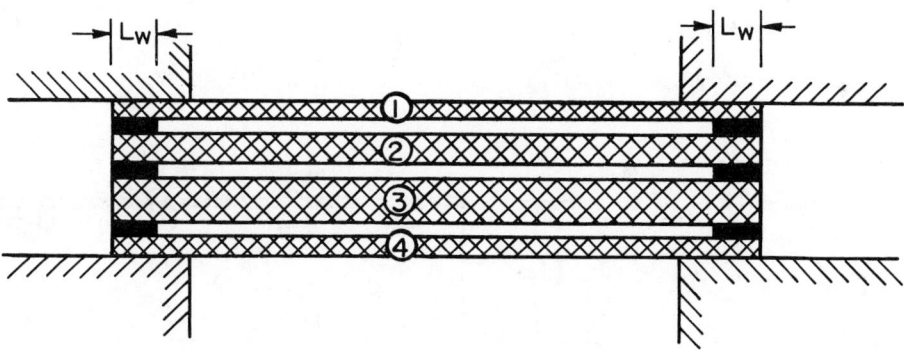

Fig. 2-5

2-6　A wing whose cross section is shown in Fig. 2-6 has an aluminum skin and stainless steel I beams. Aerodynamic heating brings the skin temperature to $250°F$ above that of the beams. If the aluminum skin (a) and stainless steel beams (s) have the properties

$$E_a = 10 \times 10^6 \text{ psi} \qquad E_s = 28 \times 10^6 \text{ psi}$$
$$\alpha_a = 10.5 \times 10^{-6}/°F \qquad \alpha_s = 9.5 \times 10^{-6}/°F$$
$$A_a = 10 \text{ in}^2 \qquad A_s = 12 \text{ in}^2 \text{ (includes both beams)}$$

find the axial normal stresses in the skin and beams. Assume one-dimensional axial extensional deformation. If the yield strengths of the skin and beams are $\sigma_{ya} = 30,000$ psi and $\sigma_{ys} = 50,000$ psi, at what temperature does yielding take place?

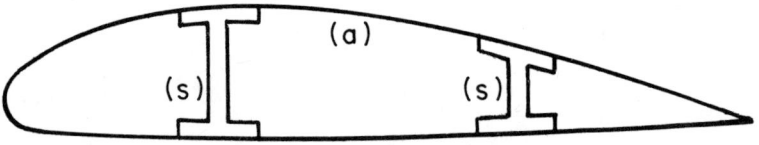

Fig. 2-6

2.7   A rocket housing structure has the form of a circular cylinder with reinforcing circular rings as shown in Fig. 2-7.

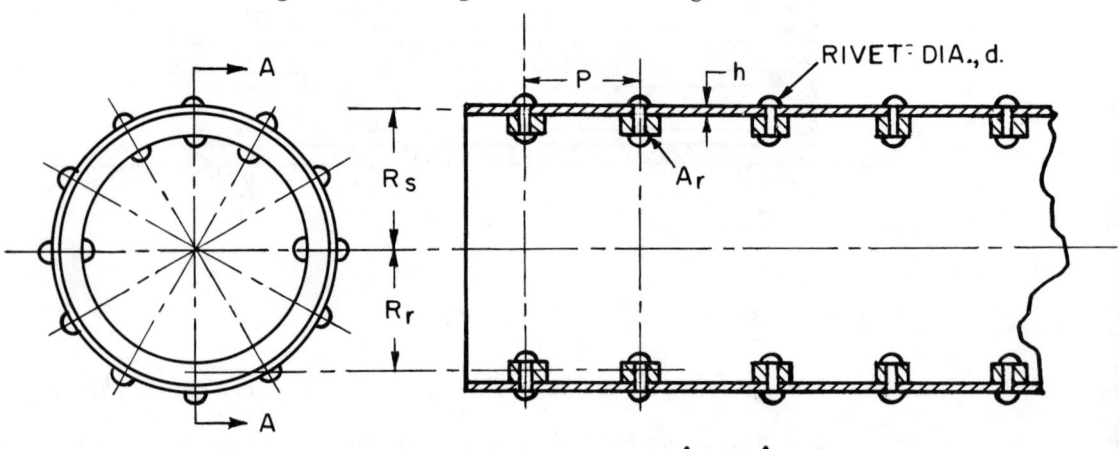

Fig. 2-7

PROBLEMS   111

The skin is heated to a temperature T_o above that of the rings. Assuming one-dimensional uniform tangential extension of the rings and shell, derive the expressions for the tangential stresses in the skin and reinforcing rings. Use E_s, α_s, h, R_s for skin, and E_r, α_r, A_r, R_r for rings. Derive an expression for the radial stress in the rivets having a diameter, d.

2-8   Two concentric rings of different materials are welded along the interface edges. What are the normal radial stresses and axial shear stresses in the weld, at $r = R_j$, (see Fig. 2.14-1), when the inner ring is at a temperature T_i and the outer ring at T_o? Assume the total length of the interfacial edge weldments to be one-tenth the axial thickness, L, of the ring assembly.

2-9   The aluminum rectangular beam shown in Fig. 2-9 has a tent-shaped temperature distribution as shown. Find the one-dimensional axial normal stresses in the central and outer fibres. Make the required correction for biaxiality of stress. Properties are: $E_a = 10^7$ psi, $\alpha_a = 10 \times 10^{-6}/°F$.

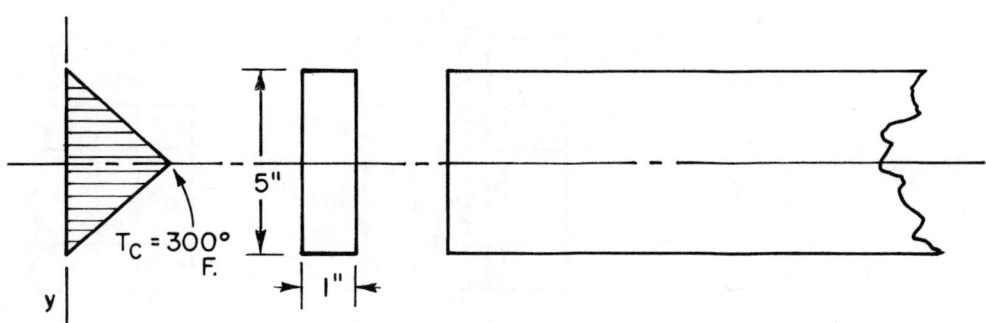

Fig. 2-9

2-10 The aluminum beam in Fig. 2-10 has a parabolic temperature distribution (°F) as shown. Find the central and outer fibre normal stresses. Make correction for biaxiality of stress. $E_a = 10^7$ psi; $\alpha_a = 10 \times 10^{-6}/°F$.

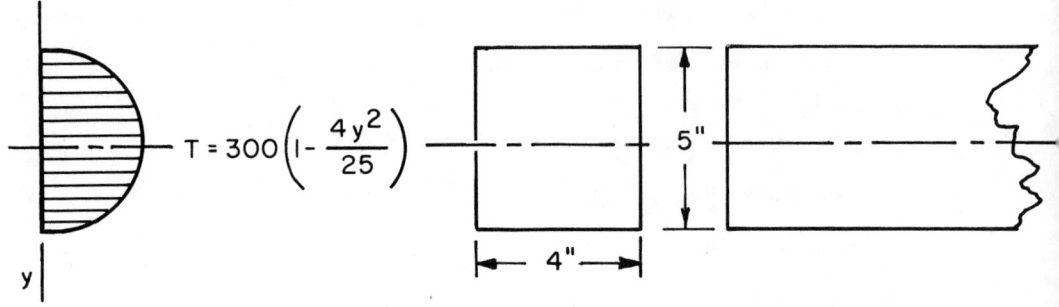

Fig. 2-10

2-11 The sandwich bar shown in Fig. 2-11 consists of stainless steel outer plates and a carbon steel inner plate. Assuming that all of the shear load is taken by the end rivet, find the shear stress in the rivet and the bearing stress in each of the rivet holes when the assembly is heated uniformly to a temperature of 200°F. Constants E and α are as in Problem 2-3.

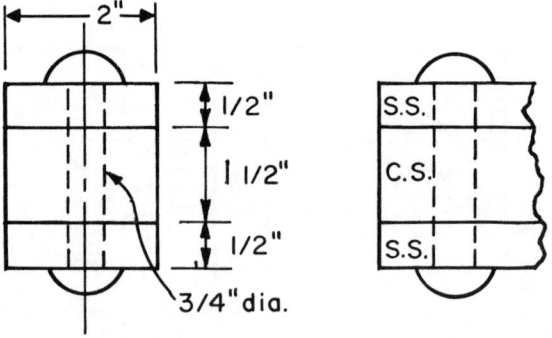

Fig. 2-11

PROBLEMS

2-12 The stainless steel rectangular bar in Fig. 2-12 is subjected to a temperature variation expressed as

$$T = T_o \left(1 - \frac{4y^2}{h^2}\right) \frac{x^2}{L^2}$$

with $T_o = 300°F$, $h = 1"$, $L = 10"$. Find the normal and shear stress distribution, and make correction for biaxiality of stress.

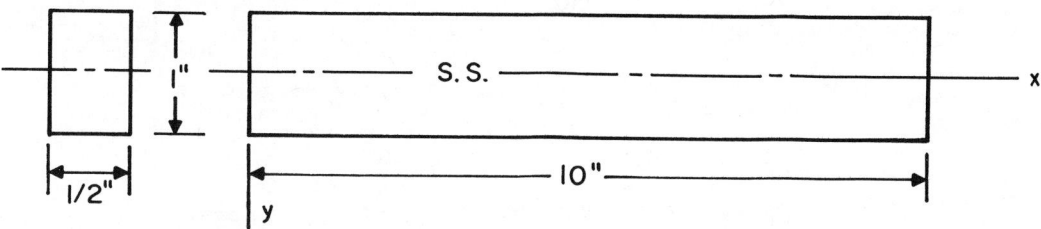

Fig. 2-12

2-13 The temperature distribution in the bimetallic sandwich strip shown in Fig. 2-13 is parabolic in the core plate and linear in the outer plates, and also varies axially. The core temperature distribution is

$$T_c = \left[T_i + T_o \left(1 - \frac{4y_c^2}{h_c^2}\right)\right] \frac{x^3}{L^3}$$

and the outer plates temperatures are expressed as

$$T_s = \frac{T_i}{2} \left(1 - \frac{2y_s}{h_s}\right) \frac{x^3}{L^3}$$

with $T_i = 100°F$, $T_o = 200°F$, $h_c = 3"$, $h_s = 1"$, and $L = 30"$. The width of the assembly is $1/2"$ and the material constants are $E_c = 30 \times 10^6$ psi, $\alpha_c = 10 \times 10^{-6}/°F$, $E_s = 20 \times 10^6$ psi, $\alpha_s = 7 \times 10^{-6}/°F$. Find the normal and shear stress distribution. Disregard end effects.

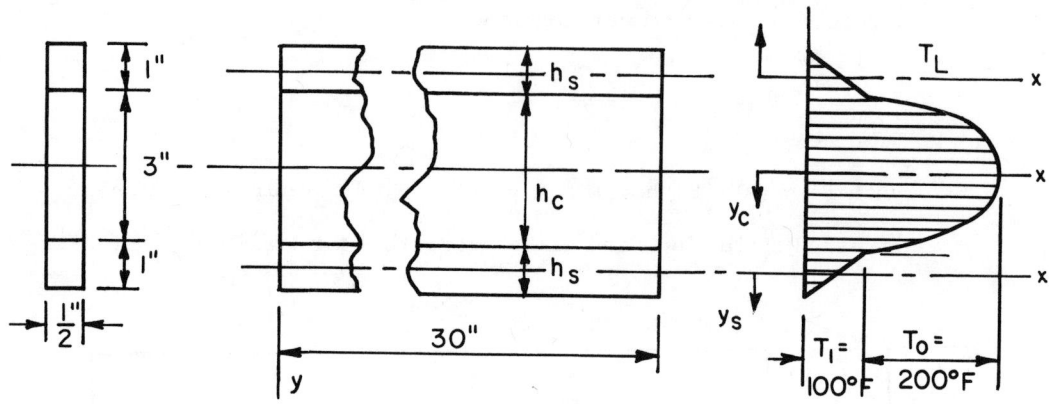

Fig. 2-13

2-14   A homogeneous rectangular beam has a three dimensional temperature distribution of the form

$$T = T_o \left(1 - \frac{4y^2}{h^2}\right)\left(1 - \frac{16z^4}{w^4}\right)\left(1 - \frac{x^2}{L^2}\right)$$

with $T_o = 250°F$, $E = 20 \times 10^6$ psi, $\alpha = 8 \times 10^{-6}/°F$. The cross section dimensions are shown in Fig. 2.14. Find the normal stress distribution and the required modification for biaxiality of stress.

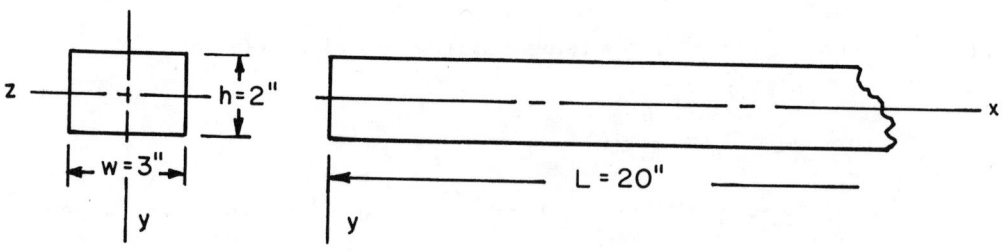

Fig. 2-14

2-15 A wide flange homogeneous beam is subjected to a temperature distribution of the form

$$T = T_o(1 - \frac{4y^2}{h^2}) \sin \frac{\pi x}{L}$$

Derive the general expression for the shear stress at the web and flange junction in the beam having dimensions as shown in Fig. 2-15. What are the normal stresses in the central and outer beam fibers? Assume $I_x = \frac{wh^3}{12} + 2wh(\frac{h}{2})^2$.

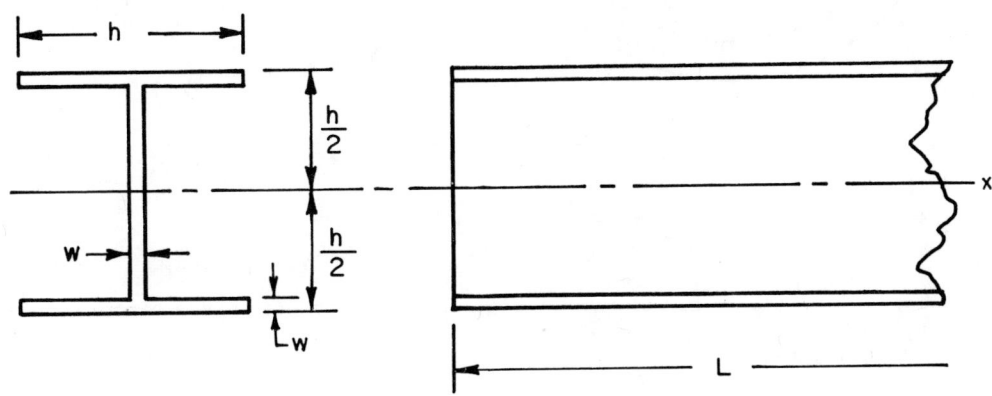

Fig. 2-15

2-16 A three ring axial assembly consists of a central aluminum ring and outer copper rings. The assembly is held together by 12 rivets as shown in Fig. 2-16. Find the tangential stresses in the rings and the shear stresses in the rivets when the assembly is heated uniformly to $180°F$. Assume the following material properties:

$E_{al} = 10 \times 10^6$ psi, $\alpha_{al} = 10 \times 10^{-6}/°F$, $E_{cop} = 14 \times 10^6$ psi, $\alpha_{cop} = 8 \times 10^{-6}/°F$.

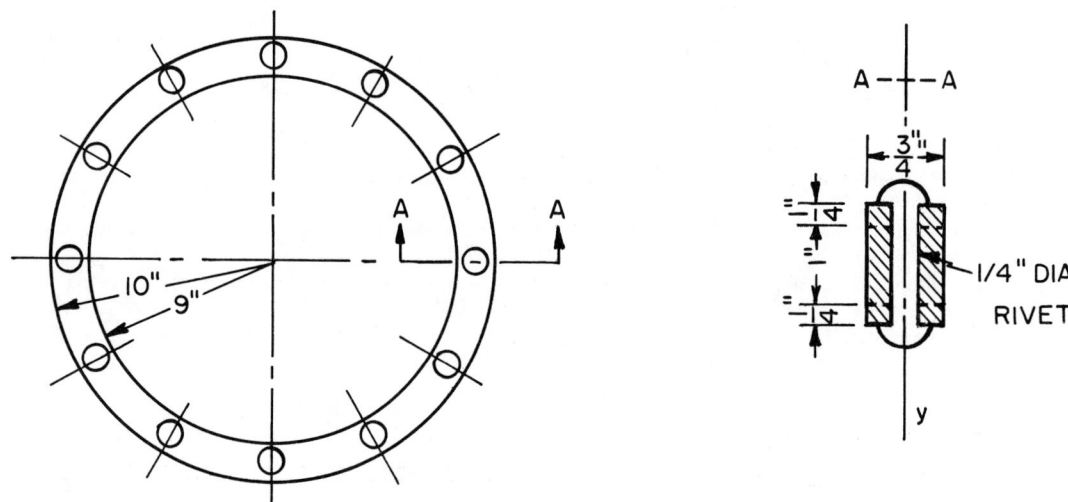

Fig. 2-16

2-17   The assembly shown in Fig. 2-16 is subjected to a temperature distribution

$$T = 200(1 - y^2)$$

with T in °F and y in inches, measured from the central plane of the assembly. Find the stress distribution in the rings and the shear stresses in the rivets.

2-18   A flat ring is held in a slot so that it can expand radially and axially but cannot twist (sliding fit). Find the tangential stresses in the ring shown in Fig. 2-18 when the axial temperature distribution is

$$T = T_o(1 - \frac{8y^3}{L^3})$$

PROBLEMS

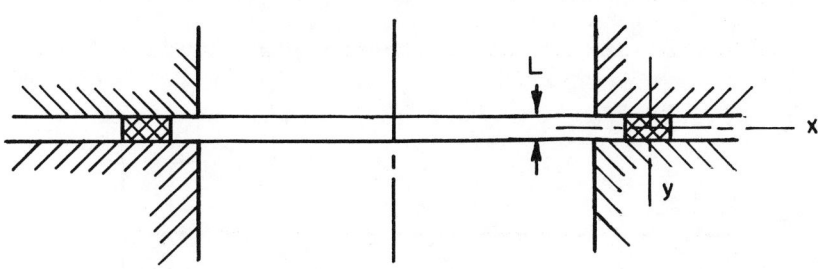

Fig. 2-18

2-19 A ring of square cross section, shown in Fig. 2-19, is subjected to a radial temperature variation of the form

$$T = T_o \left(1 - \frac{4x^2}{h^2}\right)$$

Determine the stress distribution and the radial expansion of the ring. Make correction for biaxiality of stress.

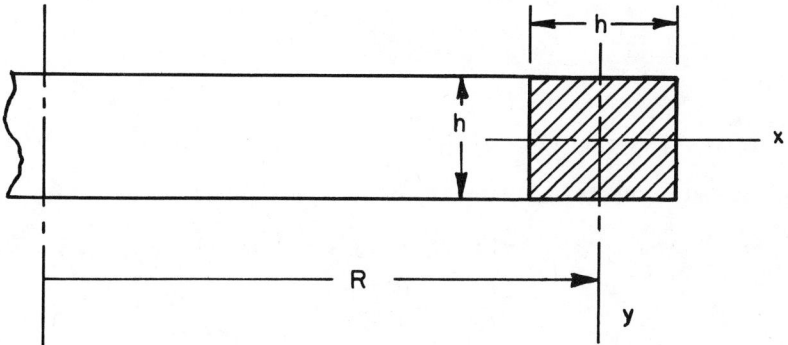

Fig. 2-19

2-20 The ring of rectangular cross section, shown in Fig. 2-20, is subjected to a radial and axial temperature variation expressed as

$$T = T_o \left(1 - \frac{4x^2}{h^2}\right)\left(1 - \frac{4y^2}{L^2}\right)$$

Determine the radial expansion of the ring, the normal stress distribution, and the biaxial stress modification for the case of $h/L = \frac{1}{4}$.

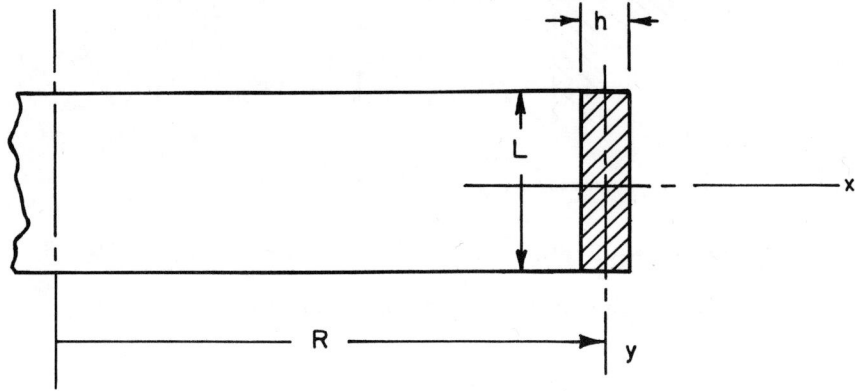

Fig. 2-20

2-21 The concentric three-ring assembly is subjected to a radial temperature variation of the form

$$T = T_o(1 - \frac{x^3}{h^3})$$

Find the radial expansion of the brazed ring assembly and the normal tangential stress distribution in the rings. What is the magnitude of the normal stress at the two interfaces? Use: $E_1, \alpha_1, E_2, \alpha_2, E_3, \alpha_3$.

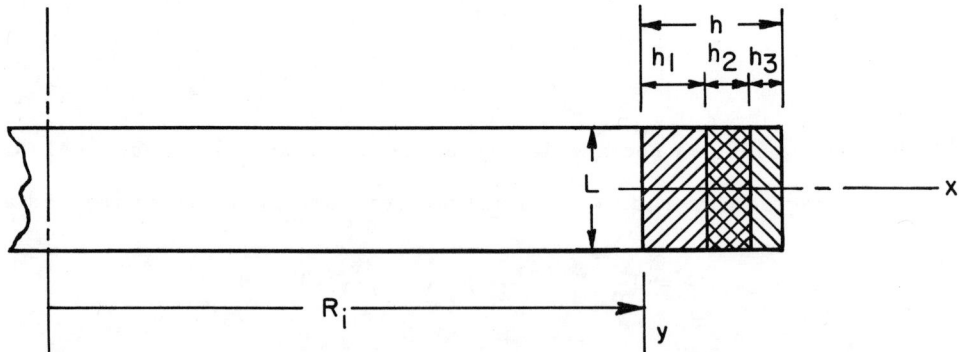

Fig. 2-21

PROBLEMS

2-22 The conical bar in Fig. 2-22 has a cross section area $A = A_o(1 + \frac{3x}{L})$ and a temperature distribution $T = T_o \cos \frac{\pi x}{2L}$. The left end of the conical bar is attached to a circular pin-ended beam of length L_b, modulus of elasticity, E_b, and moment of inertia, I_b. Derive the expression for the axial stress in the conical bar. Use E_c and α_c for the conical bar.

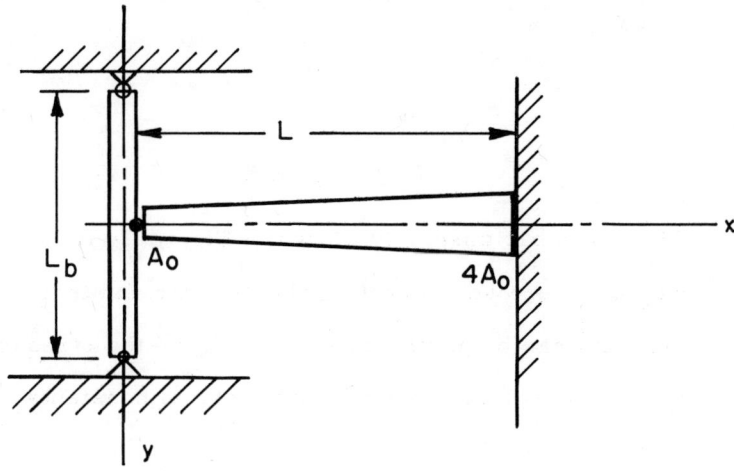

Fig. 2-22

2-23 The bar in Fig. 2-23 consists of two sections. The length L has a cross section area variation and temperature variation given as

$$A = A_j(\frac{1}{2} + \frac{x}{2L_1}) \qquad T = T_j \frac{x^2}{L_1^2}$$

The straight section of the bar, of length L_2, has a cross section area A_j and a uniform temperature T_j. Find the distributed axial stresses in each of the bar sections.

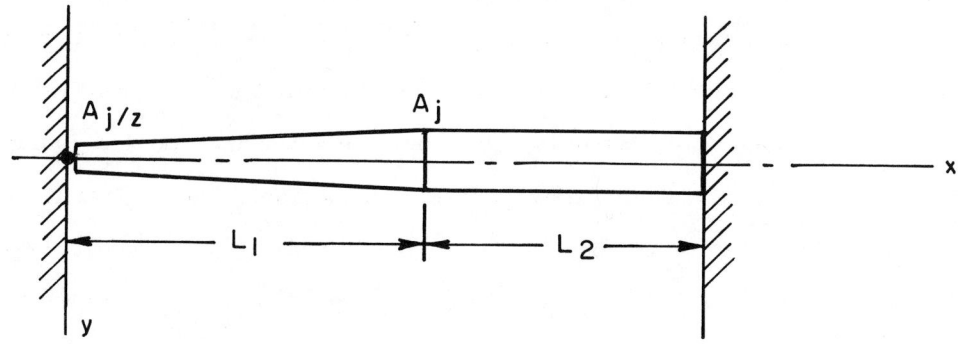

Fig. 2-23

2-24 The seven bar assembly shown in Fig. 2-24 is held between rigid walls. The yoke connecting the two-bar and three-bar sections is rigid but can be displaced axially. Find the stresses in each of the bars, having temperatures and physical constants as shown in the sketch.

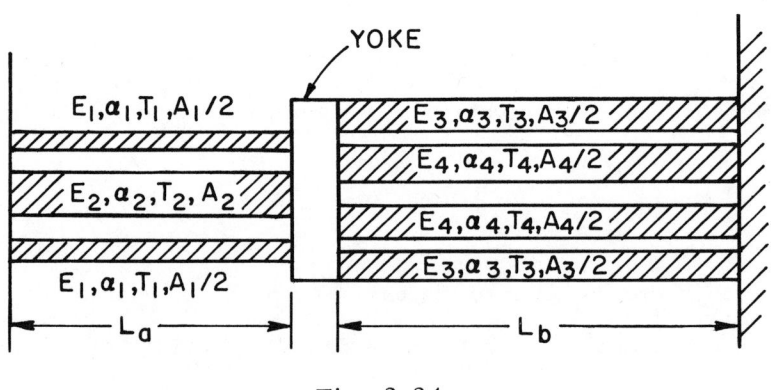

Fig. 2-24

2-25 Two circular bars of lengths L_1 and L_2 and radii R_1 and R_2, shown in Fig. 2-25, are subjected to a uniform internal heat generation which produces parabolic transverse temperature variations in the bars, expressed as

$$T_1 = T_{10}(1 - \frac{r^2}{R_1^2}) \qquad T_2 = T_{20}(1 - \frac{r^2}{R_2^2})$$

Disregard all but the axial normal stresses, and perform analysis by removing the end restraints to obtain the axial stress distribution in the unrestrained bars. Make biaxiality correction of the axial one-dimensional stress, assuming a bar of equal transverse and lateral dimensions. Add to this stress the unmodified one-dimensional stress produced by the end restraints (to which a correction for biaxiality does not apply).

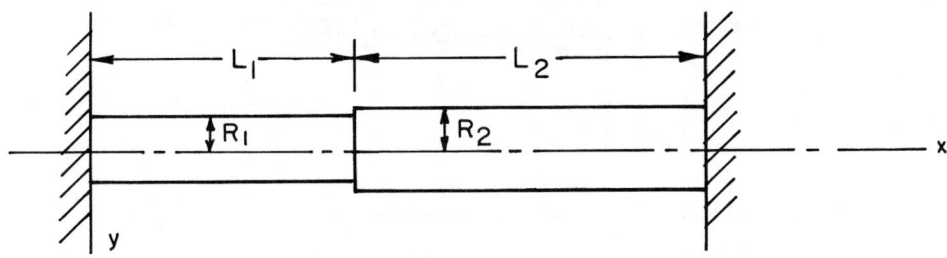

Fig. 2-25

2-26   A layered bar is supported in a flexible frame as indicated in Fig. 2-26. The temperature distribution in the core is parabolic, and linear in the outside plates, as in Fig. 2-13. The temperature distributions in the core, T_c, and in the surface plates, T_s, are

$$T_c = T_i + T_o(1 - \frac{4y_c^2}{h_c^2})$$

$$T_s = \frac{T_i}{2}(1 - \frac{2y_s}{h_s})$$

with h_c the core thickness, h_s the surface plate thickness and w, the lateral width of the assembly. The coordinate system is as shown in Fig. 2-13. Find the stresses in the core, the surface strips, and in the frame using equivalent load analysis. The loads, $P' = E_c \alpha_c \overline{T}_c h_c w + 2E_s \alpha_s \overline{T}_s h_s w$ act as shown.

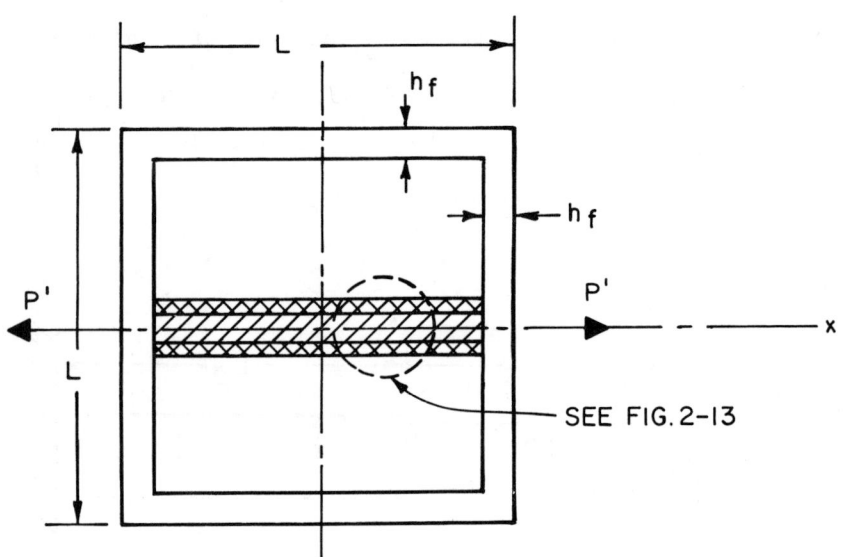

Fig. 2-26

2-27   Show that the general expression for the thermal stresses in a one-juction pinned bar assembly, Eq. (2.18-12) gives zero thermal stress for the statically determinate two-bar assembly shown in Fig. 2-27.

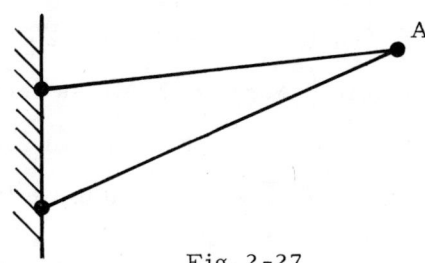

Fig. 2-27

2-28 Use the expressions derived in Section 2.18, to determine the displacement of the yoke and the stresses in each of the bars of the one-junction pinned bar assembly, shown in Prob. 2-24.

2-29 Find the horizontal and vertical displacements of point B and the stresses in each of the bars in the redundant bar assembly shown in Fig. 2-29, when all of the bars are heated uniformly to a temperature $T = 200°F$. The properties of the bars are as follows.

	Bar 1	Bar 2	Bar 3
E	10×10^6 psi	20×10^6 psi	30×10^6 psi
α	$10 \times 10^{-6}/°F$	$8 \times 10^{-6}/°F$	$6.5 \times 10^{-6}/°F$
A	0.25 in^2	0.25 in^2	0.10 in^2

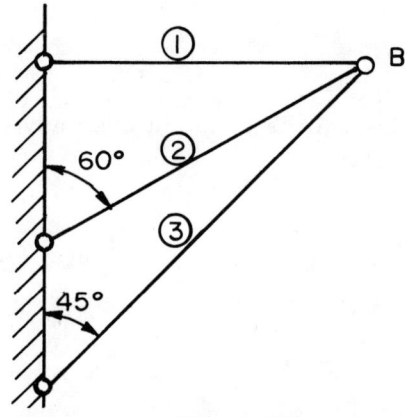

Fig. 2-29

PROBLEMS

2-30 Use equivalent load analysis to find the expressions for the stresses in the rod and beam, and the vertical deflection of the end of the beam, in the structure shown in Fig. 2-30, when the rod is heated to a temperature T_r. Assume the cross section area of the beam to be an order of magnitude larger than the rod cross section area. Use the following symbols for the rod (r) and beam (b): E_r, α_r, A_r, E_b, I_b.

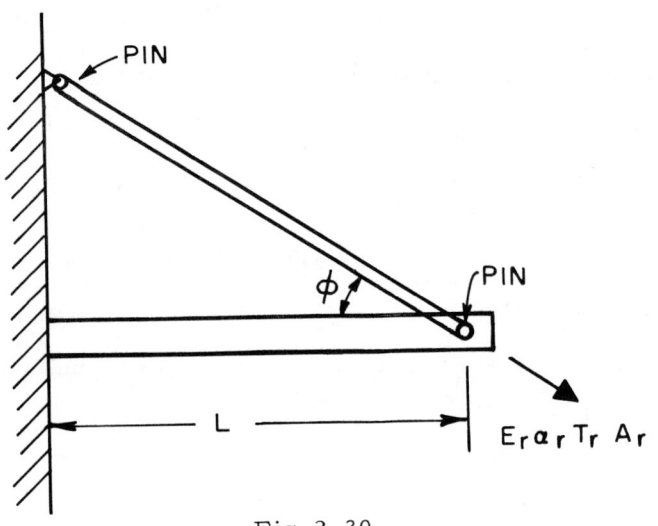

Fig. 2-30

2-31 Derive Eq. (2.5-7) by means of the classical thermoelastic equations. Apply the conditions that there is no axial load, i.e.,

$$\int_A \sigma_x dA = 0$$

and that the axial strain is constant over any cross section. Use the relationship

$$\epsilon(x) = \frac{1}{E} \sigma(x,y) + \alpha T(x,y) = \frac{1}{E} \overline{\sigma}(x) + \alpha \overline{T}(x) = \alpha \overline{T}(x)$$

PROBLEMS

2-32 Derive the expressions for the stresses in a composite bar, Eqs. (2.7-5), by means of the classical equations of thermoelasticity. Make use of the fact that there is no net axial load. That is

$$\int_A \sigma_x \, dA = \bar{\sigma}_1(x) A_1 + \bar{\sigma}_2(x) A_2 = 0$$

and that the axial strains are constant. This gives the relationship

$$\epsilon_1(x) = \epsilon_2(x) = \frac{1}{E_1} \sigma_1(x,y) + \alpha_1 T_1(x,y) = \frac{1}{E_2} \sigma_2(x,y) + \alpha_2 T_2(x,y)$$

$$= \frac{1}{E_1} \bar{\sigma}_1(x) + \alpha_1 \bar{T}_1(x) = \frac{1}{E_2} \bar{\sigma}_2(x) + \alpha_2 \bar{T}_2(x)$$

2-33 Derive Eqs. (2.12-5) and (2.12-6) for the case of the rings at uniform temperatures T_1 and T_2. Perform the analysis by means of the classical equations of thermoelasticity, making the use of the fact that there is no net tangential load, and the approximation that the tangential strain is constant throughout the ring assembly.

2-34 Derive the equation for the stress in a circular ring having a symmetrical axial temperature variation, as in Section 2.11, using the classical equations of thermoelasticity, recognizing that there is no net tangential load and assuming that the tangential strain is uniform over the ring.

2-35 Obtain the expression for the tangential strain and the distributed tangential stresses in a thin concentric two-ring assembly having an arbitrary radial temperature variation. Use the conditions that

$$\int_A \sigma \, dA = 0$$

and ϵ_t = const.

2-36 In Section 2.9 the shear stresses at the interface of composite bars with transverse and axial temperature variation are determined. Generalize the analysis to include a cross section with a variable bar width, and determine the distributed shear stresses in each of the constituents (the shear stresses at all internal locations).

3 | Plane Stress and Plane Strain

3.1 PLANE STRESS

Plane stress is a state of stress in a flat plate in which in-plane stresses σ_x, σ_y, and τ_{xy} are present which are not functions of the thickness coordinate. The stress normal to the plane of the plate, σ_z, is zero and the shear stresses in all planes perpendicular to the plate are also zero. Only the stresses σ_x, σ_y, τ_{xy}, in the plane of the plate are present, and they are functions of the x and y coordinates only. Problems of plane stress are often treated with the use of a stress function, φ. In rectangular coordinates the plane stresses are expressed in terms of the stress function as

$$\sigma_x = \frac{\partial^2 \varphi}{\partial y^2} \qquad \sigma_y = \frac{\partial^2 \varphi}{\partial x^2} \qquad \tau_{xy} = -\frac{\partial^2 \varphi}{\partial x \partial y} \qquad (3.1\text{-}1)$$

By expressing the stresses in this form they automatically satisfy the equilibrium equations. This can be verified by substituting the foregoing expressions into the equilibrium equations.

$$\frac{\partial \sigma_x}{\partial x} + \frac{\partial \tau_{xy}}{\partial y} = 0 \qquad (3.1\text{-}2)$$

and

$$\frac{\partial \sigma_y}{\partial y} + \frac{\partial \tau_{xy}}{\partial x} = 0 \qquad (3.1-3)$$

The strains in the plane of the plate are

$$\epsilon_x = \frac{\partial u}{\partial x} \qquad \epsilon_y = \frac{\partial v}{\partial y} \qquad \tau_{xy} = \frac{\partial u}{\partial y} + \frac{\partial v}{\partial x} \qquad (3.1-4)$$

Differentiation of the strains yields in the compatibility equation

$$\frac{\partial^2 \epsilon_x}{\partial y^2} + \frac{\partial^2 \epsilon_y}{\partial x^2} = \frac{\partial^2 \tau_{xy}}{\partial x \partial y} \qquad (3.1-5)$$

The two-dimensional stress-strain equations are

$$\epsilon_x = \frac{1}{E}(\sigma_x - \nu \sigma_y) + \alpha T \qquad (3.1-6)$$

$$\epsilon_y = \frac{1}{E}(\sigma_y - \nu \sigma_x) + \alpha T \qquad (3.1-7)$$

$$\gamma_{xy} = \frac{1}{G}\tau_{xy} = \frac{2(1+\nu)}{E}\tau_{xy} \qquad (3.1-8)$$

The compatibility equation (3.1-5) is expressed in terms of stress by substituting the strains from Eqs. (3.1-6), (3.1-7) and (3.1-8) into Eq. (3.1-5). The result is

$$\frac{\partial^2 \sigma_x}{\partial y^2} + \frac{\partial^2 \sigma_y}{\partial x^2} - \nu \left(\frac{\partial^2 \sigma_x}{\partial x^2} + \frac{\partial^2 \sigma_y}{\partial y^2} \right) \qquad (3.1-9)$$

$$+ E\alpha \left(\frac{\partial^2 T}{\partial x^2} + \frac{\partial^2 T}{\partial y^2} \right) = 2(1+\nu)\frac{\partial^2 \tau}{\partial x \partial y}$$

The stresses in this equation are replaced by their respective stress function derivatives, Eq. (3.1-1) and with the use of Eqs. (3.1-2) and (3.1-3) the preceding equation becomes

$$\frac{\partial^4 \varphi}{\partial x^4} + 2\frac{\partial^4 \varphi}{\partial x^2 \partial y^2} + \frac{\partial^4 \varphi}{\partial y^4} + E\alpha \left(\frac{\partial^2 T}{\partial x^2} + \frac{\partial^2 T}{\partial y^2} \right) = 0 \qquad (3.1\text{-}10)$$

The boundary conditions, for the case of plane stress, are expressed as

$$\overline{X} = \sigma_x \ell + \tau_{xy} m \qquad (3.1\text{-}11)$$

$$\overline{Y} = \tau_{xy} \ell + \sigma_y m$$

where $\overline{X}$ and $\overline{Y}$ are the vector components of the surface traction, $\overline{N}$; and ℓ and m are the direction cosines relative to the x and y axes respectively. Fig. 3.1-1 shows a surface element in equilibrium. From Fig. 3.1-1 it is clear that the boundary relations (3.1-11) can be written as

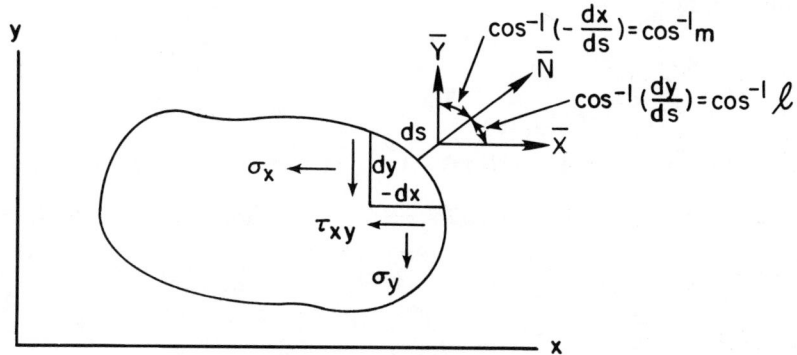

Fig. 3.1-1

$$\overline{X} = \sigma_x \frac{dy}{ds} - \tau_{xy} \frac{dx}{ds}$$

$$\overline{Y} = \tau_{xy} \frac{dy}{ds} - \sigma_y \frac{dx}{ds} \qquad (3.1\text{-}12)$$

In terms of the stress functions Eqs. (3.1-1), these equations are

$$\bar{X} = \frac{\partial^2 \varphi}{\partial y^2} \frac{dy}{ds} + \frac{\partial^2 \varphi}{\partial x \, dy} \frac{dx}{ds} = \frac{d}{ds} \left(\frac{\partial \varphi}{\partial y} \right) \qquad (3.1\text{-}13)$$

$$\bar{Y} = - \frac{\partial^2 \varphi}{\partial x \, \partial y} \frac{dy}{ds} - \frac{\partial^2 \varphi}{\partial x^2} \frac{\partial x}{\partial s} = - \frac{d}{ds} \left(\frac{\partial \varphi}{\partial x} \right)$$

When the plate is unrestrained at the boundaries, we have $\bar{X} = \bar{Y} = 0$, and from Eq. (3.1-13)

$$\frac{d}{ds} \left(\frac{\partial \varphi}{\partial x} \right) = \frac{d}{ds} \left(\frac{\partial \varphi}{\partial y} \right) = 0 \qquad (3.1\text{-}14)$$

which indicates that the first derivatives of the stress function are constants. Thus

$$\frac{\partial \varphi}{\partial x} = c_1 \qquad \frac{\partial \varphi}{\partial y} = c_2 \qquad (3.1\text{-}15)$$

Integration yields

$$\varphi = c_1 x + c_2 y + c_3 \qquad (3.1\text{-}16)$$

Since the stresses are second derivations of the stress functions the constants c_1, c_2 and c_3 can be taken as zero. The boundary conditions are then

$$\varphi = \frac{\partial \varphi}{\partial x} = \frac{\partial \varphi}{\partial y} = 0 \qquad (3.1\text{-}17)$$

on all edges. In the case of multiply connected areas such as flat rings or plates containing holes, this boundary condition is valid only at one connected boundary. At the other boundaries the stress function and its derivatives will be given as in Eqs. (3.1-15) and (3.1-16).

In the case of cylindrical geometry, Eq. (3.1-10) takes the form

$$\left(\frac{\partial^2}{\partial r^2} + \frac{1}{r}\frac{\partial}{\partial r} + \frac{\partial^2}{r^2 \partial \theta^2}\right)\left(\frac{\partial^2 \varphi}{\partial r^2} + \frac{1}{r}\frac{\partial \varphi}{\partial r} + \frac{\partial^2 \varphi}{r^2 \partial \theta^2}\right) \quad (3.1-18)$$

$$+ E\alpha\left(\frac{\partial^2 T}{\partial r^2} + \frac{1}{r}\frac{\partial T}{\partial r} + \frac{\partial^2 T}{r^2 \partial \theta^2}\right) = 0$$

The radial and tangential stresses, expressed in terms of the derivatives of the stress function, are now

$$\sigma_r = \frac{1}{r}\frac{\partial \varphi}{\partial r} + \frac{1}{r^2}\frac{\partial^2 \varphi}{\partial \theta^2} \qquad (3.1.19)$$

$$\sigma_\theta = \frac{\partial^2 \varphi}{\partial r^2} \qquad (3.1-20)$$

$$\tau_{r\theta} = \frac{1}{r^2}\frac{\partial \varphi}{\partial \theta} - \frac{1}{r}\frac{\partial^2 \varphi}{\partial r \partial \theta} = -\frac{\partial}{\partial r}\left(\frac{1}{r}\frac{\partial \varphi}{\partial \theta}\right) \qquad (3.1-21)$$

When the temperature is a function of the radius only, there will be an axially symmetrical stress and strain distribution. The stress function is then a function of the radial coordinate only and the bi-harmonic equation, Eq. (3.1-18), becomes

$$\frac{1}{r}\frac{d}{dr}\left\{r\frac{d}{dr}\left[\frac{1}{r}\frac{d}{dr}\left(r\frac{d\varphi}{dr}\right)\right]\right\} + E\alpha\frac{1}{r}\frac{d}{dr}\left(r\frac{dT}{dr}\right) = 0 \qquad (3.1-22)$$

The radial and tangential stresses are given in terms of the stress function derivatives as

$$\sigma_r = \frac{1}{r}\frac{\partial \varphi}{\partial r} \qquad \sigma_\theta = \frac{\partial^2 \varphi}{\partial r^2} \qquad \tau_{r\theta} = 0 \qquad (3.1-23)$$

With the substitution $r\frac{d\varphi}{dr} = r^2 \sigma_r$, Eq. (3.1-22) can be written as

$$\frac{1}{r}\frac{d}{dr}\left\{r\frac{d}{dr}\left[\frac{1}{r}\frac{d}{dr}(r^2\sigma_r)\right]\right\} + E\alpha\frac{1}{r}\frac{d}{dr}\left(r\frac{dT}{dr}\right) = 0 \qquad (3.1\text{-}24)$$

Integration yields

$$\sigma_r = \frac{C_1}{r^2} + C_2(-1 + 2\ln r) + C_3 - \frac{E\alpha}{r^2}\int^r Tr\,dr \qquad (3.1\text{-}25)$$

In the case of circular plate without a central hole, C_1 and C_2 are zero since otherwise the stress at $r = 0$ would be infinite. With the use of the boundary condition $\sigma_r = 0$ at the outside surface, $r = a$, we obtain

$$\sigma_r = E\alpha\left(\frac{1}{a^2}\int_0^a Tr\,dr - \frac{1}{r^2}\int_0^r Tr\,dr\right) \qquad (3.1\text{-}26)$$

This problem is treated in greater detail in Section 4.2.

When there is a central hole, the additional stress boundary condition that $\sigma_r = 0$ at the inner edge is not sufficient to determine the two additional constants C_1 and C_2, in Eq. (3.1-25). It can be shown that the requirement for single valued displacements leads to $C_2 = 0$.

Generalized Plane Stress

In plane stress, the normal and shear stresses in the plane of the plate do not vary with the thickness coordinate. That is, $\sigma_x = \sigma_x(x, y)$, $\sigma_y = \sigma_y(x, y)$ and $\tau_{xy} = \tau_{xy}(x, y)$, with σ_z, τ_{zy} and τ_{xz} taken as zero. In generalized plane stress we permit the in-plane stresses σ_x, σ_y and τ_{xy} to vary with the thickness coordinate z, such that $\sigma_x = \sigma_x(x, y, z)$, $\sigma_y = \sigma_y(x, y, z)$, $\tau_{xy} = \tau_{xy}(x, y, z)$ and impose the restriction that the flat plate remains flat; i.e., there is no bending. It is clear that there will be no bending when the temperature

PLANE STRESS

distribution and stress distribution are symmetrical about the central plane, $z = 0$. The required conditions are

$$T(x, y, z) = T(x, y, -z)$$
$$\sigma_x(x, y, z) = \sigma_x(x, y, -z)$$
$$\sigma_y(x, y, z) = \sigma_y(x, y, -z) \quad (3.1\text{-}27)$$
$$\tau_{xy}(x, y, z) = \tau_{xy}(x, y, -z)$$

In generalized plane stress the strain does not vary over the thickness of the plate. We can then integrate Eqs. (3.1-6) and (3.1-7) over the plate thickness and obtain the relationships

$$\epsilon_x = \frac{1}{E}(\bar{\sigma}_x - \nu\bar{\sigma}_y) + \alpha\bar{T} \quad (3.1\text{-}28)$$

$$\epsilon_y = \frac{1}{E}(\bar{\sigma}_y - \nu\bar{\sigma}_x) + \alpha\bar{T} \quad (3.1\text{-}29)$$

$$\gamma_{xy} = \frac{2(1+\nu)}{E}\tau_{xy} = \frac{2(1-\nu)}{E}\bar{\tau}_{xy} \quad (3.1\text{-}30)$$

where $\bar{\sigma}_x = \bar{\sigma}_x(x, y)$, $\bar{\sigma}_y = \bar{\sigma}_y(x, y)$ and $\bar{\tau}_{xy} = \bar{\tau}_{xy}(x, y)$.

The plane equilibrium equations may now be written in terms of the mean plane stresses as

$$\frac{\partial \bar{\sigma}_x}{\partial x} + \frac{\partial \bar{\tau}_{xy}}{\partial y} = 0 \qquad \frac{\partial \bar{\sigma}_y}{\partial y} + \frac{\partial \bar{\tau}_{xy}}{\partial x} = 0 \quad (3.1\text{-}31)$$

and the mean plane stresses defined in terms of the Airy stress function as

$$\bar{\sigma}_x = \frac{\partial^2 \varphi}{\partial y^2} \qquad \bar{\sigma}_y = \frac{\partial^2 \varphi}{\partial x^2} \qquad \bar{\tau}_{xy} = -\frac{\partial^2 \varphi}{dx\,dy} \quad (3.1\text{-}32)$$

The compatibility equation remains as given by Eq. (3.1-5).

The solution of a generalized plane stress problem is carried out in the same manner as a conventional plane stress problem, and the mean in-plane stresses determined. We can then obtain the distributed stresses by equating the strains in Eqs. (3.1-6) and (3.1-28), and in Eqs. (3.1-7) and (3.1-29). Thus

$$\sigma_x - \nu \sigma_y = \bar{\sigma}_x - \nu \bar{\sigma}_y + E\alpha (\bar{T} - T) \qquad (3.1\text{-}33)$$

$$\sigma_y - \nu \sigma_x = \bar{\sigma}_y - \nu \bar{\sigma}_x + E\alpha (\bar{T} - T) \qquad (3.1\text{-}34)$$

From these equations the stresses σ_x and σ_y are

$$\sigma_x = \bar{\sigma}_x + \frac{E\alpha}{1-\nu} (\bar{T} - T) \qquad (3.1\text{-}35)$$

$$\sigma_y = \bar{\sigma}_y + \frac{E\alpha}{1-\nu} (\bar{T} - T) \qquad (3.1\text{-}36)$$

and $\tau_{xy} = \bar{\tau}_{xy}$.

Pure Plane Extension

Consider the case of a free plate with a temperature distribution which is a function of the thickness coordinate only. Namely

$$T = T(z) \qquad (3.1\text{-}37)$$

This problem was discussed in Sect. 1.7 where it was indicated that the plane plate stress is uniform and omnidirectional in the plane of the plate, and varies only with the thickness coordinate z; that is, $\sigma(z) = \sigma_x(z) = \sigma_y(z)$, etc. The normal stresses in the z direction are, as before, equal to zero, but in the present problem <u>all</u> of the shear stresses are zero. As there are no shear deformations, this type of deformation is designated as <u>pure</u> extension.

We find in Sect. 1.7 that the plane strain is

$$\epsilon = \alpha \overline{T} \tag{3.1-38}$$

and that the stress distribution is

$$\sigma_z = \frac{E\alpha}{1-\nu} (\overline{T} - T(z)) \tag{3.1-39}$$

This equation is consistent with Eqs. (3.1-35) and (3.1-36), since for the case of $T = T(z)$ the mean plate stresses $\overline{\sigma}_x$ and $\overline{\sigma}_y$ are zero.

In the case of symmetrically layered plates undergoing pure extension, the strain ϵ is the same in each layer, i, and by integrating the stress-strain equations over the thickness of each layer and letting $E_i'' = \frac{E_i}{1-\nu_i}$, we obtain

$$\epsilon = \frac{\sigma_i}{E_i''} + \alpha_i T_i = \frac{\overline{\sigma}_1}{E_1''} + \alpha_1 \overline{T}_1 = \frac{\overline{\sigma}_2}{E_2''} + \alpha_2 \overline{T}_2 = \dots \text{ etc.} \tag{3.1-40}$$

In the absence of external loads we have

$$\overline{\sigma}_1 h_1 + \overline{\sigma}_2 h_2 + \dots = 0 \tag{3.1-41}$$

From Eqs. (3.1-40) and (3.1-41) we obtain

$$\epsilon = \frac{\sum\limits_1^n \frac{E_i \alpha_i T_i h_i}{1-\nu_i}}{\sum\limits_1^n \frac{E_i h_i}{1-\nu_i}} = \overline{\alpha T}' \tag{3.1-42}$$

and the distributed stress is

$$\sigma_i = E_i (\overline{\alpha T}' - \alpha_i T_i) \tag{3.1-43}$$

Eqs. (3.1-39) and (3.1-43) may be used to determine the stresses in homogeneous and layered plates when $T = T(z)$. This is done in Sect. 3.8.

3.2 PLANE STRAIN

Plane strain represents a common type of deformation of a cylindrical body (not necessarily of circular cross section) in which there is a uniform extensional strain in the axial direction and the displacements in every cross section plane are the same. The axial displacement w is then linear in z, and as u and v are functions of x and y and do not vary with z, we have the requirements for generalized plane strain that

$$\gamma_{xz} = \frac{\partial u}{\partial z} + \frac{\partial w}{\partial x} = 0$$

$$\gamma_{yz} = \frac{\partial v}{\partial z} + \frac{\partial w}{\partial y} = 0 \tag{3.2-1}$$

$$\epsilon_z = \text{const.}$$

With the use of modified elastic constants, defined as

$$E' = \frac{E}{1-\nu^2}, \quad \nu' = \frac{\nu}{1-\nu}, \quad \alpha' = \alpha(1+\nu) \tag{3.2-2}$$

the stress-strain relationships can be written as

$$\epsilon_x + \nu \epsilon_z = \frac{1}{E'} (\sigma_x - \nu' \sigma_y) + \alpha' T \qquad (3.2\text{-}3)$$

$$\epsilon_y + \nu \epsilon_z = \frac{1}{E'} (\sigma_y - \nu' \sigma_x) + \alpha' T \qquad (3.2\text{-}4)$$

$$\epsilon_z = \frac{1}{E} (\sigma_z - \nu \sigma_y - \nu \sigma_x) + \alpha T = \text{const.} \qquad (3.2\text{-}5)$$

$$\gamma_{xy} = \frac{\tau_{xy}}{G} = \frac{2(1+\nu)}{E} \tau_{xy} \qquad (3.2\text{-}6)$$

The solution of a plane strain problem is carried out in the same manner as a corresponding plane stress problem. The stresses σ_x, σ_y, and τ_{xy} are determined independently of σ_z. The appropriate axial condition with regard to axial displacement or axial load is then applied. If the elastic body under consideration is restrained from axial extension by another elastic body then the axial strain ϵ_z will be determined by the interaction of the two bodies. The more common types of plane strain problems are those in which there is complete axial restraint or no axial restraint at all. These conditions are termed zero plane strain and free plane strain respectively.

Zero Plane Strain

A cylindrical or prismatic body undergoes plane strain deformation, and is rigidly held at its extremities so that axial expansion is prevented. We then have $\epsilon_z = 0$, and Eq. (3.2-5) becomes

$$\sigma_z = \nu(\sigma_x + \sigma_y) - E\alpha T \qquad (3.2\text{-}7)$$

After the solution of the two-dimensional plane strain problem, which yields σ_x and σ_y, the axial stress, σ_z, is obtained from Eq. (3.2-7). In the solution of axisymmetric plane strain problems we find, in Sect. 4.5, that

$$\sigma_x + \sigma_y = \sigma_r + \sigma_t = \frac{E\alpha}{1-\nu}(\bar{T} - T) \tag{3.2-8}$$

The axial stress distribution, in axisymmetric problems with $\epsilon_z = 0$, is therefore

$$\sigma_z = \frac{E\alpha}{1-\nu}(\nu\bar{T} - T) \tag{3.2-9}$$

The stress-strain equations, Eqs. (3.2-3) and (3.2-4), for the case of $\epsilon_z = 0$, become

$$\epsilon_x = \frac{1}{E'}(\sigma_x - \nu'\sigma_y) + \alpha'T \tag{3.2-10}$$

$$\epsilon_y = \frac{1}{E'}(\sigma_y - \nu'\sigma_x) + \alpha'T \tag{3.2-11}$$

In the computation of strains and displacements in cross section planes, the foregoing equations show that the strains ϵ_x and ϵ_y are directly obtainable from the solution of the comparable plane stress problem by simply substituting the modified elastic constants for the true elastic constants.

Free Plane Strain

Axially unrestrained plane strain deformations are characterized by the absence of an axial load, or

$$\bar{\sigma}_z = 0 \tag{3.2-12}$$

We integrate Eq. (3.2-5) over the cross section and obtain

$$E\bar{\epsilon}_z = -\nu(\bar{\sigma}_x + \bar{\sigma}_y) + E\alpha\bar{T} \tag{3.2-13}$$

In the absence of external loads and restraints

$$\bar{\sigma}_x = \int_A \sigma_x \, dy \, dx = 0$$

$$\bar{\sigma}_y = \int_A \sigma_y \, dx \, dy = 0$$
(3.2-14)

so that Eq. (3.2-13) becomes

$$\epsilon_z = \alpha \bar{T}$$
(3.2-15)

From Eq. (3.2-5) the axial stress is

$$\sigma_z = \nu(\sigma_x + \sigma_y) + E\alpha(\bar{T} - T)$$
(3.2-16)

The solution of the plane strain problem yields σ_x and σ_y. With σ_x and σ_y known we can determine σ_z from Eq. (3.2-16). In the case of axisymmetric plane strain deformation, (from Sect. 4.5), we have

$$\sigma_x + \sigma_y = \frac{E\alpha}{1-\nu}(\bar{T} - T)$$
(3.2-17)

and the axial stress becomes

$$\sigma_z = \frac{E\alpha}{1-\nu}(\bar{T} - T)$$
(3.2-18)

The solutions of plane strain problems are carried out by using the plane stress formulation, with modified elastic constants replacing the true elastic constants. The stresses σ_x, σ_y, and τ_{xy} that are obtained are then the true stresses, and the axial stresses, σ_z, are as given either by Eq. (3.2-5), (3.2-7), (3.2-9), (3.2-16), or (3.2-18), depending upon the conditions of the problem. The cross section strains are obtained from Eqs. (2.3-3) and (2.3-4). For the case of zero plane strain we simply set

$\epsilon_z = 0$ in these equations. In the case of free plane strain, $\epsilon_z = \alpha \overline{T}$, and Eqs. (3.2-3) and (3.2-4) become

$$\epsilon_x = \frac{1}{E'}(\sigma_x - \nu' \sigma_y) + \alpha' T - \nu \alpha \overline{T} \qquad (3.2\text{-}19)$$

$$\epsilon_y = \frac{1}{E'}(\sigma_y - \nu' \sigma_x) + \alpha' T - \nu \alpha \overline{T} \qquad (3.2\text{-}20)$$

The stresses σ_x, σ_y, and τ_{xy} can be replaced by derivatives of a stress function φ. See Eqs.(3.1-1). The stresses, defined in this manner, automatically satisfy the equilibrium equations. The compatibility of strains, expressed by Eq. (3.1-5) is the same for both plane strain deformation and plane stress deformation. Eqs. (3.2-3), (3.2-4) and (3.2-6), with the stresses expressed in terms of the stress function, are set into Eq. (3.1-5). The result is the plane strain thermal biharmonic equation

$$\frac{\partial^4 \varphi}{\partial x^4} + 2 \frac{\partial^4 \varphi}{\partial x^2 \partial y^2} + \frac{\partial^4 \varphi}{\partial y^4} + \frac{E\alpha}{1-\nu}\left(\frac{\partial^2 T}{\partial x^2} + \frac{\partial^2 T}{\partial y^2}\right) = 0 \qquad (3.2\text{-}21)$$

In cylindrical coordinates, the stresses are expressed in terms of the stress function, as given by Eqs. (3.1-19), (3.1-20), and (3.1-21), and the biharmonic stress function equation takes the form

$$\left(\frac{\partial^2}{\partial r^2} + \frac{1}{r}\frac{\partial}{\partial r} + \frac{\partial^2}{r^2 \partial \theta^2}\right)\left(\frac{\partial^2 \varphi}{\partial r^2} + \frac{1}{r}\frac{\partial \varphi}{\partial r} + \frac{\partial^2 \varphi}{r^2 \partial \theta^2}\right)$$
$$+ \frac{E\alpha}{1-\nu}\left(\frac{\partial^2 T}{\partial r^2} + \frac{1}{r}\frac{\partial T}{\partial r} + \frac{\partial^2 T}{r^2 \partial \theta^2}\right) = 0 \qquad (3.2\text{-}22)$$

In the case of a circular cylindrical rod having a temperature distribution which is a function of the radius only, Eq. (3.2-22) becomes

$$\frac{1}{r}\frac{d}{dr}\left\{r\frac{d}{dr}\left[\frac{1}{r}\frac{d}{dr}\left(r\frac{d\varphi}{dr}\right)\right]\right\} + \frac{E\alpha}{1-\nu} \cdot \frac{1}{r}\frac{d}{dr}\left(r\frac{dT}{dr}\right) = 0 \qquad (3.2\text{-}23)$$

with

$$\sigma_r = \frac{1}{r}\frac{\partial \varphi}{\partial r} \qquad \sigma_\theta = \frac{\partial^2 \varphi}{dr^2} \qquad \tau_{r\theta} = 0 \qquad (3.2\text{-}24)$$

The solution, of the problem of a cylindrical rod with a radial temperature distribution, is carried out in Sect. 4.5.

3.3 STRESS-FREE TEMPERATURE DISTRIBUTIONS

Thermal stresses are produced in a body as a result of temperature distributions which do not permit the free expansion of individual elements in accordance with the local temperatures. There are certain temperature distributions which produce free expansions such that the adjacent elements do fit together without interference, and therefore without the development of stresses. To find such temperature distributions we make use of the uniqueness theorem of elasticity. This states that there is only one stress distribution that will satisfy the equations of elasticity. These equations are the stress-strain relationships, the stress equilibrium equations and the strain compatibility equations. The unique stress distribution which does not produce incompatible deformations must also satisfy the boundary conditions. The discussion which follows applies to unrestrained bodies in which boundary conditions of no surface load and no surface stress apply.

Consider a body, free from external loads and constraints, which is subjected to an unspecified temperature distribution $T = T(x, y, z)$, that will not cause stresses to be developed. When the stresses are zero, the stress-strain relationships, Eqs. (1.1-1), become

$$\epsilon_x = \alpha T \qquad \epsilon_y = \alpha T \qquad \epsilon_z = \alpha T$$
$$\gamma_{xy} = \gamma_{xz} = \gamma_{yz} = 0 \qquad (3.3\text{-}1)$$

In rectangular coordinates the strains are

$$\epsilon_x = \frac{\partial u}{\partial x} \qquad \epsilon_y = \frac{\partial v}{\partial y} \qquad \epsilon_z = \frac{\partial w}{\partial z}$$

$$\gamma_{xy} = \frac{\partial u}{\partial y} + \frac{\partial v}{\partial x} \qquad \gamma_{xz} = \frac{\partial u}{\partial z} + \frac{\partial w}{\partial x} \qquad \gamma_{yz} = \frac{\partial v}{\partial z} + \frac{\partial w}{\partial y}$$

(3.3-2)

By differentiation, the foregoing relationships yield the compatibility equations

$$\frac{\partial^2 \epsilon_y}{\partial x^2} + \frac{\partial^2 \epsilon_x}{\partial y^2} = \frac{\partial^2 \gamma_{xy}}{\partial x \partial y}$$

$$\frac{\partial^2 \epsilon_z}{\partial x^2} + \frac{\partial^2 \epsilon_x}{\partial z^2} = \frac{\partial^2 \gamma_{xz}}{\partial x \partial z} \qquad (3.3\text{-}3)$$

$$\frac{\partial^2 \epsilon_z}{\partial y^2} + \frac{\partial^2 \epsilon_y}{\partial z^2} = \frac{\partial^2 \gamma_{yz}}{\partial y \partial z}$$

Substitution of the strains from Eqs. (3.3-1) into the above compatibility equations gives

$$\frac{\partial^2 T}{\partial x^2} + \frac{\partial^2 T}{\partial y^2} = 0 \qquad \frac{\partial^2 T}{\partial x^2} + \frac{\partial^2 T}{\partial z^2} = 0 \qquad \frac{\partial^2 T}{\partial y^2} + \frac{\partial^2 T}{\partial z^2} = 0 \qquad (3.3\text{-}4)$$

A temperature distribution which satisfies these equations will therefore generate compatible determinations, and will not produce thermal stresses in an unrestrained body. The temperature must therefore be a function which satisfies the plane harmonic equations in each of the three orthogonal planes.

We consider five distinct classes of temperature distributions that produce stress-free and partially stress-free thermal deformations.

Class I: Statically Determinate Structures

Stresses in certain types of structures, notably statically determinate pin-connected trusses, can be determined by the use of load equilibrium conditions alone. This signifies that regardless of the deformation of the structure or the individual structural members, the stresses are not changed. The lengthening and shortening of the members have no effect on the stresses because the structure can deform to accommodate the dimensional changes. Strain compatibility thus always exists regardless of the temperatures of the individual members.

If we examine the two-bar frame shown in Fig. 1-8-8, it is noted that a lengthening or shortening of the bars will not cause stresses to be developed. The solution of the two-bar pinned frame, as carried out in Example (h) of Section 1.8, does indeed show that when the bars are heated to different temperatures, no stresses are produced.

The fact that thermal elongation of bars in a statically determinate truss does not produce stresses, can be demonstrated by analyzing the structure shown in Fig. 3.3-1. The analysis is performed by means of equivalent thermal loads, which in the present problem take the form of surface tractions

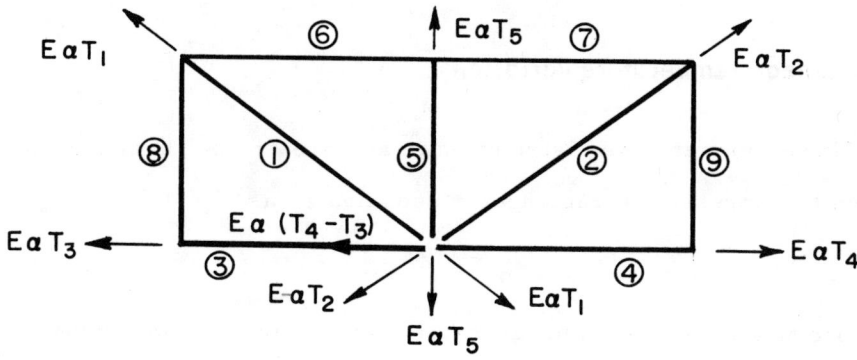

Fig. 3.3-1

applied at the joints. The diagonals are at temperatures T_1 and T_2, the lower chord members at temperatures T_3 and T_4, and the central vertical at a temperature, T_5. The surface tractions, for bars of unit cross section area, shown in Fig. 3.3-1, produce displacement stresses equal in magnitude to those surface tractions at the extremities of the bars that are in line with the bars. The displacement stresses in each of the bars are:

$$\sigma_1' = E_1 \alpha_1 T_1 \quad \sigma_2' = E_2 \alpha_2 T_2 \quad \sigma_3' = E_3 \alpha_3 T_3 \quad \sigma_4' = E_4 \alpha_4 T_4 \qquad (3.3\text{-}5)$$

$$\sigma_5' = E_5 \alpha_5 T_5 \quad \sigma_6' = \sigma_7' = \sigma_8' = \sigma_9' = 0 \qquad (3.3\text{-}6)$$

The subscripts 6, 7, 8, and 9 refer to the two outside vertical members and the two upper chord members which are not heated. The net thermal stress in a member is

$$\sigma_i = \sigma_i' - E_i \alpha_i T_i = 0 \qquad (3.3\text{-}7)$$

It is clear, from the foregoing demonstration that the thermal displacement stresses and internal pressures in the members of statically determinate frames and trusses are always self-equilibrating, so that the thermal stresses are zero.

Class 2: Linear Temperature Variation

A linear temperature distribution in an unrestrained homogeneous body, which can be expressed in rectangular coordinates as

$$T = T_0 + T_1 x + T_2 y + T_3 z \qquad (3.3\text{-}8)$$

will not produce stresses. This is readily ascertained by observing that this linear temperature variation will satisfy the three plane Laplacian equations,

Eqs. (3.3-4), which are the required conditions for the absence of thermal stresses in an unrestrained body. The linear temperature distribution, Eq. (3.3-8), will produce stress-free deformations in both simply and multiply connected homogeneous bodies. It should be recognized that in order for this condition to be fulfilled, the linear temperature distribution must be continuous over the full extent of the body. Discontinuous linear temperature distributions such as that shown in Fig. 3.5-2 will produce stresses.

One should not fail to recognize linear temperature distributions expressed in other than rectangular coordinates. A two-dimensional temperature variation can be expressed in cylindrical coordinates as

$$T = (T_a \cos \theta + T_b \sin \theta) r \qquad (3.3-9)$$

Consider, for example, a circular cylindrical tube having a cross section temperature distribution expressed as

$$T = \frac{T_o}{2} \left(1 - \frac{r \sin \theta}{R}\right) \qquad (3.3-10)$$

In rectangular coordinates this temperature distribution, shown in Fig. 3.3-2, is

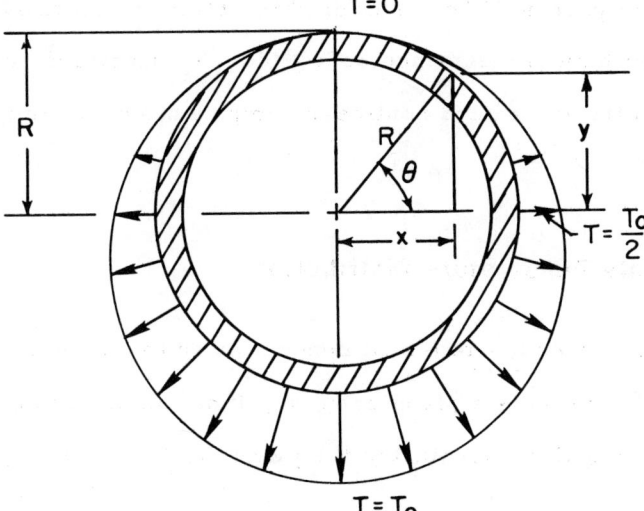

Fig. 3.3-2

$$T = \frac{T_o}{2}\left(1 - \frac{y}{R}\right) \qquad (3.3\text{-}11)$$

This is clearly recognized as a linear stress-free temperature distribution.

Linear temperature distributions are the result of steady-state heat conduction through bars, beams, and plates. For example, in the case of unrestrained bars or plates, a transverse linear temperature distribution expressed as $T = T_o y/h$, produces no stresses, but deforms the bar or the plate, giving it a constant curvature

$$K = \frac{1}{R} = \frac{\alpha T_o}{h} \qquad (3.3\text{-}12)$$

This is shown, for the case of a bar, in **Example (e)** of Section 1.8.

The fact that a linear stress distribution does not produce stresses in an unrestrained body can be used as an analytical expedient in the solution of certain types of thermal stress problems, since only the portion of the temperature that deviates from the linear is effective in producing stress. It is sometimes possible to simplify a problem by expressing a temperature distribution as a combination of a linear and nonlinear temperature variation. Only the nonlinear part need be considered to obtain the thermal stresses, if we are dealing with an unrestrained body. The deformations, however, are produced by both the linear and nonlinear components of the temperature distribution.

Class 3: Flat Plate Temperature Distributions

In the case of steady-state heat conduction in the plane of a plate the temperature distribution is a plane harmonic function satisfying the Laplacian equation. In rectangular coordinates the plane Laplacian is

$$\frac{\partial^2 T}{\partial x^2} + \frac{\partial^2 T}{\partial y^2} = 0 \tag{3.3-13}$$

and in cylindrical coordinates it is expressed as

$$\frac{\partial^2 T}{r^2 \partial \theta^2} + \frac{1}{r}\frac{\partial}{\partial r}\left(r \frac{\partial T}{\partial r}\right) = 0 \tag{3.3-14}$$

The plane thermal stress biharmonic equations, Eqs. (3.1-10) and (3.1-18) are then independent of temperature, and they become

$$\frac{\partial^4 \varphi}{\partial x^4} + 2 \frac{\partial^4 \varphi}{\partial x^2 \partial y^2} + \frac{\partial^4 \varphi}{\partial y^4} = 0 \tag{3.3-15}$$

and

$$\left[\frac{\partial^2}{r^2 \partial \theta^2} + \frac{1}{r}\frac{\partial}{\partial r}\left(r \frac{\partial}{\partial r}\right)\right]\left[\frac{\partial^2 \varphi}{r^2 \partial \theta^2} + \frac{1}{r}\frac{\partial}{\partial r}\left(r \frac{\partial \varphi}{\partial r}\right)\right] = 0 \tag{3.3-16}$$

In a simply connected body such as a plate without holes, and without external restraints, only one solution of Eq. (3.3-15) or (3.3-16) is possible; namely,

$$\sigma_x = \sigma_y = \tau_{xy} = 0 \tag{3.3-17}$$

It should also be observed that the derivation of Eq. (3.3-15) is based on the compatibility equation,

$$\frac{\partial^2 \epsilon_y}{\partial x^2} + \frac{\partial^2 \epsilon_x}{\partial y^2} = \frac{\partial^2 \gamma_{xy}}{\partial x \partial y} \tag{3.3-18}$$

For the case of zero stress, with $\epsilon_x = \epsilon_y = \alpha T$ and $\gamma_{xy} = 0$, the foregoing equation yields the first of Eqs. (3.3-4), which are the requirements for a stress-free temperature distribution. Namely,

$$\frac{\partial^2 T}{\partial x^2} + \frac{\partial^2 T}{\partial y^2} = 0 \qquad (3.3\text{-}19)$$

In order to satisfy rigorously the conditions for the absence of stress, each of the three plane temperature Laplacians must be zero, in accordance with Eqs. (3.3-4). A temperature distribution, $T = T(x, y)$, which is a plane harmonic function in x and y, satisfying Eq. (3.3-19), will also require, from the second and third of Eqs. (3.3-4) that

$$\frac{\partial^2 T}{\partial x^2} = 0 \qquad \frac{\partial^2 T}{\partial y^2} = 0 \qquad (3.3\text{-}20)$$

Thus the rigorous conditions for zero stress are:

$$\frac{\partial^2 T}{\partial x^2} = \frac{\partial^2 T}{\partial y^2} = \frac{\partial^2 T}{\partial z^2} = 0 \qquad (3.3\text{-}21)$$

and these are satisfied only by a linear temperature distribution such as given by Eq. (3.3-8).

When the temperature changes slowly with x and y, the incompatibility resulting from

$$\frac{\partial^2 T}{\partial x^2} \neq 0 \qquad \frac{\partial^2 T}{\partial y^2} \neq 0 \qquad (3.3\text{-}22)$$

results in presence of normal transverse stresses whose magnitude is of the order

$$\sigma_z \approx \frac{E\alpha}{2} (\Delta T)_h \approx \frac{E\alpha}{2} \left(\frac{\partial T}{\partial s} \right) h \qquad (3.3\text{-}23)$$

where s is any coordinate in the plane of the plate and $(\Delta T)_h$ is a monotonic temperature change in the x, y plane of the plate that takes place in a distance h, the latter being the thickness of the plate. Thus, in steady state plane heat conduction, only rapid rates of change in the temperature distribution will produce significant transverse thermal stresses.

Consider, for example, a temperature distribution in a plate of length L and width L which is expressed as

$$T = T_o \, e^{\frac{\pi}{L}(x-L)} \sin \frac{\pi y}{L} \qquad (3.3-24)$$

This distribution satisfies the plane Laplacian

$$\frac{\partial^2 T}{\partial x^2} + \frac{\partial^2 T}{\partial y^2} = 0 \qquad (3.3-25)$$

The most rapid rate of temperature change in the x direction occurs at $x = L$, $y = \frac{L}{2}$, and the most rapid rate of temperature change in the y direction occurs at $x = L$, $y = 0$.

At $x = L$ and $y = L/2$ we have

$$(\Delta T)_{h,x} = \frac{\partial T}{\partial x} \cdot h = T_o \frac{\pi h}{L} \qquad (3.3-26)$$

and at $x = L$, $y = 0$, we have

$$(\Delta T)_{h,y} = \frac{\partial T}{\partial y} \cdot h = T_o \frac{\pi h}{L} \qquad (3.3-26)(a)$$

The transverse stress, in accordance with Eq. (3.3-23) is therefore of the order

$$\sigma_z \approx \frac{\pi h}{2L} E\alpha T_o \qquad (3.3-27)$$

When h/L is small the transverse stress will be small, and the conclusion that $\sigma_x = \sigma_y = \tau_{xy} = 0$ will be reasonably accurate.

Class 4: Plane Harmonic Cross Section Temperature Distributions

It can be demonstrated that in the case of axial extensional deformations and in the bending deformations of solid bars, cylinders, or beams, a temperature distribution, $T(x, y)$, which satisfies the cross section Laplacian,

$$\frac{\partial^2 T}{\partial x^2} + \frac{\partial^2 T}{\partial y^2} = 0 \tag{3.3-28}$$

x and y being the cross section coordinates, will produce only normal axial stresses.

Assume that axial stresses, σ_z, are present and that these stresses can be computed on a one-dimensional basis, assuming that

$$\sigma_x = \sigma_y = \tau_{xy} = \tau_{xz} = \tau_{yz} = 0 \tag{3.3-29}$$

From the one-dimensional analysis of beams having a temperature distribution which is a function of the cross section coordinates, we find in Chapter 6, that when we assume the absence of all but the normal axial stresses, σ_z, we obtain

$$\sigma_z = E\alpha \left(\bar{T} + \frac{M_{tx} y}{I_x} + \frac{M_{ty} x}{I_y} - T \right) \tag{3.3-30}$$

The stress-strain equations, with only σ_z non-zero, are

$$\epsilon_x = \frac{1}{E}(-\nu \sigma_z) + \alpha T(x, y) \tag{3.3-31}$$

$$\epsilon_y = \frac{1}{E}(-\nu \sigma_z) + \alpha T(x, y) \tag{3.3-32}$$

$$\epsilon_z = \frac{1}{E}(\sigma_z) + \alpha T(x, y) \tag{3.3-33}$$

When Eq. (3.3-30) is set into Eqs. (3.3-31) (3.3-32) and (3.3-33), we obtain

$$\epsilon_x = \epsilon_y = \alpha\left[T - \nu\left(\overline{T} + \frac{M_{tx}y}{I_x} + \frac{M_{ty}x}{I_y} - T\right)\right] \tag{3.3-34}$$

$$\epsilon_z = \alpha\left(\overline{T} + \frac{M_{tx}y}{I_x} + \frac{M_{ty}x}{I_y}\right) \tag{3.3-35}$$

The strain compatibility equations, Eqs. (3.3-3), for the present case where $\gamma_{xy} = \gamma_{xz} = \gamma_{yz} = 0$ become

$$\frac{\partial^2 \epsilon_x}{\partial y^2} + \frac{\partial^2 \epsilon_y}{\partial x^2} = 0$$

$$\frac{\partial^2 \epsilon_x}{\partial z^2} + \frac{\partial^2 \epsilon_z}{\partial x^2} = 0 \tag{3.3-36}$$

$$\frac{\partial^2 \epsilon_z}{\partial y^2} + \frac{\partial^2 \epsilon_y}{\partial z^2} = 0$$

Substitution of ϵ_x and ϵ_y from Eqs. (3.3-31) and (3.3-32) into the first of the foregoing equations yields

$$\frac{\partial^2 T}{\partial x^2} + \frac{\partial^2 T}{\partial y^2} = 0 \tag{3.3-37}$$

The other two of Eqs. (3.3-36) are identically zero. Thus we have shown that when we have an axial normal stress distribution of the form given by Eq. (3.3-30), and the temperature distribution satisfies the Laplacian in the cross section plane of a bar or beam, the compatibility requirements are satisfied. And as the stress distribution, Eq. (3.3-30), satisfies the pertinent equilibrium equation

$$\frac{\partial \sigma_z}{\partial z} + \frac{\partial \tau_{xz}}{\partial x} + \frac{\partial \tau_{yz}}{\partial y} = 0 \tag{3.3-38}$$

and the non-trivial boundary condition

$$\int_A \sigma_z \, dz = 0 \tag{3.3-39}$$

it follows that in problems of extensional and bending deformation of bars, beams, and other simply-connected cylindrical structural elements, a one-dimensional analysis for the determination of the axial normal stress will represent a relatively rigorous solution of the problem, provided the temperature is a plane harmonic function of the cross section coordinates.

Class 5: Multiply Connected Plates and Cross Sections

In the discussion of Class 3 and Class 4 stress-free temperature distributions, it was shown that a temperature distribution that satisfied the two-dimensional Laplacian in the plane of an unrestrained plate, or over the cross section of an unrestrained cylinder, will produce no stresses in the plane of the plate or in a cross section plane of the cylinder. This can be shown to be valid as long as the plane or cross section is simply connected. In radial conduction through the wall of a tube, for example, $\nabla^2 T = 0$ is satisfied within the tube body, but there is a net heat flux at the surfaces. This indicates the equivalent of the presence of a heat source or heat sink within the central cavity.

In a multiply connected plate it is not sufficient that $\nabla^2 T = 0$. Discontinuities (or dislocations) can be developed in a multiply connected plate

STRESS-FREE TEMPERATURE DISTRIBUTIONS

when it is transformed into a simply-connected plate by cutting through from the inner to the outer boundaries. It can be shown that discontinuities will not exist at the adjacent surfaces, where the cut is made, if the following (Mitchell) conditions are satisfied:

$$\int_B \frac{\partial T}{\partial n} ds = 0 \qquad (3.3-40)$$

$$\int_B (y \frac{\partial T}{\partial n} - x \frac{\partial T}{\partial s}) ds = 0 \qquad (3.3-41)$$

$$\int_B (y \frac{\partial T}{\partial s} + x \frac{\partial T}{\partial n}) ds = 0 \qquad (3.3-42)$$

In the foregoing equations B represents all continuous boundaries. Eq. (3.3-40) is the condition of no heat flux in or out of a boundary surface.

Consider, for example, a circular plate with a central hole which is subjected to a plane harmonic temperature distribution

$$T = T_o r^n \cos n\theta \qquad (3.3-43)$$

which satisfies the plane Laplacian, Eq. (3.5-14). When Eq. (3.3-43) is set into Eq. (3.3-40) we obtain

$$\int_B \frac{\partial T}{\partial r} r d\theta = T_o n r^n \int_0^{2\pi} \cos n\theta \, d\theta = 0 \qquad (3.3-44)$$

We note that $x = r \cos \theta$ and $y = r \sin \theta$, so that Eqs. (3.3-41) and (3.3-42) may be written as

$$\int_0^{2\pi} (\sin \theta \frac{\partial T}{\partial r} - \cos \theta \frac{\partial T}{r \partial \theta}) r^2 d\theta = 0 \qquad (3.3-45)$$

$$\int_0^{2\pi} (\cos \theta \frac{\partial T}{\partial r} - \sin \theta \frac{\partial T}{r \partial \theta}) r^2 d\theta = 0 \qquad (3.3-46)$$

The temperature distribution, Eq. (3.3-43) is set into the above, and we obtain

$$T_o \, n \, r^{n+1} \int_0^{2\pi} (\sin\theta \cos n\theta + \cos\theta \sin n\theta) \, d\theta \equiv 0 \quad (3.3\text{-}47)$$

and

$$T_o \, n \, r^{n+1} \int_0^{2\pi} (\cos\theta \cos n\theta - \sin\theta \sin n\theta) \, d\theta \equiv 0 \quad (3.3\text{-}48)$$

Thus, we observe that a circular plate with a central hole which is subjected to a plane harmonic temperature distribution, $T = T_o \, r^n \cos n\theta$, is stress-free, as all of the required (Mitchell) boundary conditions are satisfied. Similarly, a multiply connected cylindrical rod undergoing plane strain deformation, will have stress-free cross section planes if the Mitchell conditions are satisfied, although axial stresses will be present.

Summary

(1) Thermal stresses will not be developed in statically determinate trusses.

(2) Linear temperature distributions produce stress-free deformations in both simply-connected and multiply-connected unrestrained bodies. Discontinuous linear temperatures produce stresses.

(3) Temperature distributions which satisfy the two-dimensional Laplacian do not produce plane stresses in simply connected plates.

(4) Temperature distributions which satisfy the two-dimensional cross section Laplacian do not produce stresses in the cross section plane of unrestrained simply connected bars and beams subject to plane strain or bending deformations. Only axial stresses are produced in these elements.

PLANE STRESSES IN PLATE STRIPS[1]   155

(5) Stress-free temperature distributions in multiply connected bodies, which undergo plane stress or plane strain deformation, are those which satisfy the plane Laplacian equations and also the Mitchell conditions. In multiply connected bodies subjected to plane strain, in which all of these conditions are satisfied, the stresses in the cross-sectional planes are zero, but axial stresses are present.

3.4 PLANE STRESSES IN PLATE STRIPS[1]

A long plate strip or a bar, in which the height H is much greater than

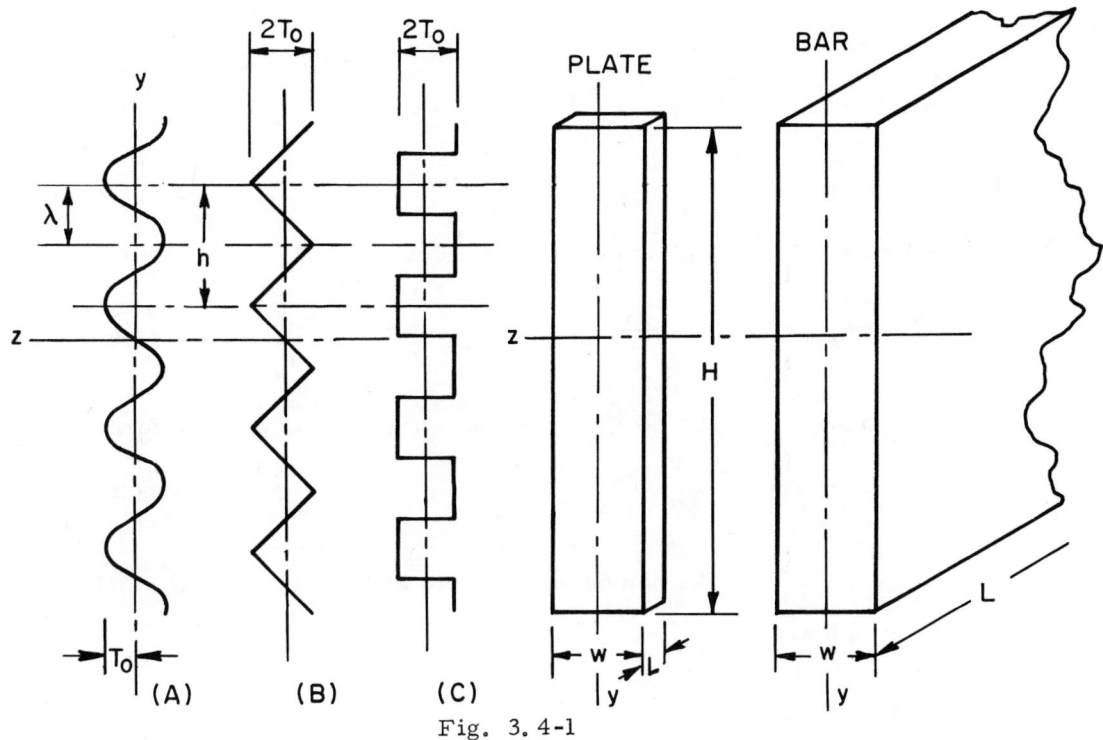

Fig. 3.4-1

---
[1] Den Hartog, J.P., J. of the Franklin Institute, Vol. 222, p. 149, (1936).

the width w, is subjected to a sinusoidal temperature variation as shown in (A) of Fig. 3.4-1. A plane stress analysis is carried out for a plate strip, $L \ll w$, but applies as well to the plane strain deformation of a bar, in which $L \gg w$. In applying the analysis which follows to a rectangular bar, the only modification that is necessary is the substitution of $E/(1-\nu)$ in each place where E appears in the plate solution.

In accordance with Eq. (3.1-10), the biharmonic equation for plane thermal stress is,

$$\frac{\partial^4 \varphi}{\partial z^4} + 2 \frac{\partial^4 \varphi}{\partial z^2 \partial y^2} + \frac{\partial^4 \varphi}{\partial y^4} + E\alpha \left(\frac{\partial^2 T}{\partial z^2} + \frac{\partial^2 T}{\partial y^2} \right) = 0 \qquad (3.4\text{-}1)$$

The temperature is specified as

$$T = T_o \sin \frac{\pi y}{\lambda} \qquad (3.4\text{-}2)$$

The particular solution of Eq. (3.4-1) is taken as

$$\varphi_p = C \sin \frac{\pi y}{\lambda} \qquad (3.4\text{-}3)$$

Substitution into Eq. (3.4-1) yields the particular solution as

$$\varphi_p = \frac{\lambda^2}{\pi^2} E\alpha T_o \sin \frac{\pi y}{\lambda} \qquad (3.4\text{-}4)$$

We take the general solution (of the homogeneous equation) in the form

$$\varphi_g = F(z) \sin \frac{\pi y}{\lambda} \qquad (3.4\text{-}5)$$

Setting Eq. (3.4-5) into the homogeneous form of Eq. (3.4-1), in which $\nabla^2 T = 0$, we obtain

$$F'''' - \frac{2\pi^2}{\lambda^2} F'' + \frac{\pi^4}{\lambda^4} F = 0 \qquad (3.4\text{-}6)$$

PLANE STRESSES IN PLATE STRIPS

As the stress distribution is symmetrical about the y axis, φ and F will be symmetrical about this axis; that is, it will be an even function of z. The solution of Eq. (3.4-6) then contains only two constants of integration, and is expressed as

$$F = C_1 z \sinh \frac{\pi z}{\lambda} + C_2 \cosh \frac{\pi z}{\lambda} \qquad (3.4\text{-}7)$$

The stress function is therefore

$$\varphi = \varphi_p + \varphi_g = (\frac{\lambda^2}{\pi^2} E\alpha T_o + C_1 z \sinh \frac{\pi z}{\lambda} + C_2 \cosh \frac{\pi z}{\lambda}) \sin \frac{\pi y}{\lambda} \qquad (3.4\text{-}8)$$

and the stresses, defined in terms of the stress function as

$$\sigma_y = \frac{\partial^2 \varphi}{\partial z^2} \qquad \sigma_z = \frac{\partial^2 \varphi}{\partial y^2} \qquad \tau_{xy} = -\frac{\partial^2 \varphi}{\partial y \partial z} \qquad (3.4\text{-}9)$$

become

$$\sigma_y = \frac{\pi^2}{\lambda^2}\left[C_1(z \sinh \frac{\pi z}{\lambda} + \frac{2\lambda}{\pi} \cosh \frac{\pi z}{\lambda}) + C_2 \cosh \frac{\pi z}{\lambda} \right] \sin \frac{\pi y}{\lambda} \qquad (3.4\text{-}10)$$

$$\sigma_z = -\frac{\pi^2}{\lambda^2}\left[\frac{\lambda^2}{\pi^2} E\alpha T_o + C_1 z \sinh \frac{\pi z}{\lambda} + C_2 \cosh \frac{\pi z}{\lambda} \right] \sin \frac{\pi y}{\lambda} \qquad (3.4\text{-}11)$$

$$\tau_{yz} = -\frac{\pi^2}{\lambda^2}\left[C_1(z \cosh \frac{\pi z}{\lambda} + \frac{\lambda}{\pi} \sinh \frac{\pi z}{\lambda}) + C_2 \sinh \frac{\pi z}{\lambda} \right] \cos \frac{\pi y}{\lambda} \qquad (3.4\text{-}12)$$

With the use of the boundary conditions, $\sigma_z = \tau_{yz} = 0$ at $z = \pm \frac{w}{2}$ we obtain the constants C_1 and C_2 as

$$C_1 = \frac{\frac{\lambda^2}{\pi^2} E\alpha T_o \sinh \frac{\pi w}{2\lambda}}{\frac{w}{2} + \frac{\lambda}{\pi}(\sinh \frac{\pi w}{2\lambda})(\cosh \frac{\pi w}{2\lambda})} \qquad (3.4\text{-}13)$$

$$C_2 = - \frac{\frac{\lambda^2}{\pi^2} E\alpha T_o (\frac{w}{2} \cosh \frac{\pi w}{2\lambda} + \sinh \frac{\pi w}{2\lambda})}{\frac{w}{2} + \frac{\lambda}{\pi} (\sinh \frac{\pi w}{2\lambda})(\cosh \frac{\pi w}{2\lambda})} \qquad (3.4\text{-}14)$$

Substitution of C_1 and C_2 into Eqs. (3.4-10), (3.4-11) and (3.4-12), and using the notation $h = 2\lambda$, we obtain

$$\sigma_y = \frac{E\alpha T_o (\sin \frac{2\pi y}{h}) \left[(\sinh \frac{\pi w}{h})(\frac{2\pi z}{h} \sinh \frac{2\pi z}{h}) + (\sinh \frac{\pi w}{h} - \frac{\pi w}{h} \cosh \frac{\pi w}{h})(\cosh \frac{2\pi z}{h}) \right]}{\frac{\pi w}{h} + (\sinh \frac{\pi w}{h})(\cosh \frac{\pi w}{h})}$$

$$(3.4\text{-}15)$$

$$\sigma_z = -E\alpha T_o (\sin \frac{2\pi y}{h}) \left[1 + \frac{(\sinh \frac{\pi w}{h})(\frac{2\pi z}{h} \sinh \frac{2\pi z}{h}) - (\sinh \frac{\pi w}{h} + \frac{\pi w}{h} \cosh \frac{\pi w}{h})(\cosh \frac{2\pi z}{h})}{\frac{\pi w}{h} + (\sinh \frac{\pi w}{h})(\cosh \frac{\pi w}{h})} \right]$$

$$(3.4\text{-}16)$$

$$\tau_{yz} = - \frac{E\alpha T_o (\cos \frac{2\pi y}{h}) \left[(\sinh \frac{\pi w}{h})(\frac{2\pi z}{h} \cosh \frac{2\pi z}{h}) - (\frac{\pi w}{h} \cosh \frac{\pi w}{h})(\sinh \frac{2\pi z}{h}) \right]}{\frac{\pi w}{h} + (\sinh \frac{\pi w}{h})(\cosh \frac{\pi w}{h})}$$

$$(3.4\text{-}17)$$

The stress distributions along the lines A, B, C, are plotted in Fig. 3.4-2 for the case of small ratios of temperature wave length to plate width, i.e., $h \ll w$. The peak normal stresses $\sigma_y = \pm E\alpha T_o$ are obtained in the lengthwise direction at the high and low temperature points, at the edges of the plate, that is, at $y = \pm \frac{h}{4}$, $z = \pm \frac{w}{2}$, and peak stresses $\sigma_z = \pm E\alpha T_o$, in the direction of the plate width are obtained along lines B and C at a distance, z, from the plate edge equal to approximately the temperature wave length, h. In this distance the shear stresses and the normal stresses in the lengthwise, y, direction have more or less disappeared. This is to be expected since the thermal surface tractions are self-equilibrating, and have little effect at distances from the edge of the plate, greater than h.

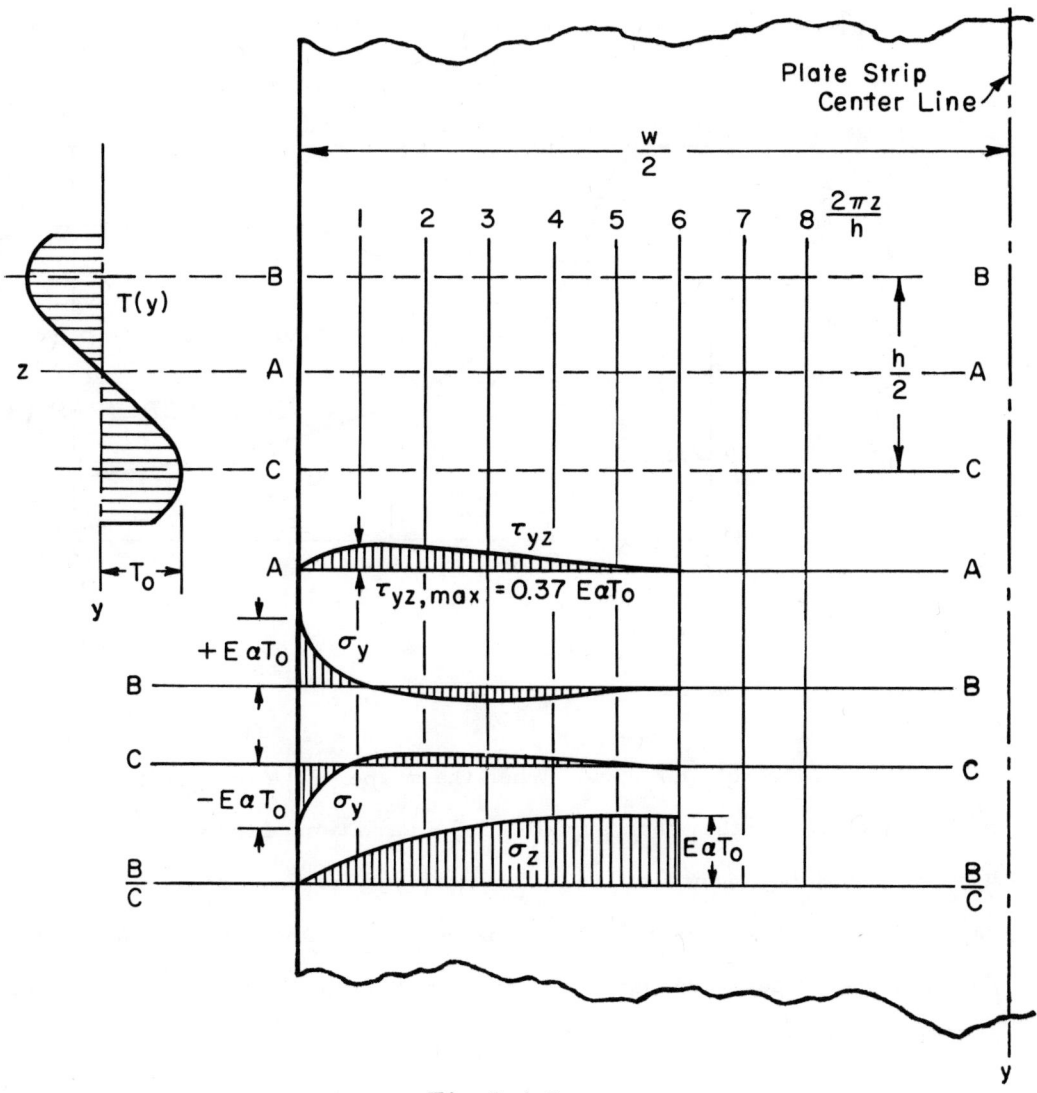

Fig. 3.4-2

The stress distributions shown in Fig. 3.4-2 apply to plate strips (or bars) having lengthwise temperature periods which are small in comparison to the plate width. Under these conditions the peak normal lengthwise stress is, $\sigma_y = E\alpha T_o$. However as the plate width w becomes small in comparison to the length of a temperature oscillation, this stress becomes small. Curve A of Fig. 3.4-3 shows the variation of peak normal stress, σ_y

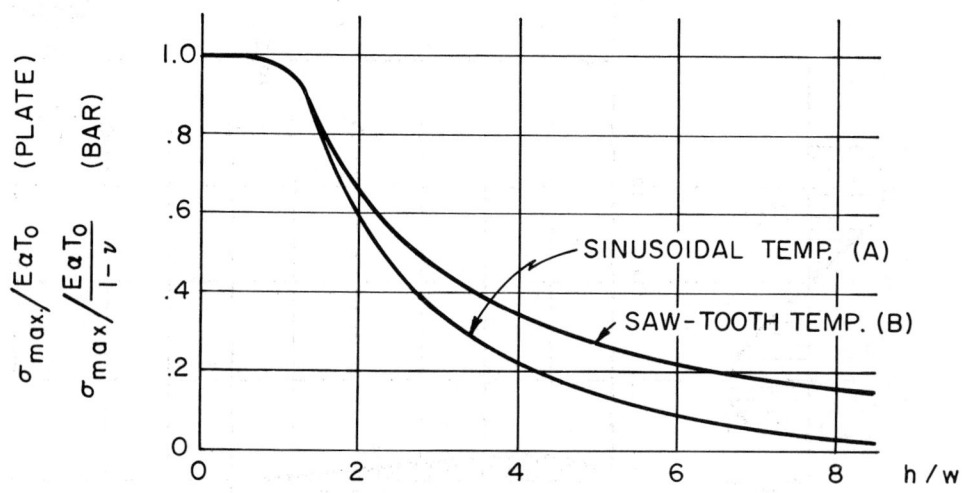

Fig. 3.4-3

with change in the ratio of h/w. When the shape of the temperature fluctuation is saw-tooth rather than sinusoidal the variation of the peak stress with h/w is seen to be fairly similar to that obtained for the sinsoidal temperature variation. At low values of h/w, denoting short temperature periods, the shape of the temperature wave does not affect the peak stress, which remains at $E\alpha T_o$ for a plate strip and $E\alpha T_o/(1-\nu)$ for a bar. It should be recognized that in the foregoing analysis the temperature T_o represents one half of the temperature range.

3.5 RECTANGULAR PLATES AND BARS

In the analysis of rectangular plates or bars with a temperature distribution in the form of a single temperature reversal such as shown in Fig. 3.5-1, we can utilize the solution given in Sect. 3.4 of the plate strip subjected to an oscillating temperature. We cut the plate shown in Fig. 3.4-2, along line C, and extend the plate above line B a distance h/2 so that we obtain plates such as those shown in Fig. 3.5-1 (a) or (b). The height of the plate

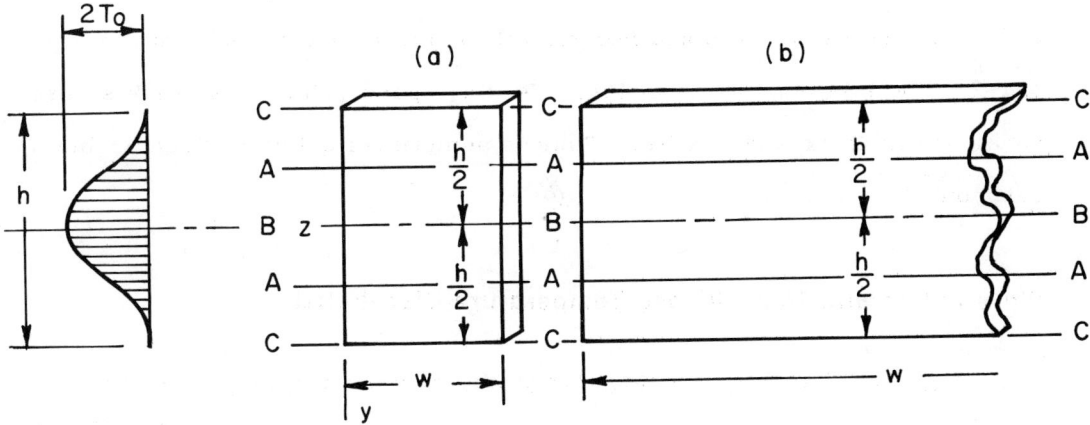

Fig. 3.5-1

and the wave length are now both equal to h.

Before the cutting of the plate along line C, transverse stresses σ_y were present along this line, such as those shown in Fig. 3.4-2. As there is now a free surface at C, the boundary conditions are satisfied by applying along line C stresses equal in magnitude to those shown in Fig. 3.4-2, but opposite in sign. These must be self-equilibrating stresses since the upper and lower boundary edges of the plate, $y = \pm h/2$, are unloaded. Their effect in generating stresses at internal points such as points B, where the normal stress would be a maximum, is small, and can be neglected. We can therefore conclude that the analysis of plate strips subjected to a periodic temperature variation, applies reasonably well to a finite plate with a single temperature reversal, such as that shown in Fig. 3.5-1.

In the case of the long plate strip, such as that shown in Fig. 3.5-1(b), it is apparent that the single temperature reversal will produce, in the vicinity of the end of plate strip, a stress distribution such as that shown in Fig. 3.4-2. We can consider the plate strip to be a thin-walled bar with a sinusoidal transverse temperature distribution, and conclude that the normal and shear stress build-up and fall-off will be more or less as shown in Fig. 3.4-2. The stress patterns in Fig. 3.4-2 are good indications of the stress development at the end of a bar. This is sometimes called an "end problem" solution.

Plate or Bar with Tent-Shaped Temperature Distribution[1]

Fig. 3.5-2 shows a rectangular plate, or rectangular bar, with a

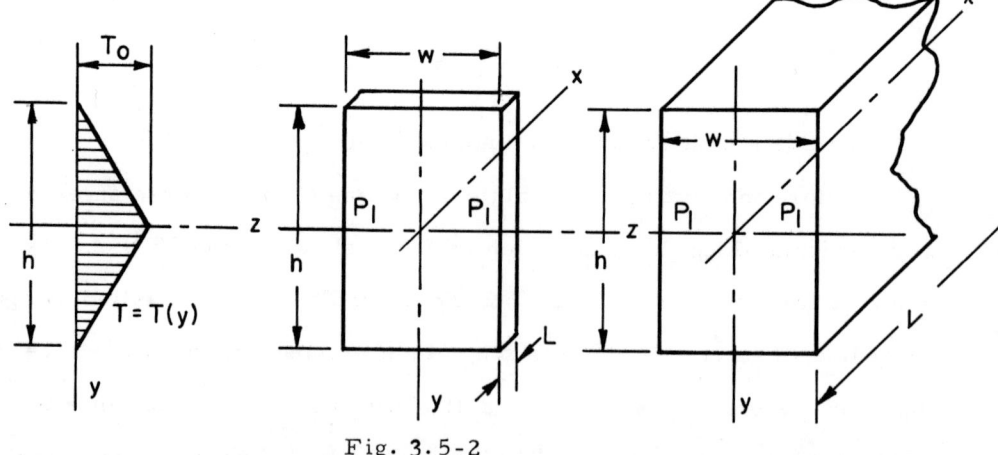

Fig. 3.5-2

---

[1] Den Hartog, J.P., Journal of the Franklin Institute, Vol. 222, p. 149, (1936).

symmetrical tent-shaped temperature variation. As in the solution of the infinite plate strip with a periodic saw-tooth temperature variation, peak transverse stresses, σ_y, are obtained at points P_1. The normal stresses, σ_y, are plotted in Fig. 3.5-3 as a function of h/w. Note that T_o is a full temperature range rather than the temperature amplitude, as was the case in the preceding problem.

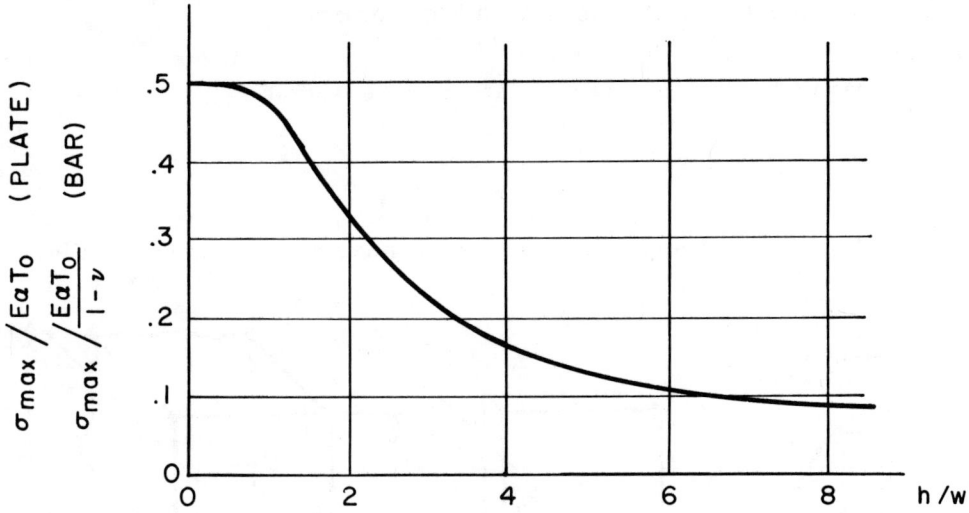

Fig. 3.5-3

The peak normal stress that is obtainable in the present problem is $E\alpha T_o/2$ for plates and $E\alpha T_o/2(1-\nu)$ for bars. In plates and bars having cross section dimensions such that $h/w < 1$ this will be the peak stress.

When $h/w \ll 1$ we can view the plate shown in Fig. 3.5-2 as a narrow bar of depth h and length w, having a symmetrical transverse temperature distribution. In such a bar it has been shown that the axial stress, σ_z, is

$$\sigma_z = E\alpha(\overline{T} - T) \tag{3.5-1}$$

As $\bar{T} = T_o/2$, the stresses at the center and at the extreme fibres are $-E\alpha T_o/2$ and $+E\alpha T_o/2$ respectively and are equal to the peak stresses σ_y at points P_1. On the other hand, when the shape of a rectangular plate is such that $\frac{h}{w} > 1$, the transverse stresses σ_y at points P_1 are less than $E\alpha T_o/2$. This is shown in Fig. 3.5-3.

Parabolic Temperature Distribution in Plate or Bar

A parabolic temperature variation expressed as

$$T = T_o \left(1 - \frac{4y^2}{h^2}\right) \tag{3.5-2}$$

is applied to a rectangular plate, or rectangular bar, as shown in Fig. 3.5-4.

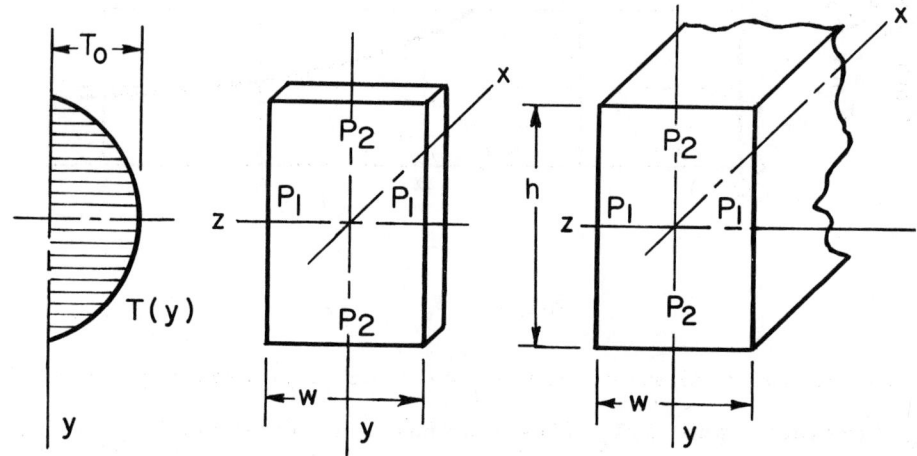

Fig. 3.5-4

The solution of this problem has been obtained by means of an analogy[1] that exists between the moments in a transversely loaded clamped plate, and an unrestrained flat plate with a plane temperature distribution. The results are

---

[1] Ross, A. L., The Clamped Plate Analogy for Thermoelasticity, ASME J. of Basic Eng., 579 Dec. 1963.

shown in Fig. 3.5-5. The maximum normal stresses occur at points P_2 when the ratio h/w is small. This peak stress is then $\sigma_z = \frac{2}{3} E\alpha T_o$ for a plate and $\frac{2E\alpha T_o}{3(1-\nu)}$ for a rectangular bar. In a bar these will also be the maximum axial stresses, σ_x, in the uppermost and lowermost surfaces of the bar. For h/w small, the peak transverse stresses occur at points P_1 and are equal to $\sigma_y = 0.457 \, E\alpha T_o$. The variation of stress with h/w is plotted in Fig. 3.5-5. When the plate width w becomes smaller than the height h, it is observed the peak stresses are the transverse stresses located at points P_1.

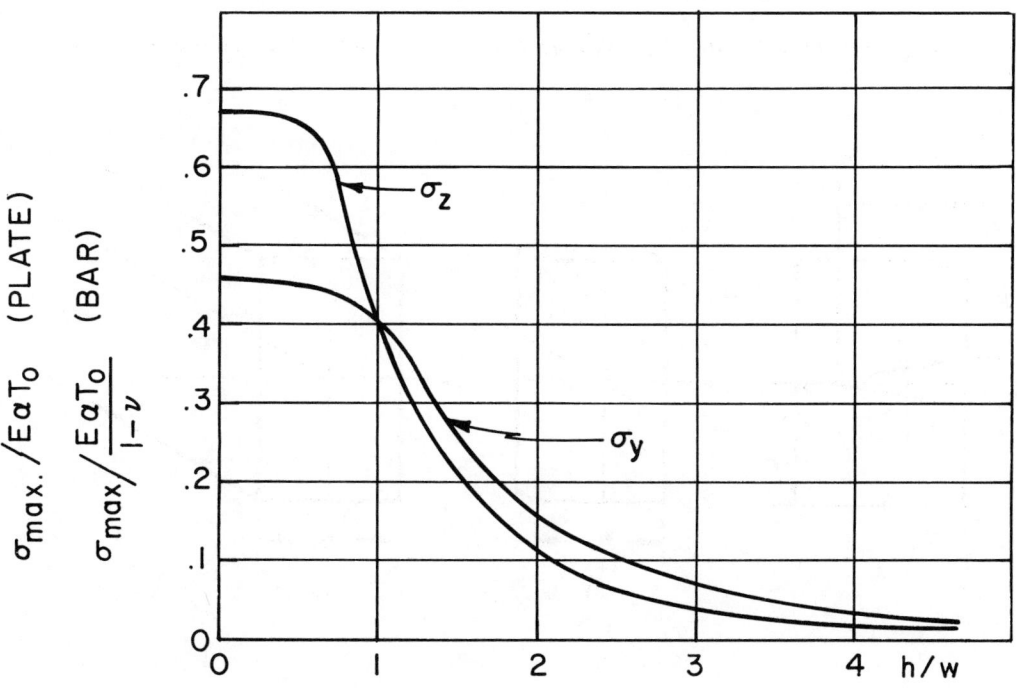

Fig. 3.5-5

Abrupt Temperature Reversals

In the problem of the finite plate with a symmetrical tent-shaped temperature variation, shown in Fig. 3.5-2 the temperature reversal takes place over the full height h, of the plate, or bar. This is a type of temperature distribution that may be found during steady state heat conduction in a plate, or bar, with an internal plane heat source. During a temperature transient the temperature reversal may be essentially confined to a short distance, h_t, shown in Fig. 3.5-6, which is considerably less than the height, h, of the plate.

When the temperature reversal distance, h_t, is quite small in comparison to the height, h, say $\frac{h}{h_t} > 10$, we can assume the presence of a

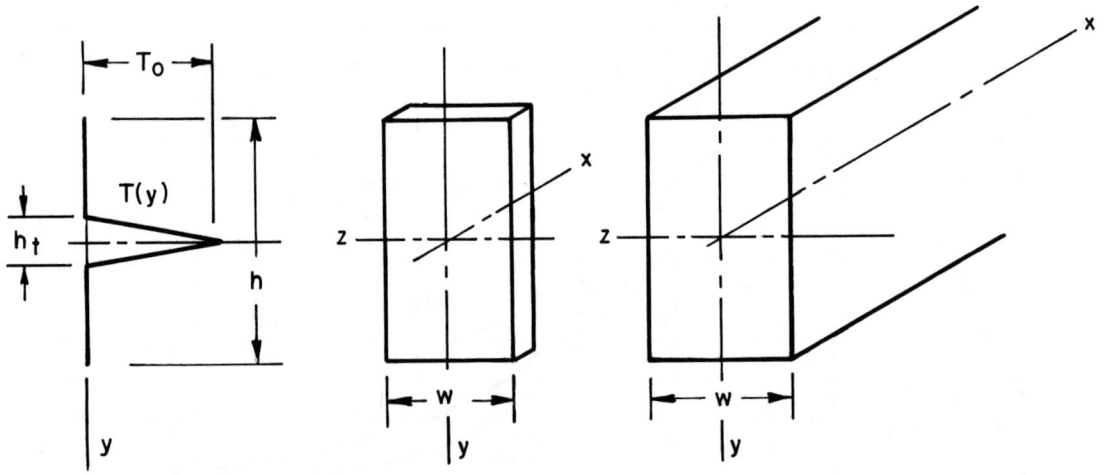

Fig. 3.5-6

narrow hot strip in a plate of infinite height. The peak normal stresses that are developed are then a function of w/h_t, the ratio of plate width to temperature reversal distance. Fig. 3.3-7 shows the variation of the peak stresses,

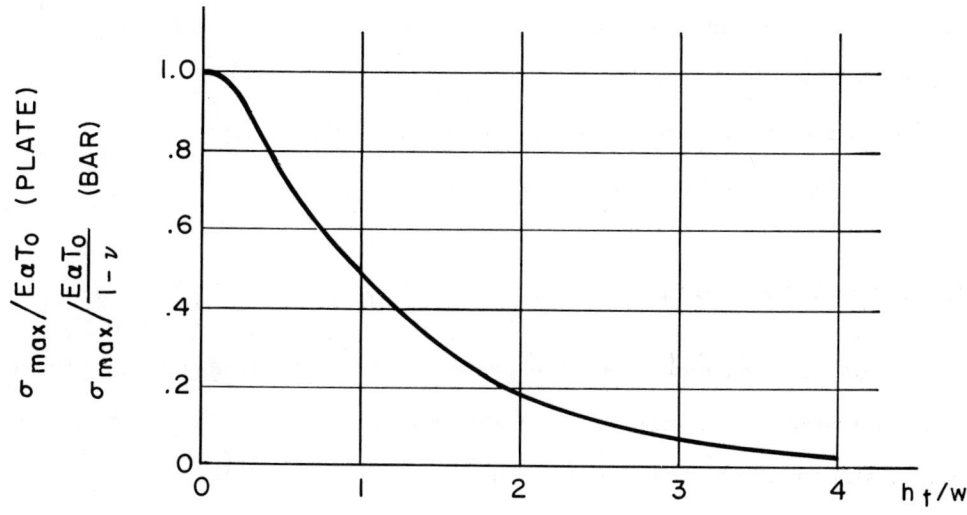

Fig. 3.5-7

in a plate or bar, with the ratio, h_t/w. Note that the peak stress is $\sigma_z = E\alpha T_o$ with T_o representing the full temperature range. When h_t occupies more of the plate, the peak stress is reduced, and when the temperature reversal, T_o, takes place over the full plate height, the peak stress becomes $E\alpha T_o/2$, as in the problem relating to Fig. 3.5-2. The reason for this is clear. In the case of the tent shaped temperature distribution in a finite plate, as in the case of an alternating temperature variation in an infinite plate, we have the equivalent of parallel hot and cold elements of essentially equal area, whereas in the case of a temperature reversal taking place over a short distance, we have the equivalent of a high temperature element of small area in parallel with a cold element (the rest of the plate) of large area. If we apply the bar formula $\sigma = E\alpha(\overline{T} - T)$ to these two cases, we obtain $\sigma = E\alpha T_o/2$ for the case of equal hot and cold areas, and $\sigma = -E\alpha T_o$ for the case of a small hot strip in parallel with a large cold area.

In thermal stress design, a detailed examination of the stress distribution in plates or in cross section planes is often required if one wishes to find the maximum stress intensity, S_{max}. The peak stress intensity which is equal to twice the maximum shear stress, is a design criterion upon which the maximum permissible stress is generally based.

Step Temperature Change

Consider, for example,[1] the problem of a plate strip in which there is a step temperature change, T_o, as shown in Fig. 3.5-8. The maximum normal stresses σ_x and σ_y are equal to $E\alpha T_o/2$, and the maximum value

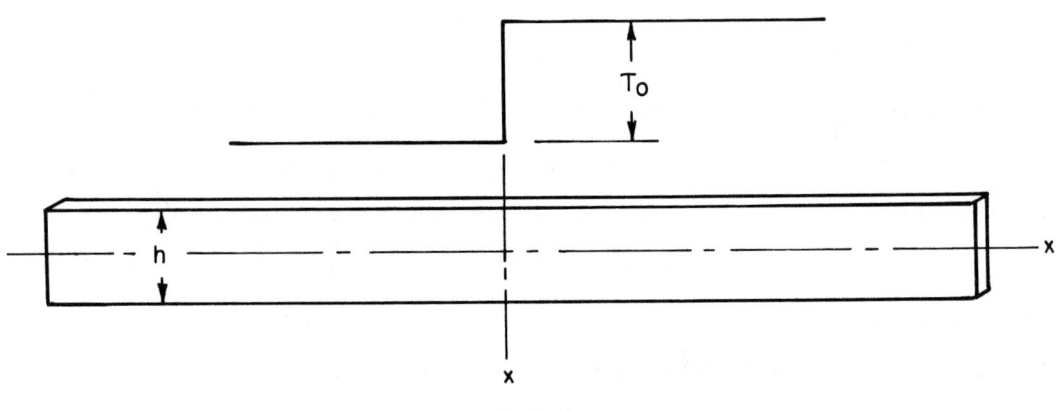

Fig. 3.5-8

of τ_{xy} is found to be $E\alpha T_o/\pi$. The maximum stress intensity due to combined stresses, S_{max}, is however found to be $1.2\, E\alpha T_o$.

Some of the approximations that have been proposed in Section 2.4 for the modification of the axial stresses in bars, to take into account the lateral and transverse stresses in the plane of the cross section, are based on the solutions presented in this section.

---

[1] Horvay G., and Born, J.S., Thermal Stresses in Rectangular Strips, KAPL - 1001, Oct. 1953.

3.6 UNIFORM HOT AREA IN LARGE PLATE

The central area of the large plate, shown in Fig. 3.6-1(a), is heated uniformly to a temperature T_o. The heated area lies within a circle of radius,

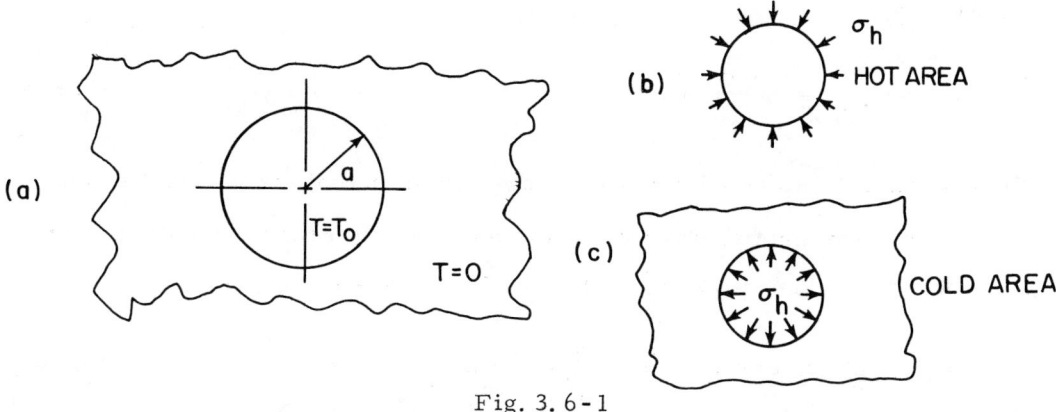

Fig. 3.6-1

a, and the remainder of the plate is unheated. It is clear that the stress in the heated area will be uniform. We designate this stress σ_h as shown in (b) above. For the cold area, (c), we begin with the general form of the solution for the stress distribution in a flat plate with radial symmetry. The general radial and tangential stress distribution is determined in the analysis preceding Eqs. (4.1-14) and (4.1-15). It shows the stress distribution to be of the form

$$\sigma_r = A - \frac{B}{r^2} \qquad (3.6-1)$$

$$\sigma_t = A + \frac{B}{r^2} \qquad (3.6-2)$$

Since the stress at $r \to \infty$ is zero, the constant A is zero, and the foregoing expressions become

$$\sigma_r = -\frac{B}{r^2} \qquad (3.6\text{-}3)$$

$$\sigma_t = \frac{B}{r^2} \qquad (3.6\text{-}4)$$

As the radial stress at $r = a$ is the same in the hot and cold area, we obtain from Eq. (3.6-3)

$$\sigma_h = \sigma_{r,a} = -\frac{B}{a^2} \qquad (3.6\text{-}5)$$

The constant, B, is determined from the condition that the radial displacements, or the tangential strains, in the hot and cold areas must be the same at $r = a$. The strain in the heated portion is uniform and equal to

$$\epsilon_h = \frac{\sigma_h}{E}(1-\nu) + \alpha T_o = -\frac{B(1-\nu)}{a^2 E} + \alpha T_o \qquad (3.6\text{-}6)$$

The tangential strain in the cold portion, at $r = a$, is, from Eqs. (3.6-3) and (3.6-4)

$$\epsilon_{t,a} = \frac{1}{E}(\sigma_{t,a} - \nu\sigma_{r,a}) = \frac{(1+\nu)}{Ea^2} B \qquad (3.6\text{-}7)$$

Since $\epsilon_{t,a}$ and ϵ_h are equal, we have from Eqs. (3.6-6) and (3.6-7)

$$B = \frac{E\alpha T_o a^2}{2} \qquad (3.6\text{-}8)$$

The stresses are therefore

$$\sigma_h = -\frac{E\alpha T_o}{2} \qquad \text{(hot area)} \qquad (3.6\text{-}9)$$

$$\sigma_r = -\frac{E\alpha T_o}{2}\left(\frac{a^2}{r^2}\right) \qquad (3.6\text{-}10)$$

(cold area)

$$\sigma_t = \frac{E\alpha T_o}{2}\left(\frac{a^2}{r^2}\right) \qquad (3.6\text{-}11)$$

3.7 PLATE WITH SYMMETRICAL TRANSVERSE TEMPERATURE VARIATION

An unrestrained plate or a thin cylindrical or spherical shell with a temperature distribution that is symmetrical about the central plane, is a problem of pure extensional deformation and can be treated as described in Sect. 1.7. Use is made of a modified modulus of elasticity, $E'' = \frac{E}{1-\nu}$. In accordance with Eq. (1.7-8)

$$\epsilon_x = \epsilon_y = \frac{\sigma_x}{E''} + \alpha T \tag{3.7-1}$$

Fig. 3.7-1 shows a plate with a symmetrical temperature distribution $T = T(z) = T(-z)$.

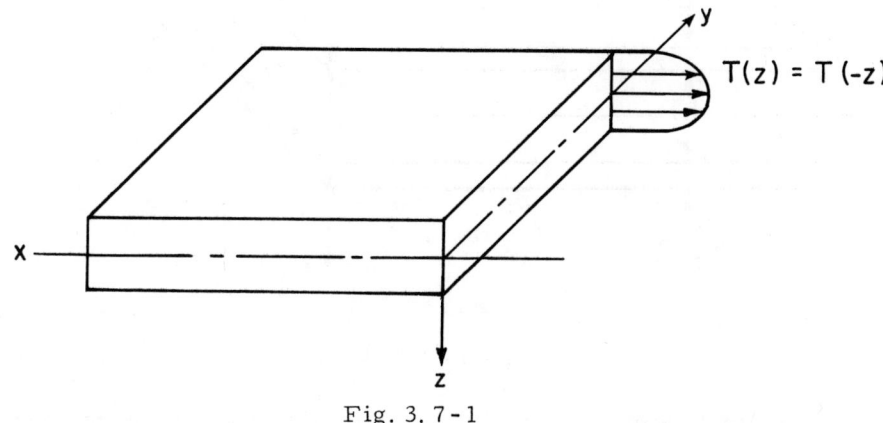

Fig. 3.7-1

From Eq. (1.7-6) we obtain the uniform strain in the plane of the plate as

$$\epsilon = \alpha \overline{T} \tag{3.7-2}$$

and the omnidirectional plane thermal stress, from Eq. (1.7-7), is

$$\sigma = \frac{E\alpha}{1-\nu} (\overline{T} - T) \tag{3.7-3}$$

3.8 EXTENSIONAL DEFORMATION OF LAYERED PLATE

A layered plate with a symmetrical distribution, of material and temperature undergoes extensional deformation. The analysis is carried out as shown in Sect. 1.7. The sandwich plate shown in Fig. 3.8-1 has a symmetrical temperature distribution, $T = T(z) = T(-z)$.

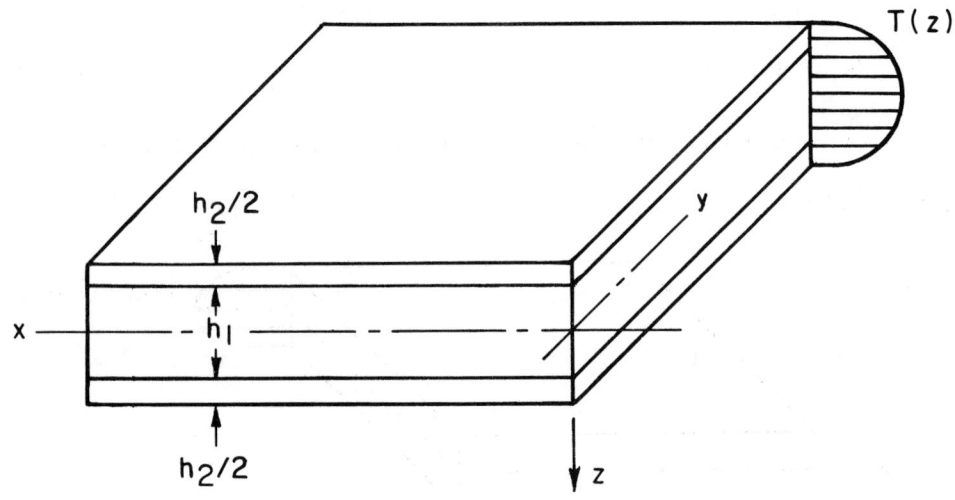

Fig. 3.8-1

As given by Eq. (1.7-10) the uniform strain in any plane parallel to the surface of the plate is

$$\epsilon = \overline{\alpha T}' = \frac{\dfrac{E_1 \alpha_1 \overline{T}_1 h_1}{1-\nu_1} + \dfrac{E_2 \alpha_2 \overline{T}_2 h_2}{1-\nu_2}}{\dfrac{E_1 h_1}{1-\nu_1} + \dfrac{E_2 h_2}{1-\nu_2}} \qquad (3.8-1)$$

and from Eqs. (1.7-8) we obtain the stresses in each of the constituents as

CYLINDRICAL SHELL WITH RADIAL TEMPERATURE   173

$$\sigma_1 = \frac{E_1}{1-\nu_1}(\overline{\alpha T}' - \alpha_1 T_1) = \frac{E_1}{1-\nu_1}\left[\frac{\frac{E_1 \alpha_1 \overline{T}_1 h_1}{1-\nu_1} + \frac{E_2 \alpha_2 \overline{T}_2 h_2}{1-\nu_2}}{\frac{E_1 h_1}{1-\nu_1} + \frac{E_2 h_2}{1-\nu_2}} - \alpha_1 T_1\right]$$

(3.8-2)

$$\sigma_2 = \frac{E_2}{1-\nu_2}(\overline{\alpha T}' - \alpha_2 T_2) = \frac{E_2}{1-\nu_2}\left[\frac{\frac{E_1 \alpha_1 \overline{T}_1 h_1}{1-\nu_1} + \frac{E_2 \alpha_2 \overline{T}_2 h_2}{1-\nu_2}}{\frac{E_1 h_1}{1-\nu_1} + \frac{E_2 h_2}{1-\nu_2}} - \alpha_2 T_2\right]$$

(3.8-3)

It should be noted, in the foregoing, that T_1 and T_2 are functions of z only. This solution is not valid when the temperature varies with the plane coordinates. A layered plate which does not have a symmetrical transverse material and temperature distribution will be subject to both bending and extension. This case is treated in Sect. 6.15.

When the temperature is a function of the plane coordinates x and y, as well as a symmetrical function of the z coordinate, the problem is treated as described on p. 132 (Generalized Plane Stress). The mean in-plane stresses are first determined and the distributed stresses then obtained.

3.9 CYLINDRICAL SHELL WITH RADIAL TEMPERATURE VARIATION

A thin cylindrical shell which is subjected to a radial temperature variation can be assumed to undergo uniform tangential extensional deformation. We treat the problem by means of two-dimensional equivalent load analysis ($\sigma_r = 0$). The cylinder, with an axial surface traction resultant, $E'\alpha'\overline{T}A$ is shown in Fig. 3.9-1(a), and the equilibrium of a shell segment is shown in Fig. 3.9-1(b).

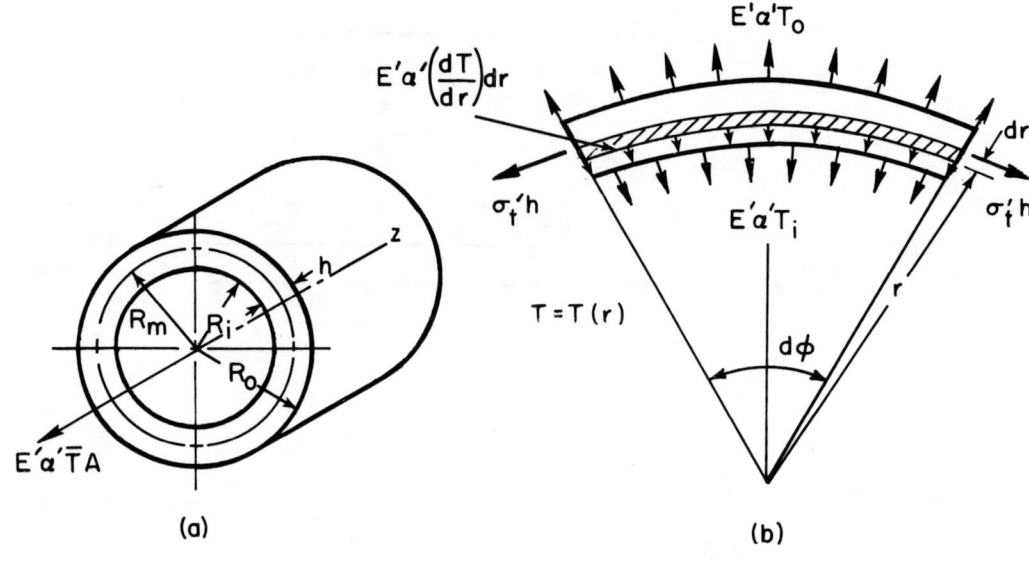

Fig. 3.9-1

The radial body forces and surface tractions produce a uniform hoop displacement stress. Radial static equilibrium of the segment yields

$$\sigma_t' h \, d\varphi = E'\alpha' T_o R_o \, d\varphi - E'\alpha' T_i R_i \, d\varphi - \int_{R_i}^{R_o} E'\alpha' \frac{dT}{dr} dr \cdot r \, d\varphi \qquad (3.9\text{-}1)$$

$$\sigma_t' h = E'\alpha'[T_o R_o - T_i R_i] - E'\alpha'[Tr]_{R_i}^{R_o} + E'\alpha' \int_{R_i}^{R_o} T \, dr \qquad (3.9\text{-}2)$$

$$\sigma_t' = E'\alpha' \overline{T} = \frac{E\alpha \overline{T}}{1-\nu} \qquad (2.9\text{-}3)$$

The net thermal stress is

$$\sigma_t = \sigma_t' - \frac{E\alpha T}{1-\nu} = \frac{E\alpha}{1-\nu}(\overline{T} - T) \qquad (3.9\text{-}4)$$

CYLINDRICAL SHELL WITH RADIAL TEMPERATURE

The equilibrium of the axial surface tractions and the displacement stresses is expressed as

$$\sigma'_z A = E' \alpha' \overline{T} A \qquad (3.9-5)$$

so that the axial displacement stress is

$$\sigma'_z = E' \alpha' \overline{T} = \frac{E \alpha \overline{T}}{1 - \nu} \qquad (3.9-6)$$

The axial thermal stress is then

$$\sigma_z = \sigma'_z - \frac{E \alpha T}{1 - \nu} = \frac{E \alpha}{1 - \nu} (\overline{T} - T) \qquad (3.9-7)$$

and is seen to be the same as the tangential stress, Eq. (3.9-4).

The derivation of Eq. (3.9-7) is based on membrane analysis which assumes a uniform displacement stress over the shell thickness. The displacement stresses σ'_t and σ'_z in Eqs. (3.9-1) and (3.9-5) are then taken to be constant in the expressions for load equilibrium in the radial and axial directions. The solution of the problem by means of the classical equations requires the assumption of a uniform strain over the shell thickness. The classical stress-strain relationships are

$$\epsilon_t = \frac{1}{E} (\sigma_t - \nu \sigma_z) + \alpha T \qquad (3.9-8)$$

$$\epsilon_z = \frac{1}{E} (\sigma_z - \nu \sigma_t) + \alpha T \qquad (3.9-9)$$

with $T = T(r)$.

Integrating the strains in Eqs. (3.9-8) and (3.9-9) over the shell thickness, and keeping the strains ϵ_t and ϵ_z constant, we obtain the relationships

$$\epsilon_t = \frac{1}{E}(\overline{\sigma}_t - \nu \overline{\sigma}_z) + \alpha \overline{T} \tag{3.9-10}$$

$$\epsilon_z = \frac{1}{E}(\overline{\sigma}_z - \nu \overline{\sigma}_t) + \alpha \overline{T} \tag{3.9-11}$$

In the absence of external loads, the mean tangential and axial stresses $\overline{\sigma}_t$ and $\overline{\sigma}_z$ are zero. Equations (3.9-10) and (3.9-11) then become

$$\epsilon = \epsilon_t = \epsilon_z = \alpha \overline{T} \tag{3.9-12}$$

and from Eqs. (3.9-8) and (3.9-9) we obtain

$$\frac{1}{E}(\sigma_t - \nu \sigma_z) + \alpha T = \alpha \overline{T} \tag{3.9-13}$$

$$\frac{1}{E}(\sigma_z - \nu \sigma_t) + \alpha T = \alpha \overline{T} \tag{3.9-14}$$

If we interchange σ_z with σ_t, in the foregoing equations we find that they remain unchanged. Therefore the stress is uniform and omnidirectional in each shell lamina, that is,

$$\sigma = \sigma_t = \sigma_z \tag{3.9-15}$$

From Eqs. (3.9-12), (3.9-13) and (3.9-14) we then obtain the extensional tangential and axial strain as

$$\epsilon = \frac{1-\nu}{E}\sigma + \alpha T = \alpha \overline{T} \tag{3.9-16}$$

from which the tangential and axial stresses are obtained as

$$\sigma = \sigma_t = \sigma_z = \frac{E\alpha}{1-\nu}(\overline{T} - T) \tag{3.9-17}$$

These are the same results as obtained by equivalent load analysis.

In a thin-walled circular cylinder having a radial temperature variation, the axial and tangential stresses are equal to $\frac{E\alpha(\bar{T} - T)}{1 - \nu}$ at all points except at the ends of the cylinder. The end surface tractions produce thermal end moments at the rim of the cylinder, and these generate axial displacement bending stresses and bending deformations which are localized and disappear at a distance from the end of the cylinder equal to approximately three or four cylinder relaxation lengths. (A relaxation length is roughly equal to the geometric mean of the cylinder radius and cylinder thickness). The axial length in which the local bending falls off to zero is considerably larger than the axial distance over which the normal axial stresses, σ_z, build up at the cylinder ends. The latter distance is comparable to the shell thickness in accordance with St. Venant's principle.

3.10 CONCENTRIC CYLINDRICAL SHELLS

The assembly shown in Fig. 3.10-1 consists of two concentric circular cylindrical shells of different materials. The inner shell has a radial

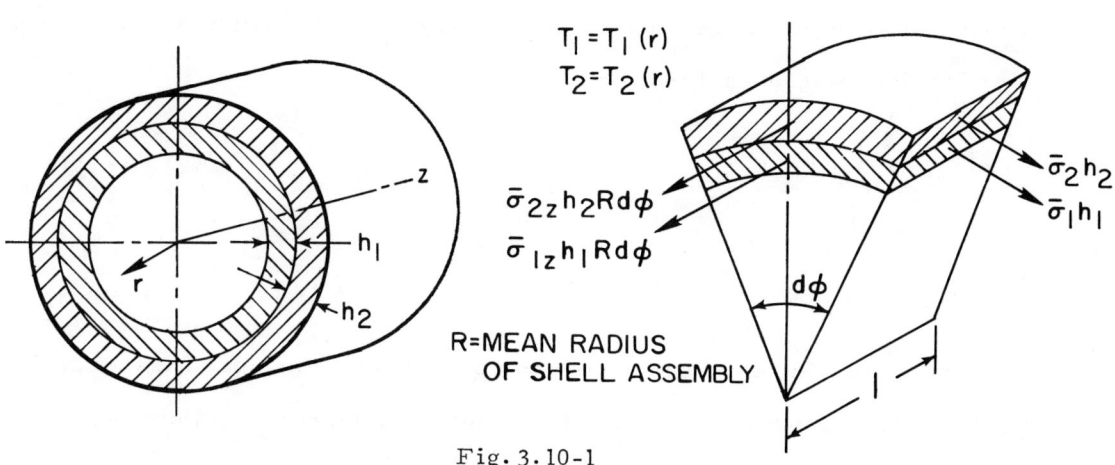

Fig. 3.10-1

temperature variation $T_1(r)$ and the outer shell a radial temperature variation $T_2(r)$. The shells are thin so that the assembly can be assumed to have a common radius, R. As in Eqs. (3.9-10) and (3.9-11) the strains are taken to be uniform over the thickness of the assembly and are expressed in terms of the mean stresses as follows:

$$\epsilon_{1t} = \frac{1}{E_1}(\bar{\sigma}_{1t} - \nu_1 \bar{\sigma}_{1z}) + \alpha_1 \bar{T}_1 \qquad (3.10\text{-}1)$$

$$\epsilon_{1z} = \frac{1}{E_1}(\bar{\sigma}_{1z} - \nu_1 \bar{\sigma}_{1t}) + \alpha_1 \bar{T}_1 \qquad (3.10\text{-}2)$$

$$\epsilon_{2t} = \frac{1}{E_2}(\bar{\sigma}_{2t} - \nu_2 \bar{\sigma}_{2z}) + \alpha_2 \bar{T}_2 \qquad (3.10\text{-}3)$$

$$\epsilon_{2z} = \frac{1}{E_2}(\bar{\sigma}_{2z} - \nu_2 \bar{\sigma}_{2t}) + \alpha_2 \bar{T}_2 \qquad (3.10\text{-}4)$$

A force balance in the radial direction yields

$$\bar{\sigma}_{1t} h_1 + \bar{\sigma}_{2t} h_2 = 0 \qquad (3.10\text{-}5)$$

and a force balance in the axial direction yields

$$\sigma_{1z} h_1 + \sigma_{2z} h_2 = 0 \qquad (3.10\text{-}6)$$

Strain compatibility in the form of uniformity of tangential and of axial strain over the thickness is expressed as

$$\epsilon_{1t} = \epsilon_{2t} = \epsilon_t \qquad (3.10\text{-}7)$$

and

$$\epsilon_{1z} = \epsilon_{2z} = \epsilon_z \qquad (3.10\text{-}8)$$

The foregoing eight equations contain eight unknown quantities, four stresses and four strains, and can be solved simultaneously. However, an examination of Eqs. (3.10-1) through (3.10-6) shows that the equations would be unaltered if the subscripts z and t were interchanged. Therefore $\bar{\sigma}_{1z} = \bar{\sigma}_{1t}$, $\bar{\sigma}_{2z} = \bar{\sigma}_{2t}$ and $\epsilon_z = \epsilon_t = \epsilon$. The equations then reduce to two equations, namely

$$\epsilon = \frac{1-\nu_1}{E_1} \bar{\sigma}_1 + \alpha_1 \bar{T}_1 = \frac{1-\nu_2}{E_2} \bar{\sigma}_2 + \alpha_2 \bar{T}_2 \qquad (3.10\text{-}9)$$

and

$$\bar{\sigma}_1 h_1 + \bar{\sigma}_2 h_2 = 0 \qquad (3.10\text{-}10)$$

From Eqs. (3.10-9) and (3.10-10) we obtain the uniform extensional strain, ϵ, and the mean stresses, $\bar{\sigma}_1$ and $\bar{\sigma}_2$ as

$$\epsilon = \frac{\dfrac{E_1 \alpha_1 \bar{T}_1 h_1}{1-\nu_1} + \dfrac{E_2 \alpha_2 \bar{T}_2 h_2}{1-\nu_2}}{\dfrac{E_1 h_1}{1-\nu_1} + \dfrac{E_2 h_2}{1-\nu_2}} = \overline{\alpha T}' \qquad (3.10\text{-}11)$$

and

$$\bar{\sigma}_1 = \frac{E_1}{1-\nu_1} \left[\frac{\dfrac{E_1 \alpha_1 \bar{T}_1 h_1}{1-\nu_1} + \dfrac{E_2 \alpha_2 \bar{T}_2 h_2}{1-\nu_2}}{\dfrac{E_1 h_1}{1-\nu_1} + \dfrac{E_2 h_2}{1-\nu_2}} - \alpha_1 \bar{T}_1 \right] = \frac{E_1}{1-\nu_1} [\overline{\alpha T}' - \alpha_1 \bar{T}_1] \qquad (3.10\text{-}12)$$

$$\bar{\sigma}_2 = \frac{E_2}{1-\nu_2} \left[\frac{\dfrac{E_1 \alpha_1 \bar{T}_1 h_1}{1-\nu_1} + \dfrac{E_2 \alpha_2 \bar{T}_2 h_2}{1-\nu_2}}{\dfrac{E_1 h_1}{1-\nu_1} + \dfrac{E_2 h_2}{1-\nu_2}} - \alpha_2 \bar{T}_2 \right] = \frac{E_2}{1-\nu_2} [\overline{\alpha T}' - \alpha_2 \bar{T}_2] \qquad (3.10\text{-}13)$$

Using Eq. (3.10-9) and the basic stress-strain equations, Eqs. (3.9-8) and (3.9-9), we find that the distributed axial and tangential stresses are related to the uniform strain ϵ, as follows:

$$\epsilon = \frac{1}{E_1}(\sigma_{1t} - \nu_1 \sigma_{1z}) + \alpha_1 T_1 = \frac{1}{E_1}(\sigma_{1z} - \nu_1 \sigma_{1t}) + \alpha_1 T_1 = \frac{1-\nu_1}{E_1}\bar{\sigma}_1 + \alpha_1 \bar{T}_1$$

(3.10-14)

$$\epsilon = \frac{1}{E_2}(\sigma_{2t} - \nu_2 \sigma_{2z}) + \alpha_2 T_2 + \frac{1}{E_2}(\sigma_{2z} - \nu_2 \sigma_{2t}) + \alpha_2 T_2 = \frac{1-\nu_2}{E_2}\bar{\sigma}_2 + \alpha_2 \bar{T}_2$$

(3.10-15)

The foregoing equations show that $\sigma_{1t} = \sigma_{1z}$ and $\sigma_{2t} = \sigma_{2z}$, and we obtain

$$\sigma_1 = \bar{\sigma}_1 + \frac{E_1 \alpha_1}{1-\nu_1}(\bar{T}_1 - T_1) \qquad (3.10\text{-}16)$$

$$\sigma_2 = \bar{\sigma}_2 + \frac{E_2 \alpha_2}{1-\nu_2}(\bar{T}_2 - T_2) \qquad (3.10\text{-}17)$$

Eqs. (3.10-12) and (3.10-13) are set into the foregoing and they become

$$\sigma_1 = \frac{E_1}{1-\nu_1}\left[\frac{\frac{E_1 \alpha_1 \bar{T}_1 h_1}{1-\nu_1} + \frac{E_2 \alpha_2 \bar{T}_2 h_2}{1-\nu_2}}{\frac{E_1 h_1}{1-\nu_1} + \frac{E_2 h_2}{1-\nu_2}} - \alpha_1 T_1\right] = \frac{E_1}{1-\nu_1}(\overline{\alpha T}' - \alpha_1 T_1)$$

(3.10-18)

$$\sigma_2 = \frac{E_2}{1-\nu_2}\left[\frac{\frac{E_1 \alpha_1 \bar{T}_1 h_1}{1-\nu_1} + \frac{E_2 \alpha_2 \bar{T}_2 h_2}{1-\nu_2}}{\frac{E_1 h_1}{1-\nu_1} + \frac{E_2 h_2}{1-\nu_2}} - \alpha_2 T_2\right] = \frac{E_2}{1-\nu_2}(\overline{\alpha T}' - \alpha_2 T_2)$$

(3.10-19)

These results are the same as those obtained for the layered plate, Sect. 3.8. It is clear that a concentric thin cylindrical shell assembly with a radial temperature distribution, is essentially the same type problem as a layered plate with a symmetrical material and temperature distribution, since both are two-dimensional problems of extensional deformation. When the problem is one in which there are more than two concentric thin shells, having radial temperature variations, Eqs. (3.10-18) and (3.10-19) can be generalized and written as

$$\sigma_i = \frac{E_i}{1-\nu_i}(\overline{\alpha T}' - \alpha_i T_i) \qquad (3.10\text{-}20)$$

with

$$\overline{\alpha T}' = \frac{\sum_{1}^{n} \frac{E_j \alpha_j \overline{T}_j h_j}{1-\nu_j}}{\sum_{1}^{n} \frac{E_j h_j}{1-\nu_j}} \qquad (3.10\text{-}21)$$

In these equations $T_i = T_i(r)$ and $\sigma_i = \sigma_i(r)$ represent the distributed temperatures and stresses in an assembly of n concentric shells. An analysis of a thick-walled cylindrical shell assembly is given in Example (b) of Sect. 4.5.

3.11 SPHERICAL SHELL WITH RADIAL TEMPERATURE VARIATION

As in the case of the cylindrical shell of Section 3.9, it is assumed that the spherical shell is thin, with all points having a common radius R, and that the extensional tangential strain over the thickness is uniform. Since the mean tangential stresses, $\overline{\sigma}$, are zero and the distributed tangential stresses, σ, are omnidirectional in tangential planes, and a function of the radius only, the relationship between the tangential stress and strain is the same as for the cylindrical shell with a radial temperature variation, given by Eq. (3.9-16). That is, for the spherical shell, a segment of which is shown in Fig. 3.11-1, we have

$$\epsilon = \frac{1-\nu}{E}\sigma + \alpha T = \frac{1-\nu}{E}\overline{\sigma} + \alpha\overline{T} = \alpha\overline{T} \qquad (3.11\text{-}1)$$

The tangential stress follows directly from Eq. (3.11-1). It is

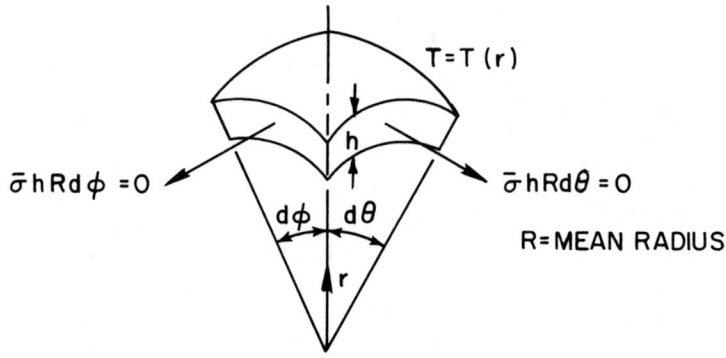

Fig. 3.11-1

$$\sigma = \frac{E\alpha}{1-\nu}(\bar{T} - T) \qquad (3.11\text{-}2)$$

This is the same form of equation as found for the stress in a homogeneous plate with a symmetrical temperature distribution, and for a cylindrical shell with a radial temperature distribution. Essentially they are all the same kinds of problems, having a thermal loading which is a function of the thickness coordinate only, and undergoing extensional deformation without bending.

3.12 CONCENTRIC SPHERICAL SHELLS

Two concentric spherical shells have radial temperature variations $T_1(r)$ and $T_2(r)$. Since the shells are assumed to be thin, a common radius, R, is used. A sketch of a section of the concentric shell assembly is shown in Fig. 3.12-1. As there is spherical symmetry and the shells are thin, $\sigma_{1\varphi} = \sigma_{1\theta} = \sigma_1$, $\sigma_{2\varphi} = \sigma_{2\theta} = \sigma_2$, $\bar{\sigma}_{1\varphi} = \bar{\sigma}_{1\theta} = \bar{\sigma}_1$, $\bar{\sigma}_{2\varphi} = \bar{\sigma}_{2\theta} = \bar{\sigma}_2$, and we assume that $\sigma_r = 0$.

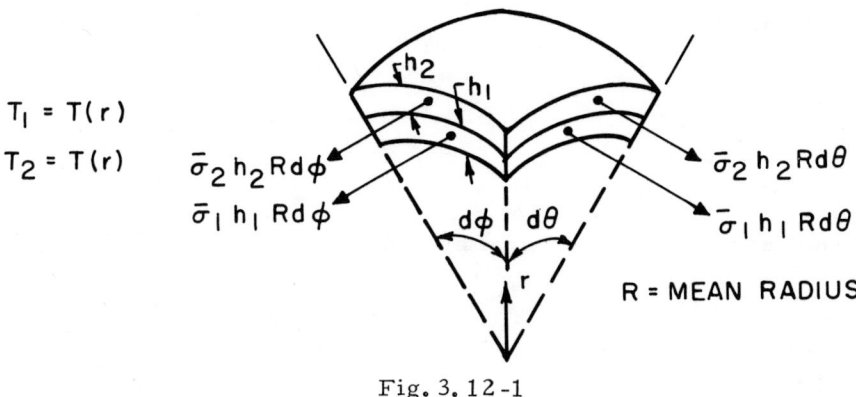

Fig. 3.12-1

The stress-strain relations gives the tangential strain ϵ, taken to be uniform over the shell assembly, as a function of the distributed stress and temperature. These equations are

$$\epsilon = \frac{1-\nu_1}{E_1}\sigma_1 + \alpha_1 T_1 \qquad (3.12\text{-}1)$$

and

$$\epsilon = \frac{1-\nu_2}{E_2}\sigma_2 + \alpha_2 T_2 \qquad (3.12\text{-}2)$$

One can also express the strain in terms of the mean stresses and temperatures, $\bar{\sigma}$ and $\bar{T}$. Eqs. (3.12-1) and (3.12-2) are integrated over the thicknesses, as shown below,

$$\int_{h_1}\epsilon\,dr = \frac{1-\nu_1}{E_1}\int_{h_1}\sigma_1(r)\,dr + \alpha_1\int_{h_1}T_1(r)\,dr$$

$$\int_{h_2}\epsilon\,dr = \frac{1-\nu_2}{E_2}\int_{h_2}\sigma_2(r)\,dr + \alpha_2\int_{h_2}T_2(r)\,dr \qquad (3.12\text{-}3)$$

and as ϵ is assumed constant, we obtain the equations

$$\epsilon = \frac{1-\nu_1}{E_1}\bar{\sigma}_1 + \alpha_1 \bar{T}_1 \tag{3.12-4}$$

$$\epsilon = \frac{1-\nu_2}{E_2}\bar{\sigma}_2 + \alpha \bar{T}_2 \tag{3.12-5}$$

Radial equilibrium of the spherical element shown in Fig. 3.12-1 requires that

$$\bar{\sigma}_1 h_1 + \bar{\sigma}_2 h_2 = 0 \tag{3.12-6}$$

From Eqs. (3.12-1), (3.12-2), and Eqs. (3.12-4), (3.12-5), we obtain the tangential strain

$$\epsilon = \frac{\dfrac{E_1 \alpha_1 \bar{T}_1 h_1}{1-\nu_1} + \dfrac{E_2 \alpha_2 \bar{T}_2 h_2}{1-\nu_2}}{\dfrac{E_1 h_1}{1-\nu_1} + \dfrac{E_2 h_2}{1-\nu_2}} = \overline{\alpha T}' \tag{3.12-7}$$

and the tangential stresses as

$$\sigma_1 = \frac{E_1}{1-\nu_1}\left[\frac{\dfrac{E_1 \alpha_1 \bar{T}_1 h_1}{1-\nu_1} + \dfrac{E_2 \alpha_2 \bar{T}_2 h_2}{1-\nu_2}}{\dfrac{E_1 h_1}{1-\nu_1} + \dfrac{E_2 h_2}{1-\nu_2}} - \alpha_1 T_1\right] = \frac{E_1}{1-\nu_1}(\overline{\alpha T}' - \alpha_1 T_1) \tag{3.12-8}$$

and

$$\sigma_2 = \frac{E_2}{1-\nu_2}\left[\frac{\dfrac{E_1 \alpha_1 \bar{T}_1 h_1}{1-\nu_1} + \dfrac{E_2 \alpha_2 \bar{T}_2 h_2}{1-\nu_2}}{\dfrac{E_1 h_1}{1-\nu_1} + \dfrac{E_2 h_2}{1-\nu_2}} - \alpha_2 T_2\right] = \frac{E_2}{1-\nu_2}(\overline{\alpha T}' - \alpha_2 T_2) \tag{3.12-9}$$

The expressions on the right hand side of Eqs. (3.12-7), (3.12-8) and (3.12-9) apply also to shell assemblies consisting of n concentric shells. The distributed tangential stress in any shell, i, is then

$$\sigma_i = \frac{E_i}{1 - \nu_i} (\overline{aT}' - a_i T_i) \qquad (3.12\text{-}10)$$

where

$$\overline{aT}' = \frac{\sum_{1}^{n} \frac{E_j a_j T_j h_j}{1 - \nu_j}}{\sum_{1}^{n} \frac{E_j h_j}{1 - \nu_j}} \qquad (3.12\text{-}11)$$

It is clear that the temperatures T_i and the tangential stresses σ_i are functions of r only. It should be observed that the analysis assumes a common radius for the shell assembly. Such an assumption may not be reasonable when an assembly of many shells is of substantial thickness. A thick wall spherical shell analysis such as given in Example (b) of Sect. 4.7 is then required.

3.13 ASSEMBLIES WITH AXIAL SYMMETRY

The discussions in the preceding text deal with homogeneous elements, or with layered elements in which the components have the same degree of dimensionality. Thus, in Chapter 1 the parallel elements in series are all one-dimensional elements, and in this chapter, the layered plates and shells are all two-dimensional. The illustrative examples which follow deal with structural assemblies in which the individual components are of different dimensionality. They are, however, problems which are in the category of axially symmetric plane stress and plane strain deformation.

The first example illustrates the analysis of a mixed one-dimensional and two-dimensional problem, and the second example illustrates the analysis of a mixed two-dimensional and three-dimensional problem.

Example (a) Circular Disc and Thin Ring

The assembly shown in Fig. 3.13-1 consists of a circular plate surrounded by a circular ring, the former being a two-dimensional element and the latter a one-dimensional element. They are of different materials and are heated to uniform temperatures T_1 and T_2 respectively. The thickness of the ring is assumed to be small, permitting the use of a radius R for both the disc and ring.

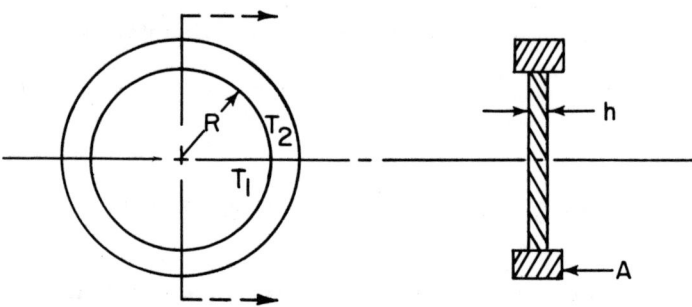

Fig. 3.13-1

Equivalent load analysis is not desirable in this problem since we are concerned with the interaction of a two-dimensional element with a one-dimensional element. It is convenient to treat the problem by the direct application of the pertinent thermoelastic equations. The solution is readily obtained when it is recognized that there is a common radial load at the ring and disc interface, and that the circumferential strain (or the radial displacement) at the surface of contact is the same in the disc and ring.

ASSEMBLIES WITH AXIAL SYMMETRY

Let the uniform plane stress in the disc be σ_1. The pressure acting radially on the inner surface of the ring is therefore also σ_1. It follows that the tangential hoop stress in the ring is

$$\sigma_2 = - \frac{\sigma_1 h R}{A} \tag{1}$$

The uniform plane strain in the disc is

$$\epsilon_1 = \frac{1 - \nu_1}{E_1} \sigma_1 + \alpha_1 T_1 \tag{2}$$

and the tangential strain in the ring, which must be equal to the disc strain, is

$$\epsilon_2 = \frac{\sigma_2}{E_2} + \alpha_2 T_2 \tag{3}$$

Substitution of Eq. (1) into Eq. (3), and equating the strains in the disc and ring we obtain

$$\epsilon = \epsilon_1 = \epsilon_2 = \frac{1 - \nu_1}{E_1} \sigma_1 + \alpha_1 T_1 = - \frac{\sigma_1 hR}{A E_2} + \alpha_2 T_2 \tag{4}$$

This yields the uniform plane stress in the disc as

$$\sigma_1 = \frac{E_1}{1 - \nu_1} \frac{\alpha_2 T_2 - \alpha_1 T_1}{\frac{E_1}{1-\nu_1} \frac{hR}{A E_2} + 1} \tag{5}$$

and the uniform tangential strain in the ring as

$$\sigma_2 = E_2 \frac{\alpha_1 T_1 - \alpha_2 T_2}{\frac{1 - \nu_1}{E_1} \frac{A E_2}{hR} + 1} \tag{6}$$

The radial growth of the assembly is clearly

$$u = \epsilon R = - \frac{\sigma_1 h R^2}{A E_2} + \alpha_2 T_2 R \tag{7}$$

Substitution of σ_1, from Eq. (5) gives

$$\epsilon = \frac{u}{R} = \frac{\dfrac{E_1 \alpha_1 T_1 h R}{1 - \nu_1} + E_2 \alpha_2 T_2 A}{\dfrac{E_1 h R}{1 - \nu_1} + E_2 A} = \overline{\alpha T}^* \qquad (8)$$

The strain has the form of a mean expansion $\overline{\alpha T}^*$, obtained by averaging the free expansions of the disc and ring using rigidities equal to $E_1 h R/(1 - \nu_1)$ and $E_2 A$ respectively.

Example (b) Concentric Rod and Thin-Walled Tube

The problem of interacting thin tube and rod is characterized by a two-dimensional state of stress in the tube and a three-dimensional state of stress in the rod. The rod and tube are assumed to be of unlike materials and at different but uniform temperatures. The rod radius and tube radius are taken as the same which is a fair approximation when the tube wall thickness is small in comparison to the radius. The configuration of the concentric rod and tube assembly is shown in Fig. 3.13-2.

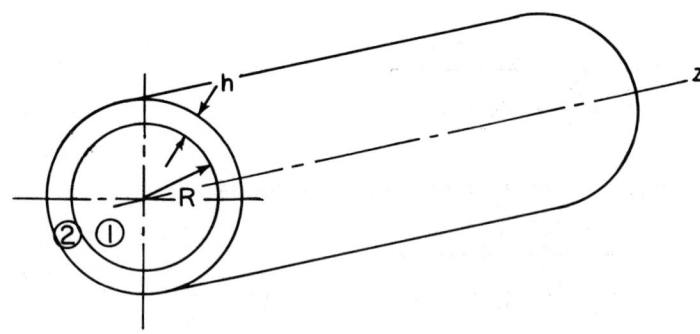

Fig. 3.13-2

ASSEMBLIES WITH AXIAL SYMMETRY

As the rod is under uniform radial compression, the stresses in the cross section planes are uniform, with the tangential and radial stresses equal; i.e., $\sigma_{1t} = \sigma_{1r}$. The tangential and axial strains in the rod (subscript 1), and tube (subscript 2) are as follows:

$$\epsilon_{1t} = \frac{1}{E_1}((1-\nu_1)\sigma_{1t} - \nu_1 \sigma_{1z}) + \alpha_1 T_1 \tag{1}$$

$$\epsilon_{1z} = \frac{1}{E_1}(\sigma_{1z} - 2\nu_1 \sigma_{1t}) + \alpha_1 T_1 \tag{2}$$

$$\epsilon_{2t} = \frac{1}{E_2}(\sigma_{2t} - \nu_2 \sigma_{2z}) + \alpha_2 T_2 \tag{3}$$

$$\epsilon_{2z} = \frac{1}{E_2}(\sigma_{2z} - \nu_2 \sigma_{2t}) + \alpha_2 T_2 \tag{4}$$

The radial balance of forces at the interface of rod and tube is expressed as

$$\sigma_{1t} R + \sigma_{2t} h = 0 \tag{5}$$

and an axial balance of the load in the tube and rod is given by

$$\sigma_{1z} R + 2 \sigma_{2z} h = 0 \tag{6}$$

The tangential strains at the interface, in the rod and tube, are the same, and the axial strains in the rod and tube are the same. Equating the tangential strains, $\epsilon_{1t} = \epsilon_{2t}$, and the axial strains, $\epsilon_{1z} = \epsilon_{2z}$, and using the load balance Equations (5) and (6) we obtain

$$\sigma_{1z}(E_2 + \frac{E_1 R}{2h}) - \sigma_{1t}(\frac{E_1 \nu_2 R}{h} + 2E_2 \nu_1) = E_1 E_2 (\alpha_2 T_2 - \alpha_1 T_1) \tag{7}$$

and

$$\sigma_{1z}(E_2 \nu_1 + \frac{E_1 \nu_2 R}{2h}) - \sigma_{1t}(E_2(1-\nu_1) + \frac{E_1 R}{h}) = -E_1 E_2 (\alpha_2 T_2 - \alpha_1 T_1) \tag{8}$$

The solution of the foregoing equations yields the axial, tangential, and radial stresses in the rod, as

$$\sigma_{1z} = \frac{\left[E_2(1+\nu_1) + \frac{E_2 R}{h}(1+\nu_2)\right]\left[E_1 E_2 (\alpha_2 T_2 - \alpha_1 T_1)\right]}{\left[E_2 + \frac{E_1 R}{2h}\right]\left[E_2(1-\nu_1) + \frac{E_1 R}{h}\right] - \left[E_2\nu_1 + \frac{E_1\nu_2 R}{2h}\right]\left[\frac{E_1\nu_2 R}{2h} + 2E_2\nu_1\right]}$$

(9)

$$\sigma_{1t} = \sigma_{1r} = \frac{\left[E_2(1+\nu_1) + \frac{E_1 R}{2h}(1+\nu_2)\right]\left[E_1 E_2 (\alpha_2 T_2 - \alpha_1 T_1)\right]}{\left[E_2 + \frac{E_1 R}{2h}\right]\left[E_2(1-\nu_1) + \frac{E_1 R}{h}\right] - \left[E_2\nu_1 + \frac{E_1\nu_2 R}{2h}\right]\left[\frac{E_1\nu_2 R}{2h} + 2E_2\nu_1\right]}$$

(10)

The stresses in the tube are written in terms of the foregoing as

$$\sigma_{2z} = -\frac{\sigma_{1z} R}{2h} \qquad \sigma_{2t} = -\frac{\sigma_{1t} R}{h} \qquad (11)$$

The strains are obtained by substituting these stresses into Eqs. (1), (2), (3) and (4).

The accurate solution of the problem of concentric thick-walled circular tubes subjected to a radial temperature distribution is given in Example (b) of Sect. 4.5. The solution in Example (b) of Sect. 4.5 applies also to an assembly consisting of a concentric rod and thick-walled tube. For the case of the solid rod it is simply required to set $R_i = 0$, in each of the equations for interfacial pressure, and for tangential, radial, and axial stress.

3.14 MAXIMUM STRESSES IN PARALLEL ELEMENTS

In reviewing the expressions that were derived for one-dimensional stresses in descrete parallel elements, in parallel fibers, and in the parallel laminae of two-dimensional structural elements, subjected to extensional deformation, it is observed that the expressions derived for the axial stresses in one-dimensional elements are of the form

$$\sigma_i = E_i (\overline{\alpha T} - \alpha_i T_i) \tag{3.14-1}$$

and for two-dimensional elements the laminar stresses are

$$\sigma_i = \frac{E_i}{1 - \nu_i} (\overline{\alpha T}' - \alpha_i T_i) \tag{3.14-2}$$

In Eqs. (3.14-1) and (3.14-2) the quantities $\overline{\alpha T}$ and $\overline{\alpha T}'$ represent mean values alternatively defined as

$$\overline{\alpha T} = \frac{\sum_{j=1}^{n} E_j \alpha_j \overline{T}_j A_j}{\sum_{j=1}^{n} E_j A_j} \quad ; \quad \overline{\alpha T} = \frac{\int_A E \alpha\, T dA}{\int_A E\, dA} \tag{3.14-3}$$

and

$$\overline{\alpha T}' = \frac{\sum_{j=1}^{n} \frac{E_j \alpha_j \overline{T}_j A_j}{1 - \nu_j}}{\sum_{j=1}^{n} \frac{E_j A_j}{1 - \nu_j}} \quad ; \quad \overline{\alpha T}' = \frac{\int_A \frac{E \alpha T\, dA}{1 - \nu}}{\int_A \frac{E\, dA}{1 - \nu}} \tag{3.14-4}$$

It is clear that for the homogeneous case $\overline{\alpha T} = \overline{\alpha T}'$. If we preclude negative temperatures, the maximum possible stress, in a one-dimensional element undergoing extensional deformation, will occur in the particular constituent or fibre in which $E_i |\overline{\alpha T} - \alpha_i T_i|$ is a maximum. Similarly, in two-dimensional homogeneous or layered plates, or thin cylindrical of spherical elements which are subjected to extensional deformation, the maximum possible stress will occur at the point where $\frac{E_i}{1-\nu} |\overline{\alpha T}' - \alpha_i T_i|$ is a maximum. These upper limits are significant in judging the magnitude of stresses that are obtainable in structural elements.

It should be recognized that stresses in parallel elements tend to be self-relieving since a thermal load does not have the character of a sustained load. When one or more parallel elements or fibres yield and undergo stress relaxation, the other elements which lie in parallel with the relaxing elements carry a reduced thermal load. Even when a number of parallel elements or fibres yield simultaneously, the hazard of structural failure is not great. Thermal yielding in a parallel element assembly is not generally accompanied by large strains or deformations, as would be the case, for example, when a sustained load hanging from a bar develops a stress equal to the ultimate strength and produces rupture. Structural failure in parallel elements may result from repeated thermal loading due to thermal cycling. As this type of loading can induce fatigue failure, it represents a type of thermal loading that is just as hazardous as positive cyclic loading.

The inherent safety that is characteristic of parallel elements subjected to thermal loading, is not present in assemblies in which the elements lie in series. In such assemblies very large deformations can be produced by a single application of a thermal load. The analysis of series elements undergoing extensional deformations, and the conditions required for the development of high stresses and large deformations in such structures, have been discussed in Section 2.16.

3.15 THERMAL SHOCK

By thermal shock is meant the generation of non-inertial thermal stresses by rapid changes in the temperature distribution in a body. Thermal shock is often produced by the sudden application of a hot or cold fluid to the surface of a structure. In fluids with large film coefficients, such as water, there is a mitigation of the thermal shock, as there is present at the fluid-solid interface, a temperature drop in a surface film. Fluids with small film coefficients and high boiling points, such as liquid metals, tend to produce severe thermal shocks; i.e., large thermal stresses are developed when there is sudden contact between the fluid and surface between which there is a large temperature differential, or when the fluid in continuous contact with the surface, undergoes a rapid change in temperature.

It can be shown that a thick homogeneous plate in which there is a temperature variation $T = T(z)$ normal to the surface, and which does not undergo bending deformation, will develop surface stresses of magnitude

$$\sigma = \frac{E\alpha}{1-\nu}(\bar{T} - T) \qquad (3.15\text{-}1)$$

In a thick-walled structure of, more or less, arbitrary shape, in which the temperature variations are confined to the region close to the surface, the state of stress near the surface will be two-dimensional, and Eq. (3.15-1) will apply. Generally the location at which the greatest difference between $\bar{T}$ and T occurs, is at the surface, so that the magnitude of the stress due to the thermal shock experienced at a surface becomes

$$\sigma = \frac{E\alpha}{1-\nu}(\bar{T} - T_s) \qquad (3.15\text{-}2)$$

where T_s is the surface temperature.

In the case of a heavy-walled structure originally at uniform temperature, the mean temperature $\bar{T}$ may be taken as zero, and the stress due to thermal shock as

$$\sigma = \frac{E\alpha}{1-\nu} T_s \qquad (3.15\text{-}3)$$

In the most severe case, in which the film coefficient is essentially zero the stress due to thermal shock becomes

$$\sigma = \frac{E\alpha}{1-\nu} T_f \qquad (3.15-4)$$

where T_f is the temperature of the fluid in contact with the surface.

As a rule the temperature change at a surface is not instantaneous but takes place over a period of time. In thick-walled structures the mean temperature, $\overline{T}$, would not change significantly as a result of the local change in the vicinity of the surface, and even a moderate rate of temperature change would produce a surface stress as given by Eq. (3.15-3) or Eq. (3.15-4). Thermal "shields" and thermal "sleeves" are sometimes installed in heavy-walled structures and in pipes which are exposed to sudden temperature changes. The thermal shields are thin-walled structural elements whose mean temperatures change rapidly and thereby keep $\overline{T} - T_s$, in the sleeves, at a low value. The installation of thermal shields on vulnerable components is good design practice since the thermal stresses are reduced, and the possible generations of fatigue cracks due to the repeated shock of thermal sleeves does not represent as critical a hazard as thermal shock and possible fracture of a major structural component.

Example (a) Stress Due to Surface Linear Temperature Ramp

A problem that is often encountered is that of a plate, tube, or shell having one surface subjected to a linear temperature ramp, with the other surface insulated. It is clear that the steeper the temperature ramp, that is, the shorter the time interval, t_o, during which a temperature increase, ΔT_o, is applied, and the lower the conductivity, k, of the structural material, the greater will be the thermal stress. Similarly, the greater the volumetric specific heat C_p, the film coefficient h, and the thickness, L, of the plate or shell, the larger will be the thermal stress. The variation of the peak stresses with these parameters[1] is shown in Fig. 3.15-1. The maximum stresses occur at a time t_o after the start of the linear transient, i.e., at the termination of the transient.

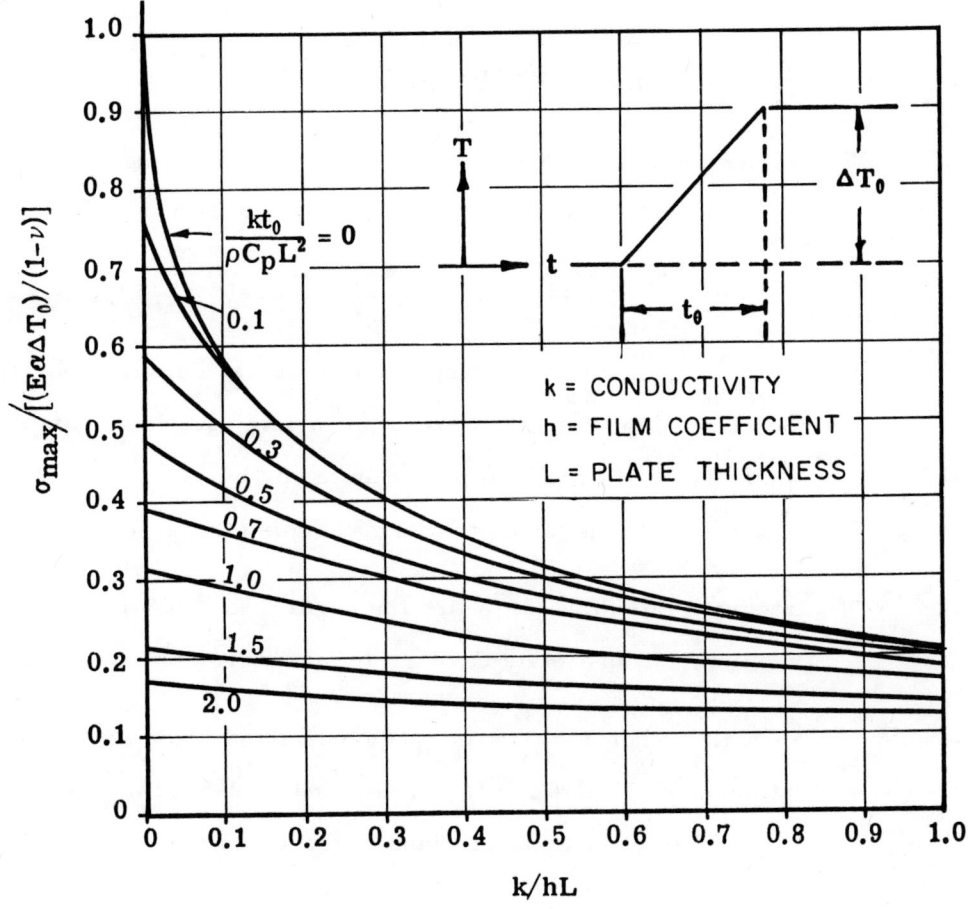

Fig. 3.15-1

---

[1] Fritz, R.J., Trans. ASME, Journal of Appl. Mech., 913-921 (Aug. 1954). Mirabel, J.A., and Van Voohris, J.K., KAPL-2000-14 (July 1961).

When the thermal shock occurs at the surface of a thick-walled structure, the parameter $kt_o/\rho C_p L^2$ is essentially equal to zero, and the uppermost curve of Fig. 3.15-1 applies. The parameter k/hL, which is the abscissa in Fig. 3.15-1 is also approximately zero. In the case of a heavy-walled structure, since L is very large, it is observed in Fig. 3.15-1 that in a very severe thermal shock, the maximum stress approaches $\dfrac{E\alpha\Delta T_o}{1-\nu}$.

PROBLEMS

3-1 Derive the biharmonic plane stress equation, using displacement stresses defined as

$$\sigma'_x = \frac{\partial^2 \varphi}{\partial y^2} + \frac{E\alpha T}{1-\nu} \qquad \sigma'_y = \frac{\partial^2 \varphi}{\partial x^2} + \frac{E\alpha T}{1-\nu} \qquad \tau'_{x,y} = -\frac{\partial^2 \varphi}{\partial x \partial y}$$

Determine whether this displacement stress definition satisfies the plane displacement stress equilibrium equations. State the boundary conditions, expressed in terms of φ, for an unrestrained plate subjected to thermal plane stress.

3-2 Integrate Eq. (3.1-22) to find the general form of the stress function, φ, in a plate having a temperature distribution $T = T(r)$.

3-3 Show that when $T = T(x, y)$, the generalized plane strain relationships, Eqs. (3.2-3), (3.2-4) and (3.2-5) always satisfy the strain compatibility Laplacians in the xz and yz planes, i.e.,

$$\frac{\partial^2 \epsilon_x}{\partial z^2} + \frac{\partial^2 \epsilon_z}{\partial x^2} = 0 \qquad \frac{\partial^2 \epsilon_y}{\partial z^2} + \frac{\partial^2 \epsilon_z}{\partial y^2} = 0$$

3-4 Explain why the temperature distribution

$$T = T_o \cos \frac{\pi x}{L} \cosh \frac{\pi y}{L}$$

will produce no stresses in a square plate of length, L, but will produce stresses in a rod with a square cross section, L by L. The cross section coordinates are x and y, with the origin at the centroid.

3-5 Which of the following temperature distributions produce stress-free deformations in the elements indicated?

(a) $T = T_o \dfrac{y^2}{h^2} \cdot \dfrac{x}{L}$, in a member of a pin-connected statically determinate truss. Transverse dimension is y, and axial dimension is x.

(b) $T = \left| \dfrac{T_o y}{L} \right|$ in a plate. The coordinate y is measured in the plane of the plate, from a central axis. The temperature is thus always positive and the temperature distribution is tent-shaped.

(c) $T = \dfrac{T_o r}{a} (\sin\theta + \cos\theta) \cdot \dfrac{x}{L}$, in an unrestrained beam having a circular cross section of radius "a" and length, L. In the coordinate system used, r is the radial coordinate, θ the angular coordinate, and x the axial coordinate. Discuss this problem in terms of the comparative magnitudes of "a" and L.

(d) $T = \dfrac{T_o r}{a} (\sin\theta + \cos\theta) + \dfrac{x}{L}$, in the beam described in (c), above.

3-6 Find the stress distribution in an unrestrained square rod with cross section dimensions L by L, having a cross section temperature distribution

$$T = T_o \cos \frac{\pi x}{L} \cosh \frac{\pi y}{L}$$

Comment on the accuracy of the analysis.

3-7 The saw-tooth temperature variation shown in Fig. 3.4-1(b) can be expressed as

$$T = \frac{8T_o}{\pi^2} \sum_1^\infty \frac{1}{n^2} \sin \frac{n\pi y}{\lambda} \qquad \text{(n odd)}$$

Derive the expression for the stresses σ_z, σ_y, and τ_{yz}, and find the σ_y at $y = 0$, $z = \frac{w}{2}$, for the case of $\lambda/w = 3$.

3-8 The temperature distribution in the form of a square wave, shown in Fig. 3.4-1(c), is expressed as

$$T = \frac{4T_o}{\pi} \sum_1^\infty \frac{1}{n} \sin \frac{n\pi y}{\lambda} \qquad \text{(n odd)}$$

Derive the expression for the stresses σ_z, σ_y, and τ_{yz}, and plot curve for σ_y at $y = 0$, $z = w/2$.

3-9 An unrestrained beryllium bar has a rectangular cross section, $h = 3''$ and $w = 2''$, with h and w as in Fig. 3.5-2. It is subjected to uniform heat generation. The bar is insulated at $z = \pm w/2$, so that the resulting steady state temperature distribution is parabolic and expressed in °F as

$$T = 200 \left(1 - \frac{4y^2}{h^2}\right)$$

Use $E = 40 \times 10^6$ psi, $\alpha = 22 \times 10^{-6}/°F$, $\nu = 0.10$. Use a Fourier series analysis, as in Problems 3-7 and 3-8, to find the cross section normal and shear stress distributions. Check results

PROBLEMS 199

against those given in Fig. 3.5-5.

3-10 In Section 3.6, we find that peak normal stresses $\pm \dfrac{E\alpha T_o}{2}$ are developed at the edge of a circular hot area in a plate. Find the maximum stress intensity $S = 2\tau_{max}$.

3-11 The stress distribution in a homogeneous plate, having a symmetrical transverse temperature distribution, is given by Eq. (3.7-2). Use equivalent load analysis to obtain this result. Repeat the problem for the case of the sandwich plate shown in Fig. 3.8-1, and obtain the stresses given by Eqs. (3.8-2) and (3.8-3).

3-12 There is a uniform heat generation of magnitude W (per unit volume) in the core of a sandwich plate. The conductivity of the core and outside plates is the same, and equal to k, and the outer surfaces are maintained at zero temperature. Using the steady state temperature distribution, find the expression for the stresses in the core and outer plates. Use the conventional symbols for the material properties with the subscript 1 for the core, and 2 for the outer plates.

3-13 Use modified elastic constants in the solution of a ring with a radial temperature variation (Sect. 2.13) to obtain the solution of a circular cylindrical shell with a radial temperature variation (Sect. 3.9).

3-14 Obtain, by means of equivalent load analysis, and also by the use of the classical equation, the solution of the problem of a circular cylindrical shell with a radial temperature variation, restrained from axial expansion, as shown in Fig. 3-14.

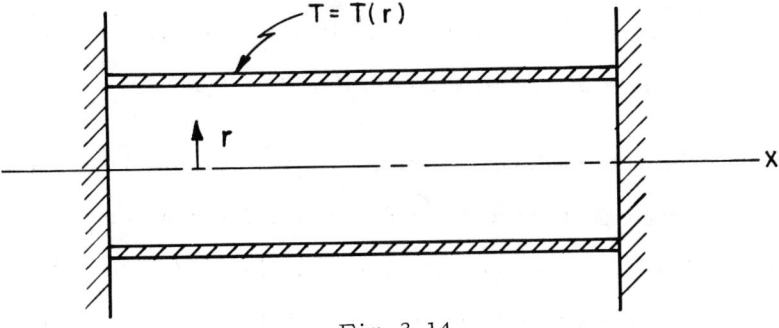

Fig. 3-14

3-15 A heated rod and tube assembly consists of a solid circular rod at a temperature T_o, and a concentric <u>thin</u> cylindrical tube at temperature $T_o(1 - \frac{x^2}{h^2})$, as shown in Fig. 3-15. Assuming that the assembly is stress-free before the application of the temperature distributions, find the expressions for the stresses in the rod and tube. Assume no axial slippage, and the maintenance of radial contact at all times. The outer radius of the rod and the mean radius of the tube can be taken as R. Use the conventional symbols with subscripts 1 for the rod and 2 for the tube.

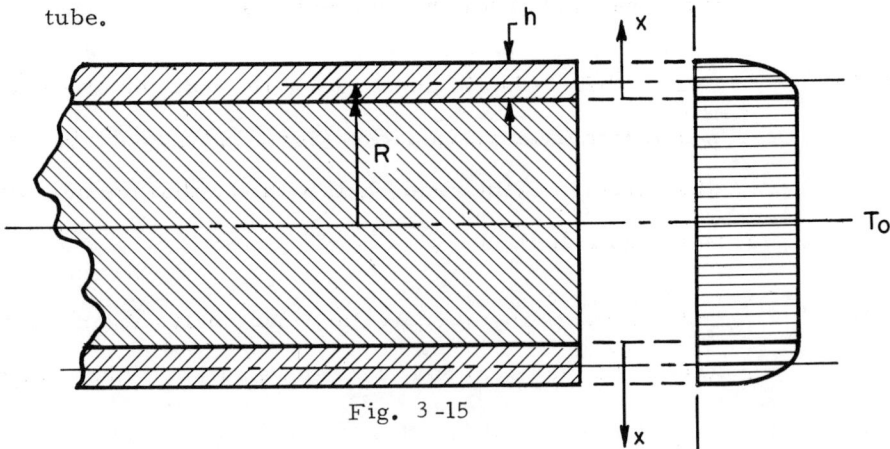

Fig. 3-15

3-16   An assembly consists of a solid sphere surrounded by a thin spherical shell. As in the preceding problem, the solid sphere is heated to a temperature T_o and the shell to a temperature $T_o(1-\frac{x^2}{h^2})$, x being measured outward from the interface. Derive the expressions for the stresses in the components. Use conventional symbols with subscripts 1 for the solid sphere and 2 for the shell.

3-17   Derive expressions for the stresses in the components of an assembly consisting of a rod and thin-walled tube held rigidly between two walls to prevent axial expansion. See Fig. 3-17.

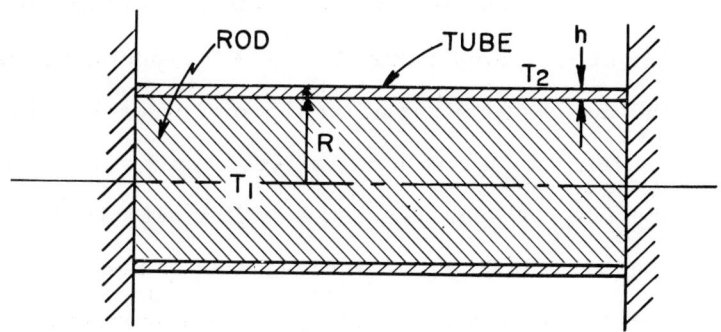

Fig. 3-17

The rod and tube are of different materials and at different but uniform temperatures T_1 and T_2 as shown. Use conventional symbols and subscripts 1 and 2.

3-18   An assembly of a rod and tube contains a lubricant at the interfacial surface which permits unrestrained axial expansion of the rod relative to the tube. See Fig. 3-18. Find the stresses in the components.

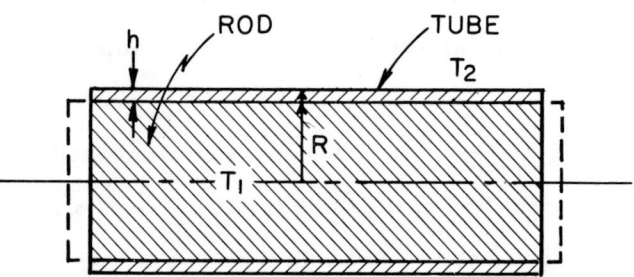

Fig. 3-18

The rod and tube are of different materials and at different but uniform temperatures, T_1 and T_2. Use E_1, α_1, for the rod and E_2, α_2, for the tube.

3-19   The inner surface of a thick-walled rocket nozzle is suddenly heated to a temperature T_o, by the propellant gases. What is the initial magnitude of the thermal stress? If the rocket is operated for a fraction of a second only, what can be said of the thermal stresses during the cooling period? If the rocket operates for a long period of time, what can be said of the changes in thermal stresses that take place with increasing time? Use E_o and α_o as the material constants. Assume the outer surface of the nozzle to be insulated.

3-20   The outer surface of a pressure vessel nozzle is insulated and the inner surface is subjected to a temperature change, $\Delta T_o = 200°F$, over a period $t_o = 2$ sec. It is found that the dimensionless parameters, shown in Fig. 3.15-1, are approximately

$$\frac{kt_o}{\rho C_p L^2} = 0.01 \qquad \frac{k}{hL} = 0$$

A cylindrical thermal sleeve, which is one tenth as thick as the nozzle wall and of the same material is placed inside the nozzle. Assuming the thermal sleeve to be insulated at the outer surface, what is the ratio of the stress in the thermal sleeve to the original stress in the nozzle, when subjected to the same temperature change in the same period of time? The inner nozzle wall experiences the same temperature change as before but the time of the temperature change is now ten seconds. What is the percent reduction in the nozzle stress?

4 | Radial Temperature Variation

4.1 CONCENTRIC CIRCULAR PLATES

The problem of an assembly of two concentric thick-walled rings or circular plates subjected to a temperature increase or decrease is in essence a ring shrink-fit problem. In the more general treatment of the problem, the two constituents may be assumed to be of unlike materials and at different but uniform temperatures. Fig. 4.1-1 shows the concentric element assembly.

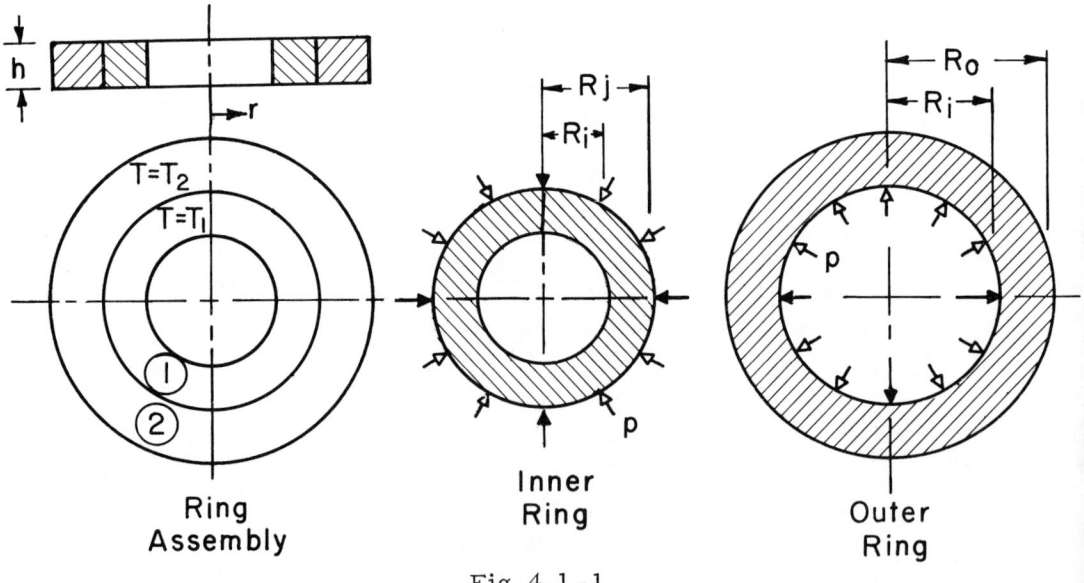

Fig. 4.1-1

In shrink-fit analysis the concentric components are taken to be stress-free at a datum temperature T_d, which is usually below room temperature, and the temperature of the assembly is raised uniformly. If Δ represents the room temperature radial interference then the datum temperature T_d is related to the room temperature, T_o as follows:

$$\Delta = -R_j(\alpha_1 - \alpha_2)(T_d - T_o) \qquad \alpha_1 > \alpha_2 \qquad (4.1\text{-}1)$$

from which we obtain

$$T_d = T_o - \frac{\Delta}{R_j(\alpha_1 - \alpha_2)} \qquad (4.1\text{-}2)$$

In thermal stress analysis the temperature T actually represents the temperature increase above room temperature; and it is tacitly assumed that at room temperature the structure is stress-free. That is, $T = T_t - T_o$ with T_t the thermometric temperature. In shrink-fit analysis the temperature rise above the no-interference stress-free, or datum temperature, is

$$T' = T_t - T_d = T_t - T_o + \frac{\Delta}{R_j(\alpha_1 - \alpha_2)} = T + \frac{\Delta}{R_j(\alpha_1 - \alpha_2)} \qquad (4.1\text{-}3)$$

so that T' should be used in lieu of T, in the analysis which follows.

Consider first either the inner or the outer ring at a uniform temperature, T, subjected to a radial pressure, p. At the contact surface, the interfacial radial pressure p increases with increasing temperature. As each of the circular plates is at a uniform temperature, the stresses that will be generated will be due to p alone. The inside plate is thus subject to an external pressure and the outside plate to an internal pressure. The general analysis which follows applies to either of the circular plates with a concentric central hole.

The two-dimensional stress-strain relationships are

$$\epsilon_r = \frac{1}{E}(\sigma_r - \nu\sigma_t) + \alpha T \tag{4.1-4}$$

$$\epsilon_t = \frac{1}{E}(\sigma_t - \nu\sigma_r) + \alpha T \tag{4.1-5}$$

In these equations the temperature T is constant. For axially symmetrical deformations we have

$$\epsilon_r = \frac{du}{dr} \qquad \epsilon_t = \frac{u}{r} \tag{4.1-6}$$

from which we obtain the compatibility relationship

$$\epsilon_t - \epsilon_r + r\frac{d\epsilon_t}{dr} = 0 \tag{4.1-7}$$

Substitution of the strains, Eqs. (4.1-4) and (4.1-5), into this equation yields the compatibility equation in terms of stress as

$$\sigma_r = \sigma_t + r\frac{d\sigma_t}{dr} - \nu(\sigma_r + r\frac{d\sigma_r}{dr} - \sigma_t) \tag{4.1-8}$$

The equilibrium of a differential element of thickness h, shown in Fig. 4.1-2, is expressed by Eq. (4.1-9).

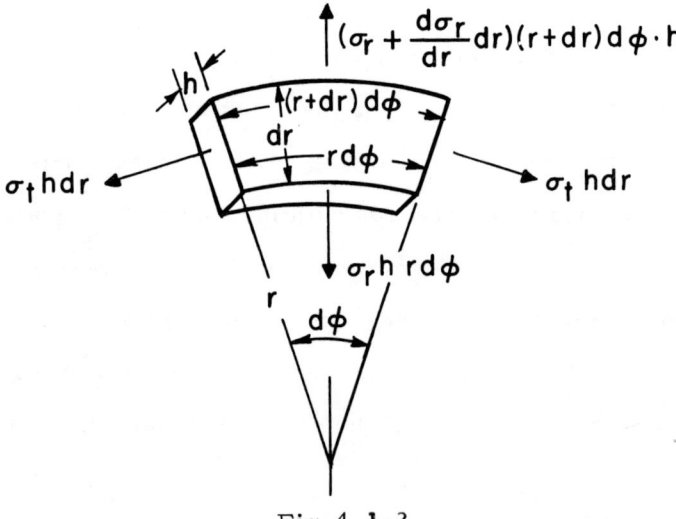

Fig. 4.1-2

CONCENTRIC CIRCULAR PLATES

$$\sigma_t dr\, d\varphi \cdot h = (\sigma_r + \frac{d\sigma_r}{dr} dr)(r + dr)\, d\varphi \cdot h - \sigma_r r\, d\varphi \cdot h = 0 \qquad (4.1\text{-}9)$$

Neglecting higher order terms, we obtain

$$\sigma_t = \sigma_r + r\frac{d\sigma_r}{dr} = \frac{d}{dr}(r\sigma_r) \qquad (4.1\text{-}10)$$

In accordance with the foregoing equation, the term in paranthesis, in Eq. (4.1-8) is zero, so that Eq. (4.1-8) can be written as

$$\sigma_r = \sigma_t + r\frac{d\sigma_t}{dr} \qquad (4.1\text{-}11)$$

Substitution of σ_t, from Eq. (4.1-10) into the above yields

$$\frac{d}{dr}(r\sigma_r) - \frac{(r\sigma_r)}{r} + r\frac{d^2(r\sigma_r)}{dr^2} = 0 \qquad (4.1\text{-}12)$$

It is easily verified that this equation can be written in the form

$$r\frac{d}{dr}\left[\frac{1}{r}\frac{d}{dr}(r^2\sigma_r)\right] = 0 \qquad (4.1\text{-}13)$$

Integration of Eq. (4.1-13) yields the radial stress as

$$\sigma_r = A - \frac{B}{r^2} \qquad (4.1\text{-}14)$$

where A and B are undetermined constants. With the use of Eq. (4.1-10) we obtain the tangential stress as

$$\sigma_t = A + \frac{B}{r^2} \qquad (4.1\text{-}15)$$

For plates of equal thickness, $h_1 = h_2$, the boundary conditions are

$$\sigma_{1r}(R_i) = 0 \qquad \sigma_{1r}(R_j) = -p$$
$$\sigma_{2r}(R_o) = 0 \qquad \sigma_{2r}(R_j) = -p \qquad (4.1\text{-}16)$$

The subscripts 1 and 2 refer to the inner and outer elements respectively, and the subscript j refers to the interface. The constants are

$$A_1 = -\frac{R_j^2 p}{R_j^2 - R_i^2} \qquad B_1 = -\frac{(R_j R_i)^2 p}{R_j^2 - R_i^2} \qquad (4.1\text{-}17)$$

$$A_2 = \frac{R_j^2 p}{R_o^2 - R_j^2} \qquad B_2 = \frac{(R_j R_o)^2 p}{R_o^2 - R_j^2} \qquad (4.1\text{-}18)$$

which substituted into Eqs. (4.1-14) and (4.1-15) yield the stresses as

$$\sigma_{1r} = -\frac{R_j^2 p}{R_j^2 - R_i^2}\left(1 - \frac{R_i^2}{r^2}\right) \qquad (4.1\text{-}19)$$

$$\sigma_{1t} = -\frac{R_j^2 p}{R_j^2 - R_i^2}\left(1 + \frac{R_i^2}{r^2}\right) \qquad (4.1\text{-}20)$$

$$\sigma_{2r} = \frac{R_j^2 p}{R_o^2 - R_j^2}\left(1 - \frac{R_o^2}{r^2}\right) \qquad (4.1\text{-}21)$$

$$\sigma_{2t} = \frac{R_j^2 p}{R_o^2 - R_j^2}\left(1 + \frac{R_o^2}{r^2}\right) \qquad (4.1\text{-}22)$$

At the interfacial surface $r = R_j$ the stresses are

$$\sigma_{1rj} = -p \qquad \sigma_{1tj} = -\frac{R_j^2 + R_i^2}{R_j^2 - R_i^2} p \qquad (4.1\text{-}23)$$

$$\sigma_{2rj} = -p \qquad \sigma_{2tj} = \frac{R_j^2 + R_o^2}{R_o^2 - R_j^2} p \qquad (4.1-24)$$

Deformation compatibility requires that the radial displacement u_j at the interface, or the tangential strain u_j/R_j be the same in both elements. From Eqs. (4.1-4), (4.1-5), (4.1-23) and (4.1-24) we have

$$\epsilon_{1tj} = \frac{1}{E_1}(\sigma_{1tj} - \nu_1 \sigma_{1rj}) + \alpha_1 T_1 = -\frac{p}{E_1}\left(\frac{R_j^2 + R_i^2}{R_j^2 - R_i^2} - \nu_1\right) + \alpha_1 T_1 \qquad (4.1-25)$$

$$\epsilon_{2tj} = \frac{1}{E_2}(\sigma_{2tj} - \nu_2 \sigma_{2rj}) + \alpha_2 T_2 = \frac{p}{E_2}\left(\frac{R_o^2 + R_j^2}{R_o^2 - R_j^2} + \nu_2\right) + \alpha_2 T_2 \qquad (4.1-26)$$

Equating the two strains and solving for p, we obtain

$$p = \frac{\alpha_1 T_1 - \alpha_2 T_2}{\frac{1}{E_1}\left(\frac{R_j^2 + R_i^2}{R_j^2 - R_i^2} - \nu_1\right) + \frac{1}{E_2}\left(\frac{R_o^2 + R_j^2}{R_o^2 - R_j^2} + \nu_2\right)} \qquad (4.1-27)$$

In a shrink-fit problem $T_1 = T_2 = T$, and T', from Eq. (4.1-3) should be used in place of T. This gives the contact pressure as

$$p = \frac{(\alpha_1 - \alpha_2)T + \frac{\Delta}{R_j}}{\frac{1}{E_1}\left(\frac{R_j^2 + R_i^2}{R_j^2 - R_i^2} - \nu_1\right) + \frac{1}{E_2}\left(\frac{R_o^2 + R_j^2}{R_o^2 - R_j^2} + \nu_2\right)} \qquad (4.1-28)$$

The radial and tangential stresses are obtained by substituting the foregoing expressions for the interfacial pressure p into Eqs. (4.1-19), (4.1-20), (4.1-21) and (4.1-22). The tangential stresses are greater than the radial stresses, with the peak tangential stresses occurring at the inner surfaces of elements 1 and 2. In the case of an initially stress-free assembly of concentric rings or circular plates, or concentric cylinders in which there is axial slippage at the interface, the peak stresses in elements 1 and 2 are

$$\sigma_{1t} = - \frac{\dfrac{2R_j^2 E_1 (\alpha_1 T_1 - \alpha_2 T_2)}{R_j^2 - R_i^2}}{\dfrac{E_1}{E_2}\left(\dfrac{R_o^2 + R_j^2}{R_o^2 - R_j^2}\right) + \dfrac{R_j^2 + R_i^2}{R_j^2 - R_i^2} + \dfrac{E_1}{E_2}\nu_2 - \nu_1} \qquad (4.1\text{-}29)$$

$$\sigma_{2t} = \frac{\dfrac{R_o^2 + R_j^2}{R_o^2 - R_j^2} E_1 (\alpha_1 T_1 - \alpha_2 T_2)}{\dfrac{E_1}{E_2}\left(\dfrac{R_o^2 + R_j^2}{R_o^2 - R_j^2}\right) + \dfrac{R_j^2 + R_i^2}{R_j^2 - R_i^2} + \dfrac{E_1}{E_2}\nu_2 - \nu_1} \qquad (4.1\text{-}30)$$

In the case of a shrink-fit with a room temperature radial interference Δ between the components, and heated uniform to a temperature T above room temperature, the peak stresses, at the inner surfaces, are

$$\sigma_{1t} = - \frac{\dfrac{2R_j^2 E_1}{R_j^2 - R_i^2}\left[(\alpha_1 - \alpha_2)T + \dfrac{\Delta}{R_j}\right]}{\dfrac{E_1}{E_2}\left(\dfrac{R_o^2 + R_j^2}{R_o^2 - R_j^2}\right) + \dfrac{R_j^2 + R_i^2}{R_j^2 - R_i^2} + \dfrac{E_1}{E_2}\nu_2 - \nu_1} \qquad (4.1\text{-}31)$$

$$\sigma_{2t} = \frac{\dfrac{R_o^2 + R_j^2}{R_o^2 - R_j^2} E_1 \left[(\alpha_1 - \alpha_2)T + \dfrac{\Delta}{R_j}\right]}{\dfrac{E_1}{E_2}\left(\dfrac{R_o^2 + R_j^2}{R_o^2 - R_j^2}\right) + \dfrac{R_j^2 + R_i^2}{R_j^2 - R_i^2} + \dfrac{E_1}{E_2}\nu_2 - \nu_1} \qquad (4.1\text{-}32)$$

When the inner element is solid, it is only required to set $R_i = 0$.

Example (b) of Sect. 4.5 contains the solution of the interference problem of two concentric cylinders having radial temperature variations.

Example (a)   Concentric Plates with Like Material Properties

Two concentric circular plates of the same material ($E_1 = E_2 = E$, $\nu_1 = \nu_2 = \nu$, $\alpha_1 = \alpha_2 = \alpha$) have inner and outer radii as shown in Fig. 4.1-1, and are heated to temperatures T_1 and T_2 respectively.

For the case of equal elastic constants and equal plate thicknesses, the contact pressure, Eq.(4.1-27), becomes

$$p = \frac{(R_o^2 - R_j^2)(R_j^2 - R_i^2)}{2R_j^2(R_o^2 - R_i^2)} E\alpha (T_1 - T_2) \tag{1}$$

This is set into Eqs. (4.1-19), (4.1-20), (4.1-21) and (4.1-22) and we obtain the stresses as

$$\sigma_{1r} = -\frac{R_o^2 - R_j^2}{2(R_o^2 - R_i^2)} \left(1 - \frac{R_i^2}{r^2}\right) E\alpha(T_1 - T_2) \tag{2}$$

$$\sigma_{1t} = -\frac{R_o^2 - R_j^2}{2(R_o^2 - R_i^2)} \left(1 + \frac{R_i^2}{r^2}\right) E\alpha(T_1 - T_2) \tag{3}$$

$$\sigma_{2r} = \frac{R_j^2 - R_i^2}{2(R_o^2 - R_i^2)} \left(1 - \frac{R_o^2}{r^2}\right) E\alpha (T_1 - T_2) \tag{4}$$

$$\sigma_{2t} = \frac{R_j^2 - R_i^2}{2(R_o^2 - R_i^2)} \left(1 + \frac{R_o^2}{r^2}\right) E\alpha(T_1 - T_2) \tag{5}$$

Note that

$$\sigma_{1r} + \sigma_{1t} = E\alpha(\overline{T} - T_1) \tag{6}$$

$$\sigma_{2r} + \sigma_{2t} = E\alpha(\overline{T} - T_2) \tag{7}$$

where

$$\overline{T} = \frac{A_1 T_1 + A_2 T_2}{A_1 + A_2} = \frac{(R_j^2 - R_i^2) T_1 + (R_o^2 - R_j^2) T_2}{R_o^2 - R_i^2} \tag{8}$$

This problem can be viewed as a circular plate with an outer radius R_o and an inner radius R_i with a temperature jump at $r = R_j$. It is thus a homogeneous plate with a radial temperature distribution for which, as we shall see in Sect. 4.2 which follows, the relationships given by Eqs. (6) and (7) always apply.

4.2 CIRCULAR PLATE WITH RADIAL TEMPERATURE VARIATION

The stresses and deformations in a circular plate with a radial temperature variation will be a function of the radius only. The loading on a differential element will be as shown in Fig. 4.1-2, and the stress-strain equations, compatibility equation, and equilibrium equation, for the case of axial symmetry given in Section 4.1, will apply to the present problem of a plate with a radial temperature variation. These equations are respectively

$$\epsilon_r = \frac{1}{E}(\sigma_r - \nu\sigma_t) + \alpha T \tag{4.2-1}$$

$$\epsilon_t = \frac{1}{E}(\sigma_t - \nu\sigma_r) + \alpha T \tag{4.2-2}$$

CIRCULAR PLATE WITH RADIAL TEMPERATURE

$$\epsilon_r = \epsilon_t + r \frac{d\epsilon_t}{dr} = \frac{d}{dr}(r\epsilon_t) \tag{4.2-3}$$

$$\sigma_t = \sigma_r + r \frac{d\sigma_r}{dr} = \frac{d}{dr}(r\sigma_r) \tag{4.2-4}$$

with $T = T(r)$.

The compatibility equation in terms of stress is obtained by setting Eqs. (4.2-1) and (4.2-2) into Eq. (4.2-3). The result is

$$\sigma_r = \sigma_t + r \frac{d\sigma_t}{dr} - \nu(\sigma_r + r\frac{d\sigma_r}{dr} - \sigma_t) + E\alpha r \frac{dT}{dr} \tag{4.2-5}$$

It follows from Eq. (4.2-4) that the expression in parenthesis in the foregoing equation is identically zero. Eq. (4.2-5) then becomes

$$\sigma_r = \frac{d}{dr}(\sigma_t r) + E\alpha r \frac{dT}{dr} \tag{4.2-6}$$

and substitution of σ_t, from Eq. (4.2-4,) transforms Eq. (4.2-6) into

$$r \frac{d\sigma_r}{dr} + r \frac{d^2(r\sigma_r)}{dr^2} = -E\alpha r \frac{dT}{dr} \tag{4.2-7}$$

The two terms on the left-hand side of the equation may be written as a single differential. The equation then becomes

$$r \frac{d}{dr}\left[\frac{1}{r} \frac{d}{dr}(r^2 \sigma_r)\right] = -E\alpha r \frac{dT}{dr} \tag{4.2-8}$$

Direct integration gives the radial stress as

$$\sigma_r = A - \frac{B}{r^2} - \frac{E\alpha}{r^2} \int^r Tr\,dr \tag{4.2-9}$$

and the tangential stress, from Eq. (4.2-4) is

$$\sigma_t = A + \frac{B}{r^2} + \frac{E\alpha}{r^2} \int^r Tr\,dr - E\alpha T \tag{4.2-10}$$

Note that the sum of the stresses $\sigma_r + \sigma_t$ is

$$\sigma_r + \sigma_t = 2A - E\alpha T \qquad (4.2-11)$$

Circular Plate with Central Hole

The boundary conditions for a plate with a central hole, having an inner radius R_i and an outer radius R_o, shown in Fig. 4.2-1, are

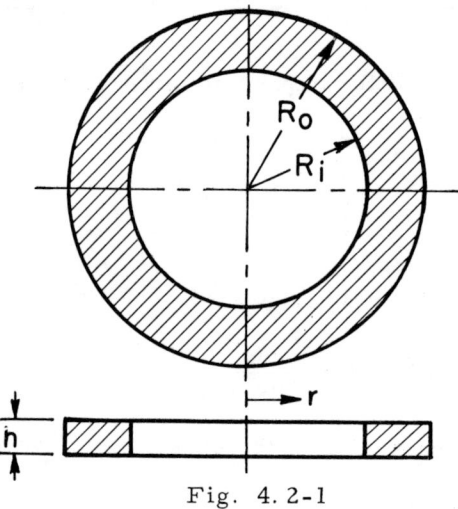

Fig. 4.2-1

$$\sigma_r(R_i) = 0 \qquad \sigma_r(R_o) = 0 \qquad (4.2-12)$$

When these boundary conditions are set into Eq. (4.2-9) we obtain the constants A and B as

$$A = \frac{E\alpha}{R_o^2 - R_i^2} \int_{R_i}^{R_o} Tr\,dr = \frac{E\alpha \overline{T}}{2} \qquad (4.2-13)$$

$$B = \frac{E\alpha R_i^2}{R_o^2 - R_i^2} \int_{R_i}^{R_o} Tr\,dr - E\alpha \int^{R_i} Tr\,dr \qquad (4.2-14)$$

With the substitution of $A = E\alpha \bar{T}/2$ into Eq. (4.2-11) the sum of the stresses $\sigma_r + \sigma_t$ is found to be

$$\sigma_r + \sigma_t = E\alpha(\bar{T} - T) \qquad (4.2\text{-}15)$$

and substitution of the values of the constants A and B into Eqs. (4.2-9) and (4.2-10) yields the stresses

$$\sigma_r = \frac{E\alpha}{r^2}\left(\frac{r^2 - R_i^2}{R_o^2 - R_i^2}\int_{R_i}^{R_o} Tr\,dr - \int_{R_i}^{r} Tr\,dr\right) \qquad (4.2\text{-}16)$$

and

$$\sigma_t = \frac{E\alpha}{r^2}\left(\frac{r^2 + R_i^2}{R_o^2 - R_i^2}\int_{R_i}^{R_o} Tr\,dr + \int_{R_i}^{r} Tr\,dr - Tr^2\right) \qquad (4.2\text{-}17)$$

The radial and tangential stresses may also be written as

$$\sigma_r = \frac{E\alpha}{2}\left(1 - \frac{R_i^2}{r^2}\right)(\bar{T} - \bar{T}_{ri}) = \frac{E\alpha}{2}\left(1 - \frac{R_o^2}{r^2}\right)(\bar{T} - \bar{T}_{ro}) \qquad (4.2\text{-}18)$$

$$\sigma_t = E\alpha(\bar{T} - T) - \frac{E\alpha}{2}\left(1 - \frac{R_i^2}{r^2}\right)(\bar{T} - \bar{T}_{ri}) \qquad (4.2\text{-}19)$$

$$= E\alpha(\bar{T} - T) - \frac{E\alpha}{2}\left(1 - \frac{R_o^2}{r^2}\right)(\bar{T} - \bar{T}_{ro})$$

where $\bar{T}$ is the mean temperature of the plate, $\bar{T}_{ri}$ the mean temperature of the inner portion of the plate, lying between R_i and r, and $\bar{T}_{ro}$ is the mean temperature of the portion of the plate between r and R_o.

In common temperature distributions, such as these due to radial heat conduction or uniform internal heat generation, the maximum stresses will be the tangential stresses σ_{ti} and σ_{to}, at the inner and outer radii of the plate.

Since $\sigma_r = 0$ at these locations we obtain these stresses directly from Eq. (4.2-15) or Eq. (4.2-19) as

$$\sigma_{ti} = E\alpha(\overline{T} - T_i)$$
$$\sigma_{to} = E\alpha(\overline{T} - T_o)$$
(4.2-20)

where the subscript i represents the inner radius and the subscript o, the outer radius.

The displacements of the inner and outer rims of the plate u_i and u_o are obtained from Eq. (4.2-2) and Eq. (4.2-20). Thus

$$u_i = \epsilon_{ti} R_i = \left(\frac{\sigma_{ti}}{E} + \alpha T_i\right) R_i = \alpha \overline{T} R_i \qquad (4.2-21)$$

$$u_o = \epsilon_{to} R_o = \left(\frac{\sigma_{to}}{E} + \alpha T_o\right) R_o = \alpha \overline{T} R_o \qquad (4.2-22)$$

Solid Circular Plate

Consider now a plate of radius, "a", without a central hole. The constant B in Eqs. (4.2-9) and (4.2-10) is zero since otherwise the stresses at the center of the plate would be infinite. The equations then become

$$\sigma_r = A - \frac{E\alpha}{r^2} \int^r T r\, dr \qquad (4.2-23)$$

$$\sigma_t = A + \frac{E\alpha}{r^2} \int^r T r\, dr - E\alpha T \qquad (4.2-24)$$

The boundary condition that $\sigma_r(R_o) = 0$ is used to find the constant of integration, A. From Eq. (4.2-23) we obtain A as

$$A = \frac{E\alpha}{a^2} \int_0^a Tr\,dr = \frac{E\alpha \overline{T}}{2} \tag{4.2-25}$$

Substitution into Eqs. (4.2-23) and (4.2-24) and adding the equations yields the expression

$$\sigma_r + \sigma_t = E\alpha(\overline{T} - T) \tag{4.2-26}$$

and we obtain the individual stress distributions as

$$\sigma_r = E\alpha \left(\frac{1}{a^2} \int_0^a Tr\,dr - \frac{1}{r^2} \int_0^r Tr\,dr \right) \tag{4.2-27}$$

$$\sigma_t = E\alpha \left(\frac{1}{a^2} \int_0^a Tr\,dr + \frac{1}{r^2} \int_0^r Tr\,dr - T \right) \tag{4.2-28}$$

These equations can also be written in terms of the mean temperatures of the plate $\overline{T}$, and the mean temperatures of the inner and outer portions of the plate. In terms of these quantities the stresses are

$$\sigma_r = \frac{E\alpha}{2}(\overline{T} - \overline{T}_{ri}) = \frac{E\alpha}{2}\left(1 - \frac{R_o^2}{r^2}\right)(\overline{T} - \overline{T}_{ra}) \tag{4.2-29}$$

$$\sigma_t = E\alpha(\overline{T} - T) - \frac{E\alpha}{2}(\overline{T} - \overline{T}_{ri}) = E\alpha(\overline{T} - T) - \frac{E\alpha}{2}\left(1 - \frac{R_o^2}{r^2}\right)(\overline{T} - \overline{T}_{ra}) \tag{4.2-30}$$

where $\overline{T}_{ri}$ is the mean temperature of the portion of the plate between $r = 0$ and r, and $\overline{T}_{ra}$ is the mean temperature of the portion of the plate between r and $r = a$. At the center of the plate we find from Eq. (4.2-27) or from Eq. (4.2-28) that the stress is

$$\sigma_c = \frac{E\alpha}{2}(\overline{T} - T_c) \tag{4.2-31}$$

where the subscript c denotes $r = 0$. This is easily verified if it is observed that

$$\frac{1}{r^2} \int_0^r T r\, dr = \frac{T_c}{2} \qquad (4.2\text{-}32)$$

$$r \to o$$

A more direct determination of σ_c may be obtained from Eq. (4.2-29), noting that at $r = o$, $\overline{T}_{ri} = T_c$. At the outer rim, at $r = a$, we have from Eq. (4.2-30), in which $\overline{T}_{ra} = \overline{T}$,

$$\sigma_{ta} = E\alpha(\overline{T} - T_a) \qquad (4.2\text{-}33)$$

The effect of a small hole at the center of a plate is found by comparing the expressions for stress given by Eqs. (4.2-20) and (4.2-31). In Eq. (4.2-20) it is seen that the tangential stress at the inner rim of a small hole at a temperature T_c is $E\alpha(\overline{T} - T_c)$, while Eq. (4.2-31) shows that in a solid plate the stress at the center would be $\frac{E\alpha}{2}(\overline{T} - T_c)$. The drilling of a small hole in the center of a solid plate thus doubles the stress at that point. This effect is similar to a stress concentration factor of two that is obtained when a hole is placed in a uniform plane stress field.

The radial displacement of the outer rim of the plate, $r = a$, is obtained from Eqs. (4.2-2) and (4.2-33) as

$$u_a = \epsilon_{ta}\, a = \left(\frac{\sigma_{ta}}{E} - \alpha T_a \right) = \alpha \overline{T} a \qquad (4.2\text{-}34)$$

Example (a) Radial Step Temperature Change in Circular Plate

Example (a) of Section 4.1 is a shrink-fit problem of two concentric circular plates with central holes, heated to temperatures T_1 and T_2. This problem is now solved assuming that a homogeneous circular plate with a central hole is subjected to a radial temperature distribution in the form of a uniform temperature T_1 from $r = R_i$ to $r = R_j$ and a uniform temperature T_2 from $r = R_j$ to $r = R_o$ so that there is a temperature jump at $r = R_j$ from T_1 to T_2. The plate is shown in Fig. 4.2-2, with the material constants of

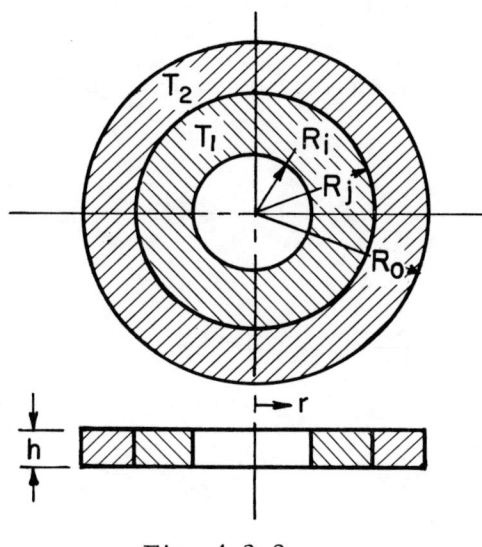

Fig. 4.2-2

the inner and outer portions taken to be the same.

The mean temperature of the plate is

$$\overline{T} = \frac{T_1 A_1 + T_2 A_2}{A_1 + A_2} = \frac{T_1(R_j^2 - R_i^2) + T_2(R_o^2 - R_j^2)}{R_o^2 - R_i^2} \tag{1}$$

so that

$$\overline{T} - T_1 = - \frac{(R_o^2 - R_j^2)(T_1 - T_2)}{R_o^2 - R_i^2} \tag{2}$$

$$\overline{T} - T_2 = \frac{(R_j^2 - R_i^2)(T_1 - T_2)}{R_o^2 - R_i^2} \tag{3}$$

From Eqs. (4.2-18) and (4.2-19), we have in terms of $\overline{T}_{ri}$

$$\sigma_{1r} = \frac{E\alpha}{2}(1-\frac{R_1^2}{r^2})(\overline{T}-T_1) = -\frac{E\alpha}{2}(1-\frac{R_i^2}{r^2})(\frac{(R_o^2-R_j^2)(T_1-T_2)}{R_o^2-R_i^2}) \quad (4)$$

$$\sigma_{1t} = \frac{E\alpha}{2}(1+\frac{R_i^2}{r^2})(\overline{T}-T_1) = -\frac{E\alpha}{2}(1+\frac{R_i^2}{r^2})(\frac{(R_o^2-R_j^2)(T_1-T_2)}{R_o^2-R_i^2}) \quad (5)$$

and from Eqs. (4.2-18) and (4.2-19), we have in terms of $\overline{T}_{ro}$

$$\sigma_{2r} = \frac{E\alpha}{2}(1-\frac{R_o^2}{r^2})(\overline{T}-T_2) = \frac{E\alpha}{2}(1-\frac{R_o^2}{r^2})(\frac{(R_j^2-R_i^2)(T_1-T_2)}{R_o^2-R_i^2}) \quad (6)$$

$$\sigma_{2t} = \frac{E\alpha}{2}(1+\frac{R_o^2}{r^2})(\overline{T}-T_2) = \frac{E\alpha}{2}(1+\frac{R_o^2}{r^2})(\frac{(R_j^2-R_i^2)(T_1-T_2)}{R_o^2-R_i^2}) \quad (7)$$

These expressions for stress coincide with the solution given in Example (a) of Section 4.1.

Example (b)   Plate with Linear Radial Temperature Distribution

The flat plate ②, shown in Fig. 4.2-3 is assumed to be infinite in extent. A linear distribution in the form

$$T = T_o(1 - \frac{r}{a}) \quad (1)$$

is applied over a circular area, ①, of radius "a" as shown.

The problem is analyzed as follows: (1), The stresses due to the interference of the surrounding cold area on the free expansion of the hot area, is computed, and (2), the stresses due to the temperature variation in the unrestrained hot area are added to the interference stresses.

Eq. (4.2-34) shows that the free expansion radial displacement at the outer edge of a circular plate with a radial temperature variation is $u = \alpha \overline{T} a$,

CIRCULAR PLATE WITH RADIAL TEMPERATURE

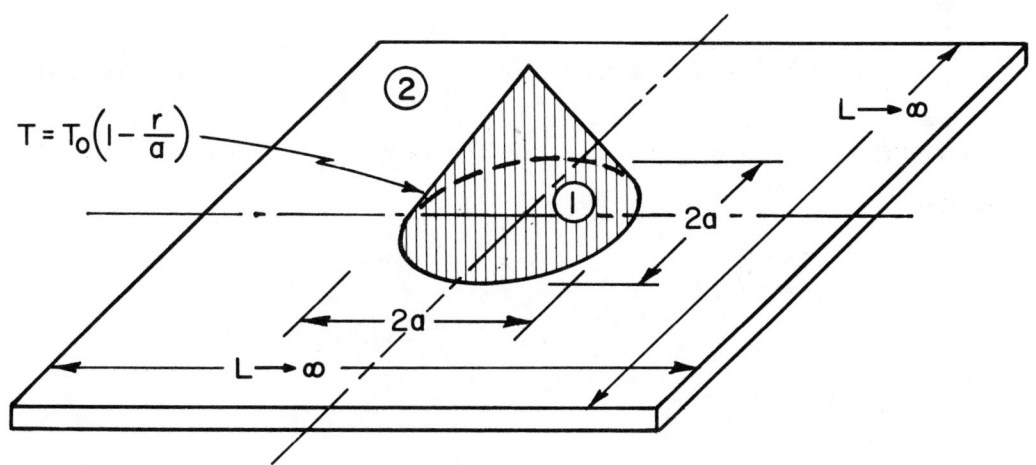

Fig. 4.2-3

or the same as that in a circular plate of radius "a" uniformly heated to a temperature $\overline{T}$. The interference stresses are therefore properly given by Eqs.(3.6-9), (3.6-10) and (3.6-11). These interference stresses are

$$\sigma_{1r} = \sigma_{1t} = -\frac{Ea\overline{T}}{2} \qquad \text{(interference)} \qquad (2)$$

$$\sigma_{2r} = -\frac{Ea\overline{T}}{2} \cdot \frac{a^2}{r^2} \qquad \text{(interference)} \qquad (3)$$

$$\sigma_{2t} = \frac{Ea\overline{T}}{2} \cdot \frac{a^2}{r^2} \qquad \text{(interference)} \qquad (4)$$

The stresses due to radial temperature variation in an unrestrained hot area are given by Eqs.(4.2-29) and (4.2-30). The unrestrained hot area stresses are

$$\sigma_{1r} = \frac{E\alpha}{2}(\overline{T} - \overline{T}_{ri}) \quad \text{(unrestrained)} \quad (5)$$

$$\sigma_{1t} = -\frac{E\alpha}{2}(\overline{T} - \overline{T}_{ri}) + E\alpha(\overline{T} - T) \quad \text{(unrestrained)} \quad (6)$$

The addition of the interference and unrestrained radial and tangential stresses yields

$$\sigma_{1r} = -\frac{E\alpha \overline{T}_{ri}}{2} \quad (7)$$

$$\sigma_{1t} = E\alpha\left(\frac{\overline{T}_{ri}}{2} - T\right) \quad (8)$$

$$\sigma_{2r} = -\frac{E\alpha \overline{T}}{2} \cdot \frac{a^2}{r^2} \quad (9)$$

$$\sigma_{2t} = \frac{E\alpha \overline{T}}{2} \cdot \frac{a^2}{r^2} \quad (10)$$

For the conical temperature distribution under consideration we have

$$\overline{T} = \frac{T_o}{3} \qquad \overline{T}_{ri} = T_o\left(1 - \frac{2r}{3a}\right) \quad (11)$$

Substitution into Eqs. (7), (8), (9) and (10) yields

$$\sigma_{1r} = -\frac{E\alpha T_o}{2}\left(1 - \frac{2r}{3a}\right) \quad (12)$$

$$\sigma_{1t} = \frac{E\alpha T_o}{2}\left(\frac{4}{3}\frac{r}{a} - 1\right) \quad (13)$$

$$\sigma_{2r} = -\frac{E\alpha T_o}{6} \frac{a^2}{r^2} \quad (14)$$

$$\sigma_{2t} = \frac{E\alpha T_o}{6} \frac{a^2}{r^2} \quad (15)$$

Note that the maximum shear stress in the plane of the plate occurs at $r = a$ and is equal to $E\alpha T_o/6$. The absolute maximum shear stress is found at $r=0$

CIRCULAR PLATE WITH RADIAL TEMPERATURE 223

in planes at right angles to the plate, at which point, the magnitude is $E\alpha T_o/4$.

Example (c)   Circular Plate with Varying Thickness

A circular plate in the form of a radially tapered disc is sometimes used as a fin on a tube to increase the rate of cooling. Fig. 4.2-4 shows such an element consisting of three concentric circular plate sections. The temperature

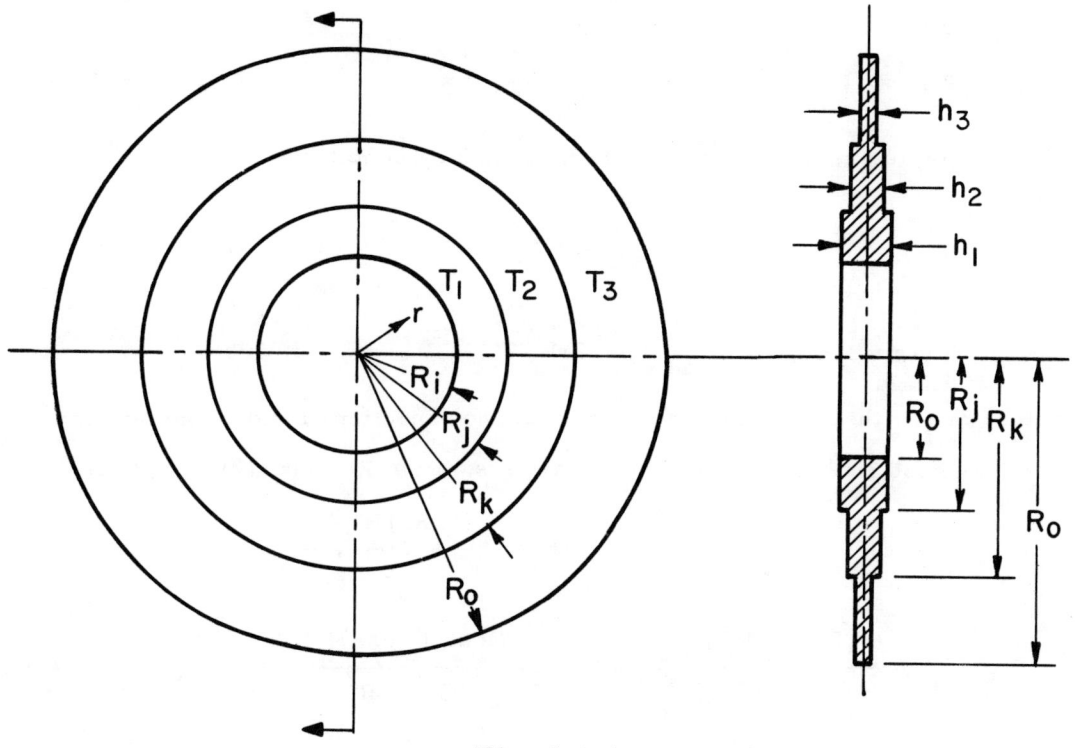

Fig. 4.2-4

$T = T(r)$ is a function of radius only.

The solution of the problem is carried out by obtaining first the interference stresses that are due to uniform heating of each of the sections to temperatures $\bar{T}_1$, $\bar{T}_2$, and $\bar{T}_3$, since, as given by Eqs.(4.2-21) and (4.2-22), such uniform heating produces the same radial displacements and interferences as the radially varying temperatures whose averages are $\bar{T}_1$, $\bar{T}_2$, and

$\overline{T}_3$. The stresses due to the radial temperature variation in the unrestrained plate sections are then added to the interference stresses.

We compute first the interference stresses, σ^*, due to the interference pressures. In accordance with Eqs. (4.1-14) and (4.1-15) the radial and tangential stresses in any of the sections are expressed as

$$\sigma^*_{1r} = A_1 - \frac{B_1}{r^2} \qquad \sigma^*_{2r} = A_2 - \frac{B_2}{r^2} \qquad \sigma^*_{3r} = A_3 - \frac{B_3}{r^2} \qquad (1)$$

$$\sigma^*_{1t} = A_1 + \frac{B_1}{r^2} \qquad \sigma^*_{2t} = A_2 + \frac{B_2}{r^2} \qquad \sigma^*_{3t} = A_3 + \frac{B_3}{r^2} \qquad (2)$$

The constants are determined from the boundary conditions

$$\sigma^*_{1ri} = 0 \qquad \sigma^*_{2rj} h_2 = -P_j \qquad \sigma^*_{3rk} h_3 = -P_k$$
$$\sigma^*_{1rj} h_1 = -P_j \qquad \sigma^*_{2rk} h_2 = -P_k \qquad \sigma^*_{3r0} = 0 \qquad (3)$$

with P_j and P_k the linear interfacial loads between sections 1 and 2, and between sections 2 and 3, respectively. The constants, from Eqs. (1), (2) and (3) are

$$A_1 = -\frac{R_j^2 P_j / h_1}{R_j^2 - R_i^2} \qquad B_1 = -\frac{(R_i R_1)^2 P_j / h_1}{R_j^2 - R_i^2}$$

$$A_2 = \frac{(P_j R_j^2 - P_k R_k^2)/h_2}{R_k^2 - R_j^2} \qquad B_2 = \frac{(R_j R_k)^2 (P_j - P_k)/h_2}{R_k^2 - R_j^2} \qquad (4)$$

$$A_3 = \frac{R_k^2 P_k / h_3}{R_0^2 - R_k^2} \qquad B_3 = \frac{(R_k R_0)^2 P_k / h_3}{R_0^2 - R_k^2}$$

The interference stresses, using Eqs. (1) and (2), are expressed in terms of P_j and P_k as follows

$$\sigma^*_{1r} = -\frac{R_j^2 P_j / h_1}{R_j^2 - R_i^2} \left(1 - \frac{R_i^2}{r^2}\right) \qquad (5)$$

$$\sigma^*_{1t} = -\frac{R_j^2 P_j / h_1}{R_j^2 - R_i^2} \left(1 + \frac{R_i^2}{r^2}\right) \tag{6}$$

$$\sigma^*_{2r} = \frac{(R_j^2 P_j - R_k^2 P_k)/h_2}{R_k^2 - R_j^2} - \frac{(R_j R_k)^2 (P_j - P_k)/h_2}{(R_k^2 - R_j^2) r^2} \tag{7}$$

$$\sigma^*_{2t} = \frac{(R_j^2 P_j - R_k^2 P_k)/h_2}{R_k^2 - R_j^2} + \frac{(R_j R_k)^2 (P_j - P_k)/h_2}{(R_k^2 - R_j^2) r^2} \tag{8}$$

$$\sigma^*_{3r} = \frac{R_k^2 P_k / h_3}{R_0^2 - R_k^2} \left(1 - \frac{R_0^2}{r^2}\right) \tag{9}$$

$$\sigma^*_{3t} = \frac{R_k^2 P_k / h_3}{R_0^2 - R_k^2} \left(1 + \frac{R_0^2}{r^2}\right) \tag{10}$$

The radial and tangential stresses at the junctions $r = R_j$ and $r = R_k$ are

$$\sigma^*_{1rj} = -\frac{P_j}{h_1} \qquad \sigma^*_{1tj} = -\frac{R_j^2 + R_i^2}{R_j^2 - R_i^2} \cdot \frac{P_j}{h_1} \tag{11}$$

$$\sigma^*_{2rj} = -\frac{P_j}{h_2} \qquad \sigma^*_{2tj} = \frac{(R_j^2 + R_k^2) P_j / h_2 - 2 R_k^2 P_k / h_2}{R_k^2 - R_j^2} \tag{12}$$

$$\sigma^*_{2rk} = -\frac{P_k}{h_2} \qquad \sigma^*_{2tk} = \frac{2 R_j^2 P_j / h_2 - (R_j^2 + R_k^2) P_k / h_2}{R_k^2 - R_j^2} \tag{13}$$

$$\sigma^*_{3rk} = -\frac{P_k}{h_3} \qquad \sigma^*_{3tk} = \frac{R_k^2 + R_0^2}{R_0^2 - R_k^2} \cdot \frac{P_k}{h_3} \tag{14}$$

Deformation compatibility at $r = R_j$ and $r = R_k$ requires that $\epsilon_{1tj} = \epsilon_{2tj}$ and $\epsilon_{2tk} = \epsilon_{3tk}$. That is

$$\sigma^*_{1tj} - \nu \sigma^*_{1rj} + E a \overline{T}_1 = \sigma^*_{2tj} - \nu \sigma^*_{2rj} + E a \overline{T}_2 \tag{15}$$

$$\sigma^*_{2tk} - \nu \sigma^*_{2rk} + E a \overline{T}_2 = \sigma^*_{3tk} - \nu \sigma^*_{3rk} + E a \overline{T}_3 \tag{16}$$

Substitution from Eqs. (11), (12), (13) and (14) into the above yields

$$P_j = \frac{E\alpha\left[(\bar{T}_2 - \bar{T}_1)H + (\bar{T}_3 - \bar{T}_2)D\right]}{DF - CH} \tag{17}$$

$$P_k = \frac{E\alpha\left[(\bar{T}_2 - \bar{T}_1)F + (\bar{T}_3 - \bar{T}_2)C\right]}{DF - CH} \tag{18}$$

wherein

$$C = \frac{R_j^2 + R_k^2}{(R_k^2 - R_j^2)h_2} + \frac{R_j^2 + R_i^2}{(R_j^2 - R_i^2)h_1} + \frac{\nu(h_1 - h_2)}{h_1 h_2} \tag{19}$$

$$D = \frac{2R_k^2}{(R_k^2 - R_j^2)h_2} \tag{20}$$

$$F = \frac{2R_j^2}{(R_k^2 - R_j^2)h_2} \tag{21}$$

$$H = \frac{R_j^2 + R_k^2}{(R_k^2 - R_j^2)h_2} + \frac{R_k^2 + R_0^2}{(R_0^2 - R_k^2)h_3} + \frac{\nu(h_2 - h_3)}{h_2 h_3} \tag{22}$$

The interference stresses, Eqs. (5), (6), (7), and (8) which are expressed in terms of P_j and P_k are therefore known.

The distributed stresses, σ_d, in the unrestrained circular plate sections, due to the radial temperature variation are given by Eqs. (4.2-16) and (4.2-17) as

$$\sigma_{1rd} = \frac{E\alpha}{r^2}\left(\frac{r^2 - R_i^2}{2}\bar{T}_1 - \int_{R_i}^{r} T_1 r\, dr\right) \tag{23}$$

$$\sigma_{2rd} = \frac{E\alpha}{r^2}\left(\frac{r^2 - R_j^2}{2}\bar{T}_2 - \int_{R_j}^{r} T_2 r\, dr\right) \tag{24}$$

$$\sigma_{3rd} = \frac{E\alpha}{r^2}\left(\frac{r^2 - R_k^2}{2}\bar{T}_3 - \int_{R_k}^{r} T_3 r\, dr\right) \tag{25}$$

$$\sigma_{1td} = \frac{E\alpha}{r^2}\left(\frac{r^2+R_i^2}{2}\overline{T}_1 + \int_{R_i}^{r} T_1 r\, dr\right) - E\alpha T_1 \tag{26}$$

$$\sigma_{2td} = \frac{E\alpha}{r^2}\left(\frac{r^2+R_j^2}{2}\overline{T}_2 + \int_{R_j}^{r} T_2 r\, dr\right) - E\alpha T_2 \tag{27}$$

$$\sigma_{3td} = \frac{E\alpha}{r^2}\left(\frac{r^2+R_k^2}{2}\overline{T}_3 + \int_{R_k}^{r} T_3 r\, dr\right) - E\alpha T_3 \tag{28}$$

The net thermal stresses are the sum of the interference stresses, σ^*, from Eqs. (5), (6), (7), (8), (9) and (10), and the foregoing distributed stresses. Thus σ_{1r} is

$$\sigma_{1r} = \sigma_{1r}^* + \sigma_{1rd} = -\frac{R_j^2}{R_j^2 - R_i^2}\left(1 - \frac{R_i^2}{r^2}\right)\frac{P_j}{h_1}$$

$$+ \frac{E\alpha}{r^2}\left(\frac{r^2 - R_i^2}{2}\right)\overline{T}_1 - \int_{R_i}^{r} T_1 r\, dr\right) \tag{29}$$

The other five stress components are obtained by making corresponding additions.

4.3 AXIALLY SYMMETRICAL TEMPERATURE VARIATION

In the preceding section the circular plate was subjected to a radial temperature variation $T(r)$. We consider now a circular plate with a radial and axial (normal to the plate) temperature variation, and impose the restriction that the axial temperature variation be symmetrical about the midplane of the plate; that is $T = T(r, z) = T(r, -z)$. It is clear that if we wish to obtain extensional deformations only, it is necessary that we have symmetry in the axial temperature distribution. This type of stress distribution is called radially symmetrical generalized plane stress.

In the absence of bending, the radial and tangential strains will be functions of the radial coordinate only. At any particular radial location the strains will not vary with the axial coordinate, z, although the stresses will be a function of axial coordinate. Fig. 4.3-1 shows a circular element with a radial and axially symmetrical temperature distribution.

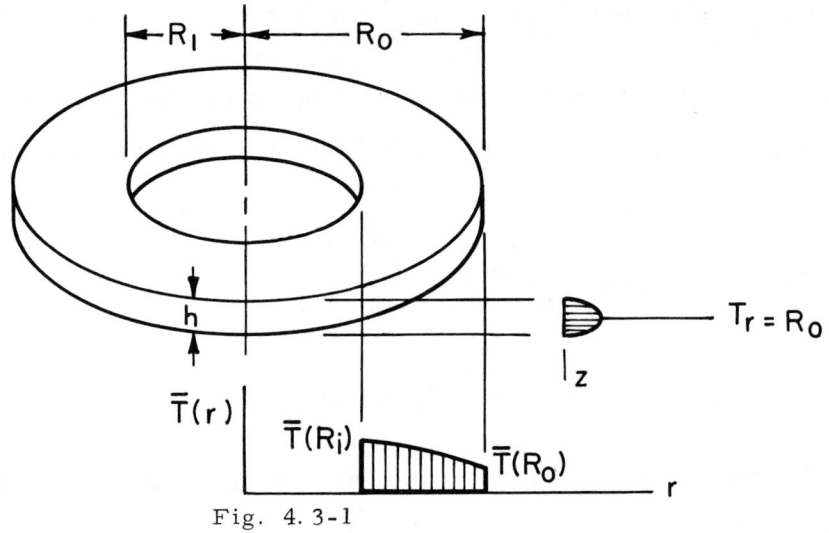

Fig. 4.3-1

The axial stress, σ_z, is assumed to be zero, so that the stress-strain relationships are

$$\epsilon_r = \frac{1}{E}(\sigma_r - \nu \sigma_t) + \alpha T \tag{4.3-1}$$

$$\epsilon_t = \frac{1}{E}(\sigma_t - \nu \sigma_r) + \alpha T \tag{4.3-2}$$

Since ϵ_r and ϵ_t are not functions of the thickness coordinate, z, Eqs. (4.3-1) and (4.3-2) when multiplied by dz and integrated over the thickness, h, yield

$$\epsilon_r = \frac{1}{E}(\bar{\sigma}_r - \nu \bar{\sigma}_t) + \alpha \bar{T}(r) \tag{4.3-3}$$

$$\epsilon_t = \frac{1}{E}(\bar{\sigma}_t - \nu \bar{\sigma}_r) + \alpha \bar{T}(r) \tag{4.3-4}$$

The barred quantities represent average values over the thickness. The equilibrium of a plate element is now established by the mean stresses. Fig. 4.3-2 shows the equilibrium of the element.

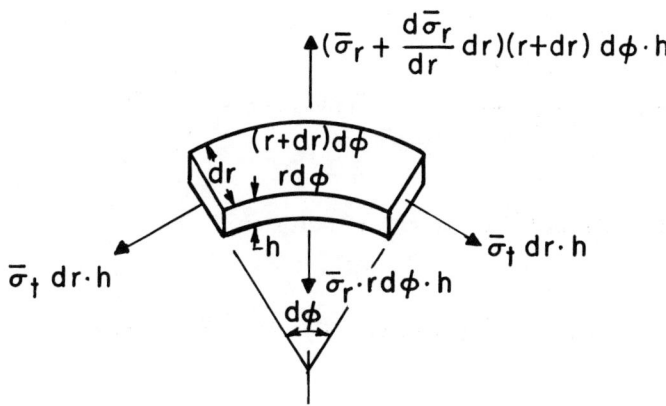

Fig. 4.3-2

The balance of radial loads gives

$$\bar{\sigma}_t = \bar{\sigma}_r + r \frac{d\bar{\sigma}_r}{dr} = \frac{d}{dr}(r\bar{\sigma}_r) \qquad (4.3-5)$$

The compatibility equation, for the case of the strains being functions of r only, is as in Section 4.1, Eq. (4.1-7)

$$\epsilon_r = \epsilon_t + r \frac{d\epsilon_t}{dr} = \frac{d}{dr}(r\epsilon_t) \qquad (4.3-6)$$

A comparison of Eq. (4.3-3), (4.3-4), and (4.3-5) with Eqs. (4.1-4), (4.1-5), and (4.1-10) shows that they are respectively identical, except for the present use of average (thickness) values of stress and temperature. The same procedure is used as in Sect. 4.2 to compute the stresses in circular plates with and without central holes. In the case of a plate with a central hole we obtain

AXIALLY SYMMETRICAL TEMPERATURE VARIATION

$$\bar{\sigma}_r = \frac{E\alpha}{r^2} \left(\frac{r^2 - R_i^2}{R_o^2 - R_i^2} \int_{R_i}^{R_o} \overline{T}(r) \cdot rdr - \int_{R_i}^{r} \overline{T}(r) \cdot rdr \right) \quad (4.3-7)$$

and

$$\bar{\sigma}_t = \frac{E\alpha}{r^2} \left(\frac{r^2 + R_i^2}{R_o^2 - R_i^2} \int_{R_i}^{R_o} \overline{T}(r) \cdot rdr + \int_{R_i}^{R_o} \overline{T}(r) \cdot rdr - \overline{T}(r) \cdot r^2 \right) \quad (4.3-8)$$

and in the case of a solid plate the results are

$$\bar{\sigma}_r = E\alpha \left(\frac{1}{a^2} \int_0^a \overline{T}(r) \cdot rdr - \frac{1}{r^2} \int_0^r \overline{T}(r) \cdot rdr \right) \quad (4.3-9)$$

and

$$\bar{\sigma}_t = E\alpha \left(\frac{1}{a^2} \int_0^a \overline{T}(r) \cdot rdr + \frac{1}{r^2} \int_0^r \overline{T}(r) \cdot rdr - \overline{T}(r) \right) \quad (4.3-10)$$

The sum of the mean radial and tangential stresses are

$$\bar{\sigma}_r + \bar{\sigma}_t = E\alpha (\overline{T} - \overline{T}(r)) \quad (4.3-11)$$

where $\overline{\overline{T}}$ represents the mean plate temperature averaged over both the axial and radial coordinates; that is, over the entire plate. $\overline{T}(r)$ represents the mean temperature over the thickness, and is function of the radial coordinate, r.

In order to obtain the distributed stresses, the radial strains, ϵ_r, from Eqs. (4.3-1) and (4.3-3) are equated, and the tangential strains, ϵ_t, from Eqs. (4.3-2) and (4.3-4) are equated. They yield

$$\sigma_r - \nu\sigma_t = \bar{\sigma}_r - \nu\bar{\sigma}_t + E\alpha(\overline{T}(r) - T) \quad (4.3-12)$$

$$\sigma_t - \nu\sigma_r = \bar{\sigma}_t - \nu\bar{\sigma}_r + E\alpha(\overline{T}(r) - T) \quad (4.3-13)$$

AXIALLY SYMMETRICAL TEMPERATURE VARIATION

Simultaneous solution of the foregoing gives the radial and tangential stresses as

$$\sigma_r = \bar{\sigma}_r + \frac{E\alpha}{1-\nu}(\bar{T}(r) - T) \qquad (4.3\text{-}14)$$

and

$$\sigma_t = \bar{\sigma}_t + \frac{E\alpha}{1-\nu}(\bar{T}(r) - T) \qquad (4.3\text{-}15)$$

Substitution of Eqs. (4.3-7) and (4.3-8) into the above equations yields the stresses in a plate with a central hole as

$$\sigma_r = \frac{E\alpha}{r^2}\left[\frac{r^2 - R_i^2}{R_o^2 - R_i^2}\int_{R_i}^{R_o} \bar{T}(r)\cdot r\,dr - \int_{R_i}^{r} \bar{T}(r)\cdot r\,dr + \frac{r^2}{1-\nu}(\bar{T}(r) - T)\right] \qquad (4.3\text{-}16)$$

and

$$\sigma_t = \frac{E\alpha}{r^2}\left[\frac{r^2 - R_i^2}{R_o^2 - R_i^2}\int_{R_i}^{R_o} \bar{T}(r)\cdot r\,dr + \int_{R_i}^{r} \bar{T}(r)\cdot r\,dr - \bar{T}(r)\cdot r^2 + \frac{r^2}{1-\nu}(\bar{T}(r) - T)\right] \qquad (4.3\text{-}17)$$

For a plate without a central hole, we obtain, with the use of Eqs. (4.3-9) and (4.3-10),

$$\sigma_r = E\alpha\left[\frac{1}{a^2}\int_0^a \bar{T}(r)\cdot r\,dr - \frac{1}{r^2}\int_0^r \bar{T}(r)\cdot r\,dr + \frac{\bar{T}(r) - T}{1-\nu}\right] \qquad (4.3\text{-}18)$$

and

$$\sigma_t = E\alpha\left[\frac{1}{a^2}\int_0^a \bar{T}(r)\cdot r\,dr + \frac{1}{r^2}\int_0^r \bar{T}(r)\cdot r\,dr - \bar{T}(r) + \frac{\bar{T}(r) - T}{1-\nu}\right] \qquad (4.3\text{-}19)$$

When the temperature is a function of r only these results are the same as those obtained in Sect. 4.2, and when the temperature is a function of the thickness coordinate only, these four equations give the common result,

$$\sigma = \sigma_r = \sigma_t = \frac{E\alpha}{1-\nu}(\bar{T} - T) \qquad (4.3\text{-}20)$$

which was found in Sect. 3.7 to be the proper expression for stress in a plate with a symmetrical transverse temperature variation.

The form of the foregoing solution is seen to be, to a large extent, similar to that obtained in Section 4.2 for the case of the temperature being a function of r only. If we substitute mean thickness values of temperature, the only additional modification that is required, in the equations for stress, is the addition of the term

$$E\alpha \left(\frac{\overline{T}(r)-T}{1-\nu}\right) \tag{4.3-21}$$

Eqs. (4.3-16) through (4.3-19) may also be written for a plate with a central hole in terms of mean values, as follows

$$\sigma_r = \frac{E\alpha}{2}\left(1-\frac{R_i^2}{r^2}\right)(\overline{T}-\overline{T}_i(r)) + E\alpha\,\frac{\overline{T}(r)-T}{1-\nu} \tag{4.3-22}$$

$$\sigma_t = E\alpha(\overline{T}-\overline{T}(r)) - \frac{E\alpha}{2}\left(1-\frac{R_i^2}{r^2}\right)(\overline{T}-\overline{T}_i(r)) + E\alpha\,\frac{\overline{T}(r)-T}{1-\nu} \tag{4.3-23}$$

and for a circular plate without a central hole the equations are written as

$$\sigma_r = \frac{E\alpha}{2}(\overline{T}-\overline{T}_i(r)) + E\alpha\,\frac{\overline{T}(r)-T}{1-\nu} \tag{4.3-24}$$

$$\sigma_t = E\alpha(\overline{T}-\overline{T}(r)) - \frac{E\alpha}{2}(\overline{T}-\overline{T}_i(r)) + E\alpha\,\frac{\overline{T}(r)-T}{1-\nu} \tag{4.3-25}$$

The temperature parameters in these equations are defined as

$$T = T(r,z) \qquad \overline{T}(r) = \frac{1}{h}\int_{-\frac{h}{2}}^{\frac{h}{2}} T\,dz$$

$$\tag{4.3-26}$$

$$\overline{T}_i(r) = \frac{2}{(r^2-R_i^2)}\int_{R_i}^{r} \overline{T}(r)\cdot r\,dr \qquad \overline{T} = \frac{2}{(R_o^2-R_i^2)}\int_{R_i}^{R_o} \overline{T}(r)\,r\,dr$$

AXIALLY SYMMETRICAL TEMPERATURE VARIATION   233

For a solid plate without a central hole, Equations (4.3-22) through (4.3-25) also apply, with $R_i = 0$ and $R_o = a$.

The displacements at the inner and outer radii are obtained from Eqs. (4.3-4) and (4.3-11). Setting $\bar{\sigma}_r = 0$ we have

$$u_{r=R_i} = \alpha \bar{T} R_i \qquad u_{r=R_o} = \alpha \bar{T} R_o \qquad (4.3-27)$$

Example (a)   Circular Slab with Axially Symmetrical Temperature

The circular slab shown in Fig. 4.3-3 has a temperature distribution

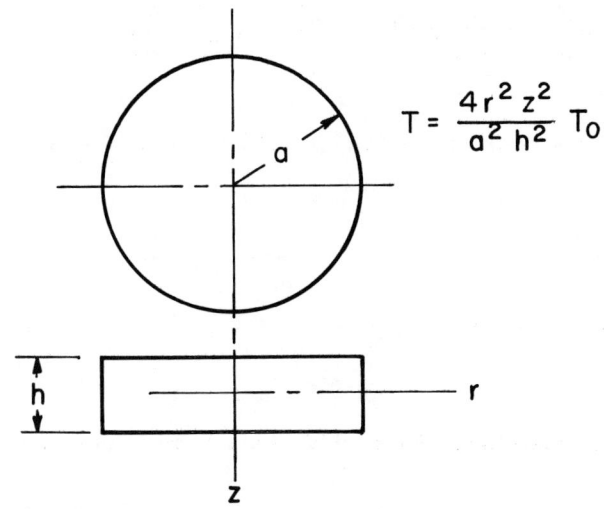

Fig. 4.3-3

$$T = \frac{4r^2 z^2}{a^2 h^2} T_o \qquad (1)$$

The mean temperature parameters, from Eqs. (4.3-26) are

$$\bar{T}(r) = \frac{1}{h} \int_{-\frac{h}{2}}^{\frac{h}{2}} T\, dz = \frac{r^2}{3a^2} T_o \qquad (2)$$

$$\overline{T}_i(r) = \frac{2}{r^2} \int_o^r \overline{T}(r) \, rdr = \frac{r^2}{6a^2} T_o \tag{3}$$

$$\overline{T} = \frac{2}{a^2} \int_o^a \overline{T}(r) \cdot rdr = \frac{T_o}{6} \tag{4}$$

These values are set into Eqs. (4.3-24) and (4.3-25), which yield

$$\sigma_r = E\alpha T_o \left[\frac{a^2 - r^2}{12a^2} + \frac{r^2}{(1-\nu)a^2} \left(\frac{1}{3} - \frac{4z^2}{h^2} \right) \right] \tag{5}$$

$$\sigma_t = E\alpha T_o \left[\frac{a^2 - 3r^2}{12a^2} + \frac{r^2}{(1-\nu)a^2} \left(\frac{1}{3} - \frac{4z^2}{h^2} \right) \right] \tag{6}$$

The maximum stress is the tangential stress at $r = a$, $z = \pm \frac{h}{2}$.

For $\nu = .3$ it is

$$\sigma_t (\max) = -1.12 \, E\alpha T_o \tag{7}$$

Example (b) Sandwich Plate with Axially Symmetrical Temperature Variation

Fig. 4.3-4 shows a segment of a circular sandwich plate having an axially symmetrical material distribution, and a radially varying and axially symmetrical temperature distribution, $T = T(r, z) = T(r, -z)$. We shall assume that the Poisson's ratio, ν, is the same in both materials. As ϵ_r and ϵ_t are not functions of the thickness coordinate z, Eqs. (4.3-1) and (4.3-2) when integrated over the thicknesses yield

$$\epsilon_r E_1 h_1 = \overline{\sigma}_{1r} h_1 - \nu \overline{\sigma}_{1t} h_1 + E_1 \alpha_1 \overline{T}_1(r) h_1 \tag{1}$$

$$\epsilon_r E_2 h_2 = \overline{\sigma}_{2r} h_2 - \nu \overline{\sigma}_{2t} h_2 + E_2 \alpha_2 \overline{T}_2(r) h_2 \tag{2}$$

$$\epsilon_t E_1 h_1 = \overline{\sigma}_{1t} h_1 - \nu \overline{\sigma}_{1r} h_1 + E_1 \alpha_1 \overline{T}_1(r) h_1 \qquad (3)$$

$$\epsilon_t E_2 h_2 = \overline{\sigma}_{2t} h_2 - \nu \overline{\sigma}_{2r} h_2 + E_2 \alpha_2 T_2(r) h_2 \qquad (4)$$

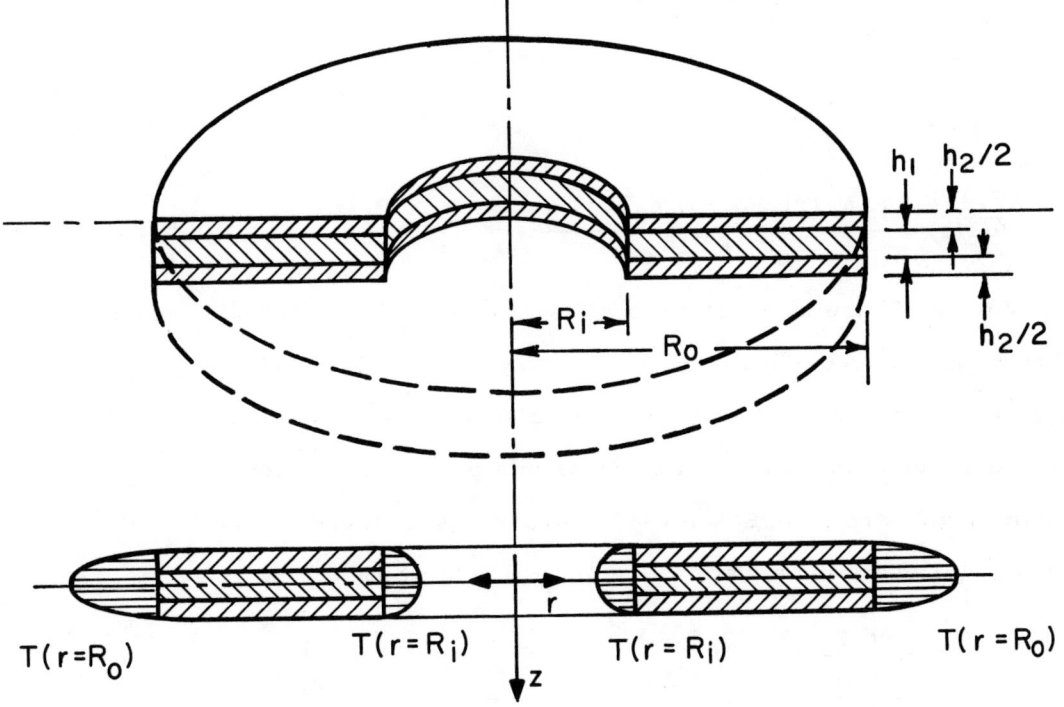

Fig. 4.3-4

The addition of Eqs. (1) and (2) yields

$$\epsilon_r = \frac{(\overline{\sigma}_{1r} h_1 + \overline{\sigma}_{2r} h_2) - \nu(\overline{\sigma}_{1t} h_1 + \overline{\sigma}_{2t} h_2) + E_1 \alpha_1 \overline{T}_1(r) h_1 + E_2 \alpha_2 T_2(r) h_2}{E_1 h_1 + E_2 h_2} \qquad (5)$$

or in terms of parameters defined by Eqs. (8), (9), (10) and (11)

$$\epsilon_r = \frac{1}{\overline{E}} (\overline{\sigma}_r - \nu \overline{\sigma}_t) + \overline{\alpha T}(r) \qquad (6)$$

and from Eqs. (3) and (4) we obtain

$$\epsilon_t = \frac{1}{\overline{E}} (\overline{\sigma}_t - \nu \overline{\sigma}_r) + \overline{\alpha T}(r) \qquad (7)$$

The mean quantities in the foregoing equations are defined as

$$\bar{E} = \frac{E_1 h_1 + E_2 h_2}{h_1 + h_2} \tag{8}$$

$$\bar{\sigma}_r = \frac{\bar{\sigma}_{1r} h_1 + \bar{\sigma}_{2r} h_2}{h_1 + h_2} \tag{9}$$

$$\bar{\sigma}_t = \frac{\bar{\sigma}_{1t} h_1 + \bar{\sigma}_{2t} h_2}{h_1 + h_2} \tag{10}$$

$$\overline{aT}(r) = \frac{E_1 a_1 \bar{T}_1(r) h_1 + E_2 a_2 \bar{T}_2(r) h_2}{E_1 h_1 + E_2 h_2} \tag{11}$$

Eqs. (6) and (7) are in the same form as Eqs. (4.3-3) and (4.3-4), except a and $\bar{T}$ now appear together as a lumped quantity. Eqs. (4.3-5) and (4.3-6), the equilibrium and compatibility equations also apply as before. Accordingly the mean radial and tangential stresses in a plate with a central hole, and in a solid plate, are as in Eqs. (4.3-7), (4.3-8), (4.3-9) and (4.3-10), with $a\bar{T} \rightarrow \overline{aT}(r)$. Thus for a plate with a central hole we have

$$\bar{\sigma}_r = \frac{\bar{E}}{r^2} \left[\frac{r^2 - R_i^2}{R_0^2 - R_i^2} \int_{R_i}^{R_0} \overline{aT}(r) \cdot r\, dr - \int_{R_i}^{r} \overline{aT}(r) \cdot r\, dr \right] \tag{12}$$

$$\bar{\sigma}_t = \frac{\bar{E}}{r^2} \left[\frac{r^2 + R_i^2}{R_0^2 - R_i^2} \int_{R_i}^{R_0} \overline{aT}(r) \cdot r\, dr + \int_{R_i}^{r} \overline{aT}(r) \cdot r\, dr - \overline{aT}(r) \cdot r^2 \right] \tag{13}$$

and for a solid plate of radius "a" the mean stresses are

$$\bar{\sigma}_r = \bar{E} \left[\frac{1}{a^2} \int_0^a \overline{aT}(r) \cdot r\, dr - \frac{1}{r^2} \int_0^r \overline{aT}(r) \cdot r\, dr \right] \tag{14}$$

$$\bar{\sigma}_t = \bar{E} \left[\frac{1}{a^2} \int_0^a \overline{aT}(r) \cdot r\, dr + \frac{1}{r^2} \int_0^r \overline{aT}(r) \cdot r\, dr - \overline{aT}(r) \right] \tag{15}$$

The radial and tangential strains in the central layer, ①, are expressed in terms of the distributed and mean stress as

$$\epsilon_r = \frac{\bar{\sigma}_r}{\bar{E}} - \nu \frac{\bar{\sigma}_t}{\bar{E}} + \overline{aT}(r) = \frac{\sigma_{1r}}{E_1} - \nu \frac{\sigma_{1t}}{E_1} + a_1 T_1 \tag{16}$$

$$\epsilon_t = \frac{\bar{\sigma}_t}{\bar{E}} - \nu \frac{\bar{\sigma}_r}{\bar{E}} + \overline{aT}(r) = \frac{\sigma_{1t}}{E_1} - \nu \frac{\sigma_{1r}}{E_1} + a_1 T_1 \tag{17}$$

From these equations we obtain

$$\sigma_{1r} = \frac{E_1}{\bar{E}} \bar{\sigma}_r + \frac{E_1}{1-\nu} (\overline{aT}(r) - a_1 T_1) \tag{18}$$

$$\sigma_{1t} = \frac{E_1}{\bar{E}} \bar{\sigma}_t + \frac{E_1}{1-\nu} (\overline{aT}(r) - a_1 T_1) \tag{19}$$

Similarly, the stresses in the surface plates are obtained as

$$\sigma_{2r} = \frac{E_2}{\bar{E}} \bar{\sigma}_r + \frac{E_2}{1-\nu} (\overline{aT}(r) - a_2 T_2) \tag{20}$$

$$\sigma_{2t} = \frac{E_2}{\bar{E}} \bar{\sigma}_t + \frac{E_2}{1-\nu} (\overline{aT}(r) - a_2 T_2) \tag{21}$$

Substitution $\bar{\sigma}_r$ and $\bar{\sigma}_t$ from Eqs. (12) and (13) into the foregoing equations gives the stresses in a sandwich plate with a central hole as

$$\sigma_{1r} = \frac{E_1}{r^2} \left[\frac{r^2 - R_i^2}{R_0^2 - R_i^2} \int_{R_i}^{R_0} \overline{aT}(r) \cdot r\, dr - \int_{R_i}^{r} \overline{aT}(r) \cdot r\, dr \right.$$
$$\left. + \frac{r^2}{1-\nu} (\overline{aT}(r) - a_1 T_1) \right] \tag{22}$$

$$\sigma_{1t} = \frac{E_1}{r^2} \left[\frac{r^2 + R_i^2}{R_0^2 - R_i^2} \int_{R_i}^{R_0} \overline{aT}(r) \cdot r\, dr + \int_{R_i}^{r} \overline{aT}(r) \cdot r\, dr \right.$$
$$\left. - \overline{aT}(r) \cdot r^2 + \frac{r^2}{1-\nu} (\overline{aT}(r) - a_1 T_1) \right] \tag{23}$$

$$\sigma_{2r} = \frac{E_2}{r^2}\left[\frac{r^2-R_i^2}{R_0^2-R_i^2}\int_{R_i}^{R}\overline{aT}(r)\cdot r\,dr - \int_{R_i}^{r}\overline{aT}(r)\cdot r\,dr\right.$$
$$\left. + \frac{r^2}{1-\nu}(\overline{aT}(r) - a_2 T_2)\right] \qquad (24)$$

$$\sigma_{2t} = \frac{E_2}{r^2}\left[\frac{r^2+R_i^2}{R_0^2-R_i^2}\int_{R_i}^{R_0}\overline{aT}(r)\cdot r\,dr + \int_{R_i}^{r}\overline{aT}(r)\cdot r\,dr\right.$$
$$\left. - \overline{aT}(r)\cdot r^2 + \frac{r^2}{1-\nu}(\overline{aT}(r) - a_2 T_2)\right] \qquad (25)$$

For a solid plate of radius "a" the stresses become

$$\sigma_{1r} = E_1\left(\frac{1}{a^2}\int_0^a \overline{aT}(r)\cdot r\,dr - \frac{1}{r^2}\int_0^r \overline{aT}(r)\cdot r\,dr\right.$$
$$\left. + \frac{\overline{aT}(r) - a_1 T_1}{1-\nu}\right) \qquad (26)$$

$$\sigma_{1t} = E_1\left(\frac{1}{a^2}\int_0^a \overline{aT}(r)\cdot r\,dr + \frac{1}{r^2}\int_0^r \overline{aT}(r)\cdot r\,dr\right.$$
$$\left. - \overline{aT}(r) + \frac{\overline{aT}(r) - a_1 T_1}{1-\nu}\right) \qquad (27)$$

$$\sigma_{2r} = E_2\left(\frac{1}{a^2}\int_0^a \overline{aT}(r)\cdot r\,dr - \frac{1}{r^2}\int_0^r \overline{aT}(r)\cdot r\,dr\right.$$
$$\left. + \frac{\overline{aT}(r) - a_2 T_2}{1-\nu}\right) \qquad (28)$$

$$\sigma_{2t} = E_2\left(\frac{1}{a^2}\int_0^a \overline{aT}(r)\cdot r\,dr + \frac{1}{r^2}\int_0^r \overline{aT}(r)\cdot r\,dr\right.$$
$$\left. - \overline{aT}(r) + \frac{\overline{aT}(r) - a_2 T_2}{1-\nu}\right) \qquad (29)$$

Note that in the event that T_1 and T_2 are functions of the thickness coordinate only, the foregoing equation gives the result obtained in Sect. 3.8; namely that

$$\sigma_{1r} = \sigma_{1t} = \frac{E_1}{1-\nu}(\overline{\alpha T} - \alpha_1 T_1) \tag{30}$$

$$\sigma_{2r} = \sigma_{2t} = \frac{E_1}{1-\nu}(\overline{\alpha T} - \alpha_2 T_2) \tag{31}$$

In these equations the mean free expansion is $\overline{\alpha T}$ rather than $\overline{\alpha T}'$ as in Eqs. (3.8-2) and (3.8-3). This is proper, since in the present problem $\nu_1 = \nu_2 = \nu$, so that $\overline{\alpha T}' = \overline{\alpha T}$.

4.4 CONCENTRIC CIRCULAR CYLINDERS

Fig. 4.4-1 shows an assembly of concentric cylinders of dissimilar

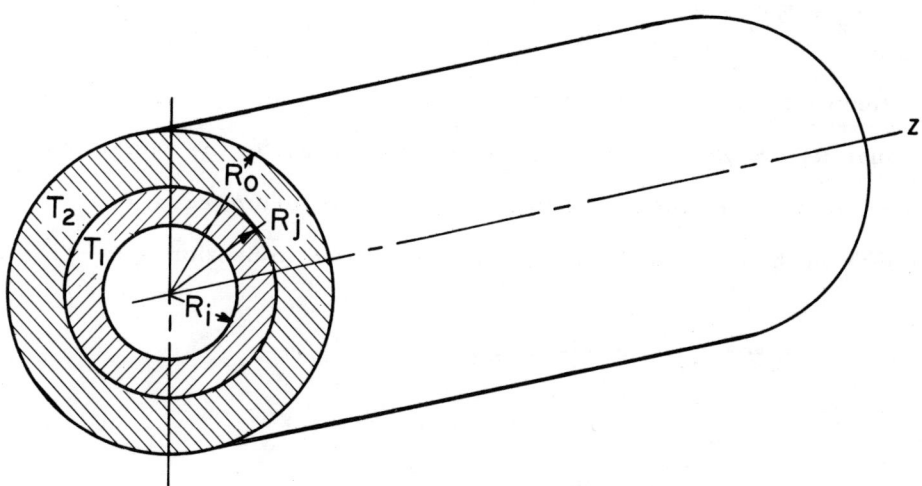

Fig. 4.4-1

materials, heated to different but uniform temperatures T_1 and T_2. They are attached at their interface $r = R_j$ so that the interfacial radial displacements of the two cylinders is the same, and the axial strains are uniform over the entire assembly. This is the typical concentric cylinder shrink-fit problem.

In Section 1.3 it is shown that a three-dimensional plane strain problem such as the present problem, can be simplified by a transformation of elastic constants. Modified (primed) elastic constants defined as

$$E' = \frac{E}{1-\nu^2} \qquad \nu' = \frac{\nu}{1-\nu} \qquad \alpha' = \alpha(1+\nu) \qquad (4.4-1)$$

are used in the analysis which follows. The stress-strain relationships, expressed in terms of modified and unmodified elastic constants, are, as given by Eqs. (1.7-2) and (1.7-5),

$$\epsilon_r + \nu \epsilon_z = \frac{1}{E'}(\sigma_r - \nu' \sigma_t) + \alpha' T \qquad (4.4-2)$$

$$\epsilon_t + \nu \epsilon_z = \frac{1}{E'}(\sigma_t - \nu' \sigma_r) + \alpha' T \qquad (4.4-3)$$

$$\sigma_z = E \epsilon_z + \nu (\sigma_r + \sigma_t) - E \alpha T \qquad (4.4-4)$$

The term $E\epsilon_z$ is not a function of the coordinates, and as will be shown later the quantity $\nu(\sigma_r + \sigma_t)$ is also invariant with respect to the radial and axial coordinates. The compatibility and radial equilibrium equations are the same as in the concentric ring problems treated in Section 4.1. They are

$$\epsilon_r = \epsilon_t + r \frac{d\epsilon_t}{dr} = \frac{d}{dr}(r \epsilon_t) \qquad (4.4-5)$$

and

$$\sigma_t = \sigma_r + r \frac{d\sigma_r}{dr} = \frac{d}{dr}(r \sigma_r) \qquad (4.4-6)$$

In the absence of external loads or expansion restraints, the equilibrium of forces in the axial direction requires that the integrated stresses over the assembly cross section be zero, or

$$\int_{A_1} \sigma_{1z} \, dA_1 + \int_{A_2} \sigma_{2z} \, dA_2 = 0 \qquad (4.4-7)$$

wherein σ_z is given by Eq.(4.4-4) with the appropriate subscript.

When the strains from Eqs. (4.4-2) and (4.4-3) are set in Eq. (4.4-5), the results are the same whether the term $\nu \epsilon_z$ is present or absent. Furthermore since T is constant in each of the cylinders, the terms containing αT disappear. The result is therefore the same as in the case of the ring analysis of Section 4.1. Namely,

$$\sigma_t - \sigma_r + r \frac{d\sigma_t}{dr} - \nu' [\sigma_r + r \frac{d\sigma_r}{dr} - \sigma_t] = 0 \qquad (4.4-8)$$

The quantity in brackets is zero in accordance with Eq. (4.4-6), and substitution for σ_t from Eq. (4.4-6) into the foregoing equation yields, as in the case of concentric rings, the integrable differential

$$r \frac{d}{dr} [\frac{1}{r} \frac{d}{dr} (r^2 \sigma_r)] = 0 \qquad (4.4-9)$$

Integration yields the radial stress in terms of undetermined constants A and B, as

$$\sigma_r = A - \frac{B}{r^2} \qquad (4.4-10)$$

and from Eq. (4.4-6) the tangential stress is

$$\sigma_t = A + \frac{B}{r^2} \qquad (4.4-11)$$

It is seen that $\sigma_r + \sigma_t$ is constant, and as the temperature is constant in each of the cylinders, the stress σ_z, expressed by Eq. (4.4-4), is also constant in each of the cylinders. Eq. (4.4-7) becomes

$$\sigma_{1z} A_1 + \sigma_{2z} A_2 = 0 \qquad (4.4-12)$$

The boundary conditions are

$$\sigma_{1r}(R_i) = 0 \qquad \sigma_{1r}(R_j) = -p$$

$$\sigma_{2r}(R_o) = 0 \qquad \sigma_{2r}(R_j) = -p \qquad (4.4\text{-}13)$$

with p the radial contact pressure. Setting these boundary conditions into Eqs (4.4-10) and (4.4-11) we obtain the radial and tangential stresses in cylinders 1 and 2 expressed in terms of the still undetermined interfacial pressure p. They are

$$\sigma_{1r} = -\frac{R_j^2 p}{R_j^2 - R_i^2}\left(1 - \frac{R_i^2}{r^2}\right) \qquad (4.4\text{-}14)$$

$$\sigma_{1t} = -\frac{R_j^2 p}{R_j^2 - R_i^2}\left(1 + \frac{R_i^2}{r^2}\right) \qquad (4.4\text{-}15)$$

$$\sigma_{2r} = \frac{R_j^2 p}{R_o^2 - R_j^2}\left(1 - \frac{R_o^2}{r^2}\right) \qquad (4.4\text{-}16)$$

$$\sigma_{2t} = \frac{R_j^2 p}{R_o^2 - R_j^2}\left(1 + \frac{R_o^2}{r^2}\right) \qquad (4.4\text{-}17)$$

Interfacial tangential strain compatibility requires that $\epsilon_{1tj} = \epsilon_{2tj}$. From Eq. (4.4-3)

$$\frac{1}{E_1'}(\sigma_{1tj} - \nu_1' \sigma_{1rj}) + \alpha_1' T_1 - \nu_1' \epsilon_z = \frac{1}{E_2'}(\sigma_{2tj} - \nu_2' \sigma_{2rj}) + \alpha_2' T_2 - \nu_2' \epsilon_z \qquad (4.4\text{-}18)$$

With the use of Eqs. (4.4-14) through (4.4-17), wherein we set r equal to R_j, the foregoing equation becomes

$$\alpha'_1 T_1 - \alpha'_2 T_2 + \epsilon_z(\nu_2 - \nu_1) = p\left[\frac{1}{E'_2}\left(\frac{R_o^2 + R_j^2}{R_o^2 - R_j^2} + \nu'_2\right) + \frac{1}{E'_1}\left(\frac{R_j^2 + R_i^2}{R_j^2 - R_i^2} - \nu'_1\right)\right] \quad (4.4\text{-}19)$$

Substitution of σ_{1z} and σ_{2z}, from Eq. (4.4-4) into the axial equilibrium equation, Eq. (4.4-12) yields

$$(E_1 A_1 + E_2 A_2)\epsilon_z + \nu_1 A_1(\sigma_{1r} + \sigma_{1t}) + \nu_2 A_2(\sigma_{2r} + \sigma_{2t}) = E_1 \alpha_1 T_1 A_1 \quad (4.4\text{-}20)$$
$$+ E_2 \alpha_2 T_2 A_2$$

Also from Eqs. (4.4-14) and (4.4-15), we have

$$\nu_1 A_1(\sigma_{1r} + \sigma_{1t}) = -2\nu_1 A_j p \quad (4.4\text{-}21)$$

in which $A_j = \pi R_j^2$, and from Eqs. (4.4-16) and (4.4-17) we have

$$\nu_2 A_2(\sigma_{1r} + \sigma_{1t}) = 2\nu_2 A_j p \quad (4.4\text{-}22)$$

From the three preceding equations the axial strain can be expressed in terms of the contact pressure, p, as

$$\epsilon_z = \overline{\alpha T} + \frac{2A_j(\nu_1 - \nu_2)}{E_1 A_1 + E_2 A_2} \cdot p \quad (4.4\text{-}23)$$

with

$$\overline{\alpha T} = \frac{E_1 \alpha_1 T_1 A_1 + E_2 \alpha_2 T_2 A_2}{E_1 A_1 + E_2 A_2}$$

Setting ϵ_z, from the foregoing equation, into Eq. (4.4-19) gives the interfacial pressure p as

$$p = \frac{\alpha'_1 T_1 - \alpha'_2 T_2 + (\nu_2 - \nu_1)\overline{\alpha T}}{\frac{1}{E'_2}\left(\frac{R_o^2 + R_j^2}{R_o^2 - R_j^2} + \nu'_2\right) + \frac{1}{E'_1}\left(\frac{R_j^2 + R_i^2}{R_j^2 - R_i^2} - \nu'_1\right) + \frac{2A_j(\nu_2 - \nu_1)^2}{E_1 A_1 + E_2 A_2}} \qquad (4.4\text{-}24)$$

When $\nu_1 = \nu_2$ the interfacial pressure becomes

$$p = \frac{\alpha'_1 T_1 - \alpha'_2 T_2}{\frac{1}{E'_2}\left(\frac{R_o^2 + R_j^2}{R_o^2 - R_j^2} + \nu'\right) + \frac{1}{E'_1}\left(\frac{R_j^2 + R_i^2}{R_j^2 - R_i^2} - \nu'\right)} \qquad (4.4\text{-}25)$$

In the case of shrink-fitting two concentric tubes or a cylindrical jacket on a rod, the assembly is subcooled to a base temperature, at which there is a stress-free contact of the two components. If Δ is the room temperature radial interference, the temperature T in Eq. (4.4-24) should be replaced by

$$T + \frac{\Delta}{R_j(\alpha_1 - \alpha_2)}$$

as shown earlier (Eq. (4.1-3)) in the ring shrink-fit problem. When the assembly is at a uniform temperature T the interference radial pressure at the interface is

$$p = \frac{[(\alpha_1 - \alpha_2) + (\nu_2 - \nu_1)\overline{\alpha}]\left[T + \frac{\Delta}{R_j(\alpha_1 - \alpha_2)}\right]}{\frac{1}{E'_2}\left(\frac{R_o^2 + R_j^2}{R_o^2 - R_j^2} + \nu'_2\right) + \frac{1}{E'_1}\left(\frac{R_j^2 + R_i^2}{R_j^2 - R_i^2} - \nu'_1\right) + \frac{2A_j(\nu_2 - \nu_1)^2}{E_1 A_1 + E_2 A_2}} \qquad (4.4\text{-}26)$$

The quantity $\overline{\alpha}$ is the mean coefficient of expansion; i.e., $\overline{\alpha} = \frac{\alpha_1 + \alpha_2}{2}$. In the shrink-fit problem the assumption is implicit that axial slippage between the two cylinders does not take place after the initial contact of the surfaces. The radial and tangential stresses in each of the cylinders is obtained explicitly when the pressure p is substituted in Eqs. (4.4-14) through (4.4-17). Substitution of p from Eq. (4.4-24) into Eq. (4.4-23) gives the axial strain as

$$\epsilon_z = \overline{\alpha T} + \frac{\left(\frac{2A_j(\nu_1-\nu_2)}{E_1 A_1 + E_2 A_2}\right)(\alpha'_1 T_1 - \alpha'_2 T_2 + (\nu_2-\nu_1)\overline{\alpha T})}{\frac{1}{E'_2}\left(\frac{R_o^2+R_j^2}{R_o^2-R_j^2}+\nu'_2\right) + \frac{1}{E'_1}\left(\frac{R_j^2+R_i^2}{R_j^2-R_i^2}-\nu'_1\right) + \frac{2A_j(\nu_2-\nu_1)^2}{E_1 A_1 + E_2 A_2}} \qquad (4.4\text{-}27)$$

Note that when $\nu_1 = \nu_2$, the axial strain is

$$\epsilon_z = \overline{\alpha T} \qquad (4.4\text{-}28)$$

The axial stresses are, from Eq. (4.4-4),

$$\sigma_{1z} = E_1 \epsilon_z + \nu_1(\sigma_{1r} + \sigma_{1t}) - E_1 \alpha_1 T_1 \qquad (4.4\text{-}29)$$

$$\sigma_{2z} = E_2 \epsilon_z + \nu_2(\sigma_{2r} + \sigma_{2t}) - E_2 \alpha_2 T_2 \qquad (4.4\text{-}30)$$

With the use of Eqs. (4.4-21), (4.4-22), (4.4-23) the foregoing become

$$\sigma_{1z} = E_1\left(\overline{\alpha T} + \frac{2A_j(\nu_1-\nu_2)}{E_1 A_1 + E_2 A_2}p\right) - \frac{2\nu_1 A_j}{A_1}p - E_1 \alpha_1 T_1 \qquad (4.4\text{-}31)$$

$$\sigma_{2z} = E_2\left(\overline{\alpha T} + \frac{2A_j(\nu_1-\nu_2)}{E_1 A_1 + E_2 A_2}p\right) + \frac{2\nu_2 A_j}{A_2}p - E_2 \alpha_2 T_2 \qquad (4.4\text{-}32)$$

with p given by Eq. (4.4-24). When the materials of both cylinders are the same, $E_1 = E_2 = E$, $\nu_1 = \nu_2 = \nu$, and $\alpha_1 = \alpha_2 = \alpha$, but $T_1 \neq T_2$, it can be verified that the axial stress becomes

$$\sigma_z = \frac{E\alpha}{1-\nu}(\overline{T} - T) \qquad (4.4\text{-}33)$$

with T equal to either T_1 or T_2. $\qquad (4.4\text{-}34)$

This is recognized as the proper expression for the axial stress in an unrestrained homogeneous cylinder with a radial temperature variation.

4.5 CIRCULAR ROD WITH RADIAL TEMPERATURE VARIATION

The analysis which follows is applicable to a circular rod or a hollow thick-walled cylinder, such as those shown in Fig. 4.5-1, having a radial temperature variation. The axial strain, ϵ_z, produced by the radial temperature

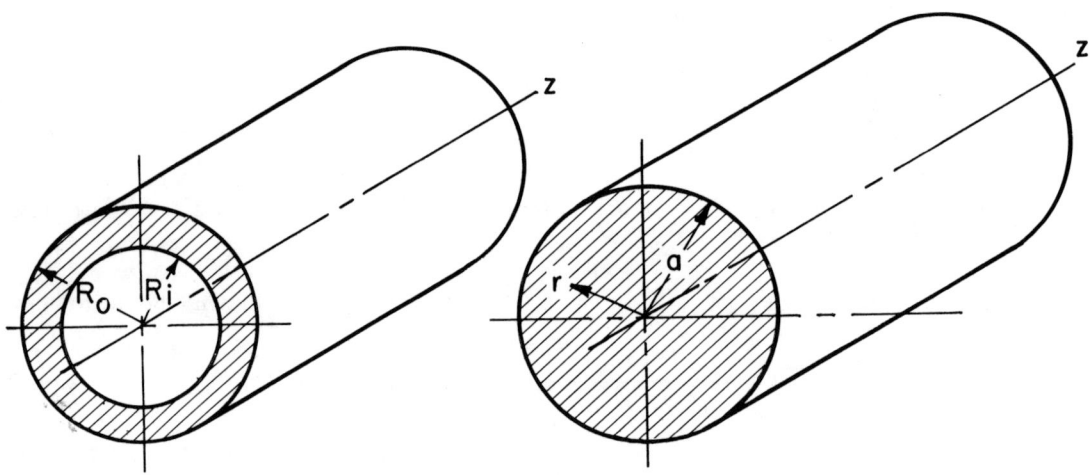

Fig. 4.5-1

distribution, is uniform throughout the body so that we are dealing with a plane strain problem. In these problems it is expedient to use modified elastic constants defined as

$$E' = \frac{E}{1-\nu^2} \qquad \nu' = \frac{\nu}{1-\nu} \qquad \alpha' = \alpha(1+\nu) \qquad (4.5\text{-}1)$$

The stress-strain relationships then take the form

$$\epsilon_r + \nu \epsilon_z = \frac{1}{E'}(\sigma_r - \nu' \sigma_t) + \alpha' T \tag{4.5-2}$$

$$\epsilon_t + \nu \epsilon_z = \frac{1}{E'}(\sigma_t - \nu' \sigma_r) + \alpha' T \tag{4.5-3}$$

$$\sigma_z = E\epsilon_z + \nu(\sigma_r + \sigma_t) - E\alpha T \tag{4.5-4}$$

The temperature is a function of the radius only, $T = T(r)$, and as in the preceding section, the equations of compatibility and equilibrium for plane axially symmetrical deformations, are

$$\epsilon_r = \epsilon_t + r\frac{d\epsilon_t}{dr} = \frac{d}{dr}(r\epsilon_t) \tag{4.5-5}$$

$$\sigma_t = \sigma_r + r\frac{d\sigma_r}{dr} = \frac{d}{dr}(r\sigma_r) \tag{4.5-6}$$

The compatibility equation is expressed in terms of stress by substitution of ϵ_r and ϵ_t from Eqs. (4.5-2) and (4.5-3) into Eq. (4.5-5). The result is

$$\sigma_r = \sigma_t + r\frac{d\sigma_t}{dr} - \nu'(\sigma_r + r\frac{d\sigma_r}{dr} - \sigma_t) + E'\alpha' r\frac{dT}{dr} \tag{4.5-7}$$

With the use of Eq. (4.5-6), σ_t is eliminated and the foregoing equation becomes

$$\frac{r d\sigma_r}{dr} + r\frac{d^2(r\sigma_r)}{dr^2} = -E'\alpha'\frac{dT}{dr} \tag{4.5-8}$$

This equation can be written as

$$r\frac{d}{dr}\left[\frac{1}{r}\frac{d}{dr}(r^2 \sigma_r)\right] = -E'\alpha' r\frac{dT}{dr} \tag{4.5-9}$$

and when integrated gives the radial stress as

$$\sigma_r = A - \frac{B}{r^2} - \frac{E'\alpha'}{r^2} \int^r Tr\, dr \qquad (4.5\text{-}10)$$

Substitution of σ_r into Eq. (4.5-6) gives the tangential stress as

$$\sigma_t = A + \frac{B}{r^2} + \frac{E'\alpha'}{r^2} \int^r Tr\, dr \qquad (4.5\text{-}11)$$

and the sum of the radial and tangential stresses is

$$\sigma_r + \sigma_t = 2A - E'\alpha' T \qquad (4.5\text{-}12)$$

Thick-Walled Circular Cylinder

In the thick-walled circular cylinder the boundary conditions needed to determine the constants A and B in Eqs. (4.5-10) and (4.5-11) are

$$\sigma_r(r = R_i) = 0 \qquad \sigma_r(r = R_o) = 0 \qquad (4.5\text{-}13)$$

Applying these boundary conditions to Eq. (4.5-10) we obtain

$$A = \frac{E'\alpha'}{R_o^2 - R_i^2} \int_{R_i}^{R_o} Tr\, dr = \frac{E'\alpha \overline{T}}{2} \qquad (4.5\text{-}14)$$

$$B = \frac{E'\alpha' R_i^2}{R_o^2 - R_i^2} \int_{R_i}^{R_o} Tr\, dr - E'\alpha' \int^{R_i} Tr\, dr \qquad (4.5\text{-}15)$$

These values are set into Eqs. (4.5-10) and (4.5-11), and we obtain the stresses

$$\sigma_r = \frac{E'\alpha'}{r^2} \left(\frac{r^2 - R_i^2}{R_o^2 - R_i^2} \int_{R_i}^{R_o} Tr\,dr - \int_{R_i}^{r} Tr\,dr \right) \tag{4.5-16}$$

$$\sigma_t = \frac{E'\alpha'}{r^2} \left(\frac{r^2 + R_i^2}{R_o^2 - R_i^2} \int_{R_i}^{R_o} Tr\,dr + \int_{R_i}^{R_o} Tr\,dr - Tr^2 \right) \tag{4.5-17}$$

The foregoing equations may also be expressed in terms of mean thermal parameters in the form

$$\sigma_r = \frac{E'\alpha'}{2}\left(1 - \frac{R_i^2}{r^2}\right)(\overline{T} - \overline{T}_{ri}) = \frac{E'\alpha'}{2}\left(1 - \frac{R_o^2}{r^2}\right)(\overline{T} - \overline{T}_{ro}) \tag{4.5-18}$$

$$\sigma_t = E'\alpha'(\overline{T} - T) - \frac{E'\alpha'}{2}\left(1 - \frac{R_i^2}{r^2}\right)(\overline{T} - \overline{T}_{ri})$$

$$= E'\alpha'(\overline{T} - T) - \frac{E'\alpha'}{2}\left(1 - \frac{R_o^2}{r^2}\right)(\overline{T} - \overline{T}_{ro}) \tag{4.5-19}$$

where $\overline{T}$ is the mean temperature of the cylinder, $\overline{T}_{ri}$ the mean temperature of the inner portion of the cylinder lying between R_i and r, and $\overline{T}_{ro}$ the mean temperature of the outer portion of the cylinder lying between r and R_o. That is,

$$\overline{T} = \frac{2}{R_o^2 - R_i^2}\int_{R_i}^{R_o} Tr\,dr \qquad \overline{T}_{ri} = \frac{2}{r^2 - R_i^2}\int_{R_i}^{r} Tr\,dr \qquad \overline{T}_{ro} = \frac{2}{R_o^2 - r^2}\int_{r}^{R_o} Tr\,dr$$
$$\tag{4.5-20}$$

Substitution of $A = E'\alpha'\overline{T}/2$ into Eq. (4.5-12) gives the sum of the radial and tangential stresses as

$$\sigma_r + \sigma_t = E'\alpha'(\overline{T} - T) = \frac{E\alpha}{1-\nu}(\overline{T} - T) \tag{4.5-21}$$

This is set into Eq. (4.5-4) and we obtain the axial stress

$$\sigma_z = E\epsilon_z + \frac{E\alpha}{1-\nu}(\nu\overline{T} - T) \tag{4.5-22}$$

When the cylinder is completely restrained from axial expansion $\epsilon_z = 0$ and the axial stress is

$$\sigma_z = \frac{E\alpha}{1-\nu}(\nu\overline{T} - T) \tag{4.5-23}$$

In the event that there is no axial restraint, the axial boundary condition requires that the net load over cross section be zero, or

$$\int_{R_i}^{R_o} \sigma_z r\, dr = 0 \tag{4.5-24}$$

The expression for σ_z, Eq. (4.5-22) is set into the above and we obtain

$$\epsilon_z = \alpha\overline{T} \tag{4.5-25}$$

so that for the case of unrestrained axial expansion

$$\sigma_z = \frac{E\alpha}{1-\nu}(\overline{T} - T) \tag{4.5-26}$$

In the more commonly encountered temperature distributions the maximum stress will be found at either the inner or outer surfaces of the cylinder. We find from Eqs. (4.5-19) and (4.5-26) that the tangential and axial stresses at the surface are equal. Thus

$$\sigma_{ts} = \sigma_{zs} = \frac{E\alpha}{1-\nu}(\overline{T} - T_s) \tag{4.5-27}$$

where the subscript s signifies either the inner or the outer surface.

The radial displacements of the inner and outer surfaces are obtained from Eqs. (4.5-3) and (4.5-27). Thus

$$u_i = \epsilon_{ti} R_i = R_i [\alpha'(\overline{T} - T_i) + \alpha' T_i - \nu\alpha\overline{T}] = \alpha\overline{T} R_i \tag{4.5-28}$$

$$u_o = \epsilon_{to} R_o = R_o [\alpha'(\overline{T} - T_o) + \alpha' T_o - \nu\alpha\overline{T}] = \alpha\overline{T} R_o \tag{4.5-29}$$

Solid Rod

The general equations for radial and tangential stresses, Eqs. (4.5-10), (4.5-11) and (4.5-12) are applicable to both hollow and solid circular cylinders. In the case of a solid cylinder the constant B is zero since otherwise the stress at the center of the cylinder would become infinite. The constant A is determined from the boundary condition

$$\sigma_r (r = a) = 0 \tag{4.5-30}$$

where a is the outer radius of the cylinder. With this boundary condition, we obtain A from Eq. (4.5-10) as

$$A = \frac{E'\alpha'}{a^2} \int_0^a T r\, dr = \frac{E'\alpha' \overline{T}}{2} \tag{4.5-31}$$

The values of A and B are set into Eqs. (4.5-10) and (4.5-11), and we obtain the radial and tangential stresses

$$\sigma_r = E'\alpha' \left(\frac{1}{a^2} \int_0^a T r\, dr - \frac{1}{r^2} \int_0^r T r\, dr \right) \tag{4.5-32}$$

$$\sigma_t = E'\alpha' \left(\frac{1}{a^2} \int_0^a T r\, dr + \frac{1}{r^2} \int_0^r T r\, dr - T \right) \tag{4.5-33}$$

and the sum of the stresses $\sigma_r + \sigma_t$ is as found previously

$$\sigma_r + \sigma_t = \frac{E\alpha}{1-\nu} (\overline{T} - T) \tag{4.5-34}$$

The radial and tangential stresses may also be written in terms of mean thermal parameters as follows.

$$\sigma_r = \frac{E'\alpha'}{2} (\overline{T} - \overline{T}_{ri}) = \frac{E'\alpha'}{2}\left(1 - \frac{a^2}{r^2}\right)(\overline{T} - \overline{T}_{ro}) \tag{4.5-35}$$

$$\sigma_t = E'\alpha'(\bar{T}-T) - \frac{E'\alpha'}{2}(\bar{T}-\bar{T}_{ri}) = E'\alpha'(\bar{T}-T) - \frac{E'\alpha'}{2}\left(1-\frac{a^2}{r^2}\right)(\bar{T}-\bar{T}_{ro})$$

(4.5-36)

In these equations the mean temperatures are defined as

$$\bar{T} = \frac{2}{a^2}\int_0^a Tr\,dr \qquad \bar{T}_{ri} = \frac{2}{r^2}\int_0^r Tr\,dr \qquad \bar{T}_{ro} = \frac{2}{a^2-r^2}\int_r^a Tr\,dr$$

(4.5-37)

The axial stress, σ_z, is given by the same equations as derived for the hollow cylinder. In the latter analysis, Eq. (4.5-23) gives the axial stress for the case of completely restrained axial expansion as

$$\sigma_z = \frac{E\alpha}{1-\nu}(\nu\bar{T}-T)$$

(4.5-38)

which applies also for the solid rod. When there is no axial restraint, Eq. (4.5-26) gives the axial stress as

$$\sigma_z = \frac{E\alpha}{1-\nu}(\bar{T}-T)$$

(4.5-39)

and this expression also applies to the current problem of the solid rod.

At the center of the rod $\sigma_s = \sigma_t$ so that we have, from Eq. (4.5-34)

$$\sigma_r + \sigma_t = 2\sigma_{r=0} = \frac{E\alpha}{1-\nu}(\bar{T}-T_{r=0})$$

(4.5-40)

or

$$\sigma_{r=0} = \frac{E\alpha}{2(1-\nu)}(\bar{T}-T_{r=0})$$

(4.5-41)

On the other hand, if a small hole were drilled at the center of a solid rod, σ_r would become zero and Eq. (4.5-34) (or Eq. (4.5-21)) would become

$$\sigma_{r=0} = \frac{E\alpha}{1-\nu}(\bar{T}-T_{r=0})$$

(4.5-42)

We thus introduce a stress concentration factor of two by placing a small hole at the center of a rod. This was also found to be true when a small hole was placed at the center of a circular plate (See Section 4.2).

At the surface of the solid rod, where $\sigma_r = 0$, Eq. (4.5-34) gives the tangential stress as

$$\sigma_{t, r=a} = \frac{E\alpha}{1-\nu} (\overline{T} - T_{r=a}) \tag{4.5-43}$$

and in an axially unrestrained rod the axial stress at $r = a$, from Eq. (4.5-39), is

$$\sigma_{z, r=a} = \frac{E\alpha}{1-\nu} (\overline{T} - T_{r=a}) \tag{4.5-44}$$

The tangential and axial thermal stresses at the surface of the solid rod are therefore equal.

The radial displacement u_a at $r = a$ is obtained from Eqs. (4.5-3) and (4.5-34). It is

$$u_a = \epsilon_{ta} a = a[\alpha'(\overline{T}-T_a) + \alpha' T_a - \nu\alpha\overline{T}] = \alpha \overline{T} a \tag{4.5-45}$$

Some examples of the application of the foregoing results follow..

Example (a)   Solid Circular Rod with Uniform Heat Generation

A solid circular rod of radius a, and conductivity, k, has a uniform heat generation density, W. The temperature distribution obtained for the case of uniform heat generation, is

$$T = \frac{Wa^2}{4k} \left(1 - \frac{r^2}{a^2}\right) \tag{1}$$

The stresses in the rod are determined with the use of mean temperature parameters. As defined by Eq. (4.5-37) they are

$$\overline{T} = \frac{2}{a^2} \int_o^a Tr\,dr = \frac{W}{2k} \int_o^a \left(r - \frac{r^3}{a^2}\right) dr = \frac{Wa^2}{8k} \tag{2}$$

$$\bar{T}_{ri} = \frac{2}{r^2} \int_0^r Tr\,dr = \frac{Wa^2}{2r^2} \int_0^r (r - \frac{r^3}{a^2})\,dr = \frac{Wa^2}{4}(1 - \frac{r^2}{2a^2}) \qquad (3)$$

From Eqs. (4.5-35) and (4.5-36) we have

$$\sigma_r = \frac{E\alpha}{2(1-\nu)}(\bar{T} - \bar{T}_{ri}) = -\frac{E\alpha Wa^2}{16(1-\nu)k}(1 - \frac{r^2}{a^2}) \qquad (4)$$

and

$$\sigma_t = \frac{E\alpha}{1-\nu}(\bar{T}-T) - \sigma_r = \frac{E\alpha Wa^2}{16(1-\nu)k}(\frac{3r^2}{a^2} - 1) \qquad (5)$$

The axial stress is obtained from Eq. (4.5-39). It is

$$\sigma_z = \frac{E\alpha}{1-\nu}(\bar{T}-T) = \frac{E\alpha Wa^2}{8(1-\nu)k}(\frac{2r^2}{a^2} - 1) \qquad (6)$$

Example (b) Two Concentric Thick-Walled Cylinders

The two concentric cylinders shown in Fig. 4.5-2 have radial tempera-

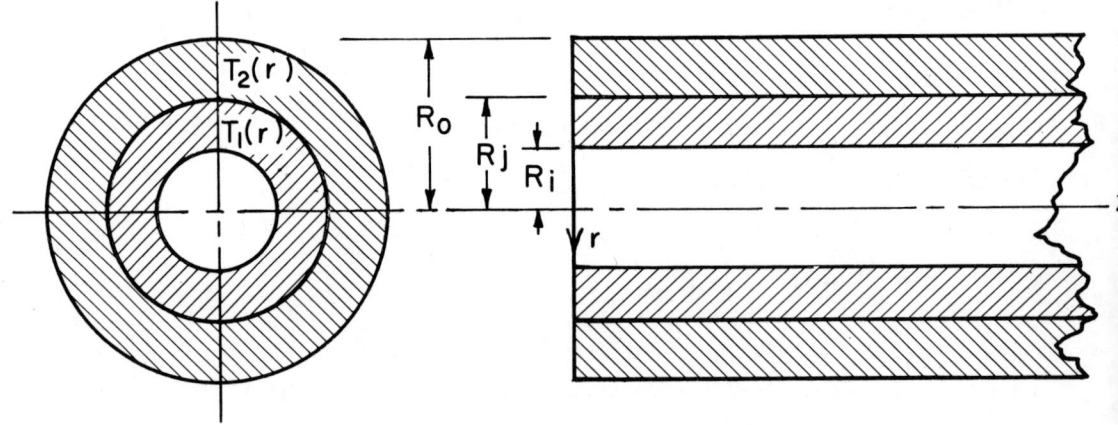

Fig. 4.5-2

ture distributions $T_1(r)$ and $T_2(r)$.

The stresses in the cylinders can be considered as due to two separate effects; first, the stresses generated in an unrestrained cylinder with a radial temperature variation and the second, the stresses due to the radial interference. Eqs. (4.5-25) and (4.5-28) show that the axial strains and interfacial tangential strains in a cylinder with a radial temperature variation, are

$$\epsilon_{z1} = \epsilon_{tj1} = \alpha_1 \overline{T}_1 \qquad \epsilon_{z2} = \epsilon_{tj2} = \alpha_2 \overline{T}_2 \qquad (1)$$

The stresses due to the interference or interaction of the two cylinders is therefore the same as that obtained for concentric cylinders uniformly heated to temperatures $\overline{T}_1$ and $\overline{T}_2$, since in these cylinders the axial and tangential strains are as given by Eqs. (1). We can therefore superpose the solutions of the free cylinders with radial temperature variation and a concentric cylinder assembly with the cylinders uniformly heated to temperatures $\overline{T}_1$ and $\overline{T}_2$ respectively.

The interfacial pressure, p, will be as given by Eq. (4.4-24), with $\overline{T}_1$ and $\overline{T}_2$ replacing T_1 and T_2. Thus

$$p = \frac{\alpha_1' \overline{T}_1 - \alpha_2' \overline{T}_2 + (\nu_2 - \nu_1) \overline{\alpha T}}{\frac{1}{E_2'}(\frac{R_o^2 + R_j^2}{R_o^2 - R_j^2} + \nu_2') + \frac{1}{E_1'}(\frac{R_j^2 + R_i^2}{R_j^2 - R_i^2} - \nu_1') + \frac{2 A_j (\nu_2 - \nu_1)^2}{E_1 A_1 + E_2 A_2}} \qquad (2)$$

The addition of the stresses from Eqs. (4.4-14) through (4.4-17) with the stresses from Eqs. (4.5-18) and (4.5-19) yields

$$\sigma_{1r} = (\frac{E_1' \alpha_1' (\overline{T}_1 - \overline{T}_{1ri})}{2} - \frac{R_j^2 p}{R_j^2 - R_i^2})(1 - \frac{R_i^2}{r^2}) \qquad (3)$$

$$\sigma_{1t} = E_1' \alpha_1' (\overline{T}_1 - T_1) - \frac{2A_j}{A_1} p - \sigma_{1r} \tag{4}$$

$$\sigma_{2r} = \left(\frac{E_2' \alpha_2' (\overline{T}_2 - \overline{T}_{2ro})}{2} + \frac{R_j^2 p}{R_o^2 - R_j^2} \right) \left(1 - \frac{R_o^2}{r^2} \right) \tag{5}$$

$$\sigma_{2t} = E_2' \alpha_2' (\overline{T}_2 - T_2) + \frac{2A_j}{A_2} p - \sigma_{2r} \tag{6}$$

With the use of Eqs. (4.4-31), (4.4-32), and (4.5-26) we obtain the axial stresses

$$\sigma_{1z} = E_1 (\overline{\alpha T} + \frac{2A_j(\nu_1 - \nu_2)}{E_1 A_1 + E_2 A_2} p) - \frac{2\nu_1 A_j}{A_1} p + \frac{E_1 \alpha_1}{1 - \nu_1} (\nu_1 \overline{T}_1 - T_1) \tag{7}$$

$$\sigma_{2z} = E_2 (\overline{\alpha T} + \frac{2A_j(\nu_1 - \nu_2)}{E_1 A_1 + E_2 A_2} p) + \frac{2\nu_2 A_j}{A_2} p + \frac{E_2 \alpha_2}{1 - \nu_2} (\nu_2 \overline{T}_2 - \overline{T}_2) \tag{8}$$

4.6 CONCENTRIC HOLLOW SPHERES

The two concentric hollow spheres shown in Fig. 4.6-1 are of dissimilar

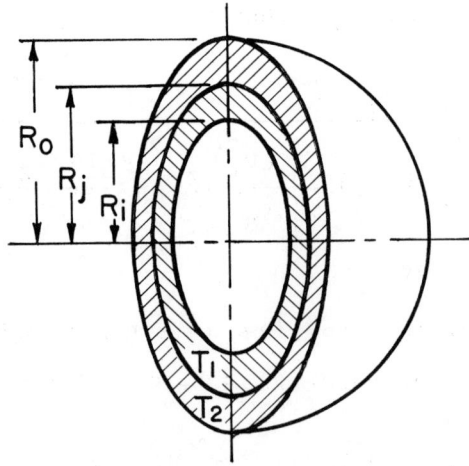

Fig. 4.6-1

CONCENTRIC HOLLOW SPHERES

materials and at different but uniform temperatures T_1 and T_2. It is clear that there will be spherical symmetry, with the stresses, strains, and displacements, functions of the radial coordinate only. The normal stresses and strains in any two orthogonal tangential coordinate directions are identified by the subscript, t.

In each of the hollow spheres, the stress-strain relationships, for spherical symmetry, are written as

$$\epsilon_t = \frac{1}{E}[(1-\nu)\sigma_t - \nu\sigma_r] + \alpha T \tag{4.6-1}$$

$$\epsilon_r = \frac{1}{E}[\sigma_r - 2\nu\sigma_t] + \alpha T \tag{4.6-2}$$

with T constant. The radial strain is $\epsilon_r = \frac{du}{dr}$, and the tangential strain is $\epsilon_t = \frac{u}{r}$, so that the strain compatibility relationship is

$$\epsilon_r = \epsilon_t + r\frac{d\epsilon_t}{dr} = \frac{d}{dr}(r\epsilon_t) \tag{4.6-3}$$

The spherical element shown in Fig. 4.6-2 is in static equilibrium under the

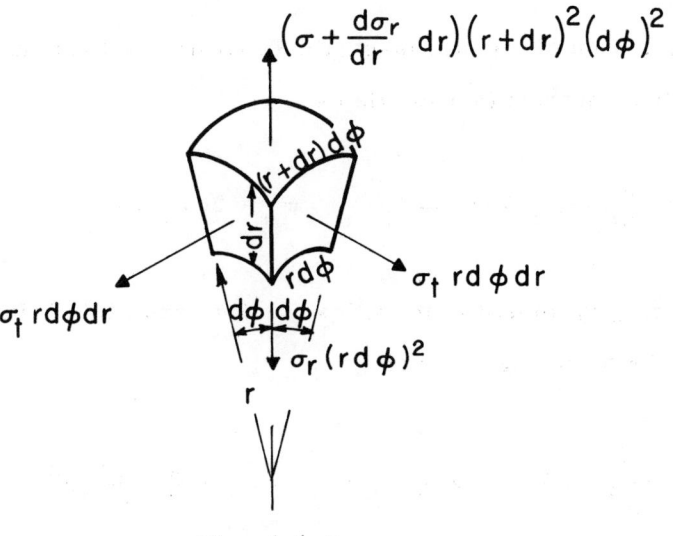

Fig. 4.6-2

radial and tangential loads acting on the element. The radial balance of loads is expressed as

$$4(\sigma_t r \cdot dr \cdot d\varphi) \frac{d\varphi}{2} = (\sigma_r + \frac{d\sigma_r}{dr} dr)(r + dr)^2 (d\varphi)^2 - \sigma_r (r\, d\varphi)^2 \qquad (4.6\text{-}4)$$

Neglecting higher order quantities, the foregoing equation becomes

$$\sigma_t = \sigma_r + \frac{r}{2} \frac{d\sigma_r}{dr} \qquad (4.6\text{-}5)$$

This equation may be written as

$$r\sigma_t = \frac{1}{2} \frac{d}{dr}(r^2 \sigma_r) \qquad (4.6\text{-}6)$$

and upon differentiation, we have

$$\frac{d}{dr}(r\sigma_t) = \frac{1}{2} \frac{d^2}{dr^2}(r^2 \sigma_r) \qquad (4.6\text{-}7)$$

Substitution of the stress-strain Equations (4.6-1) and (4.6-2) into the compatibility Equation (4.6-3) yields

$$(1-\nu) \frac{d}{dr}(r\sigma_t) - \nu \frac{d}{dr}(r\sigma_r) - \sigma_r + 2\nu\sigma_t = 0 \qquad (4.6\text{-}8)$$

Eliminating σ_t by substitution of Eqs. (4.6-6) and (4.6-7) into Eq. (4.6-8) results in the following equation.

$$\frac{1-\nu}{2} \frac{d^2}{dr^2}(r^2 \sigma_r) - \nu \frac{d}{dr}(r\sigma_r) - \sigma_r + \frac{\nu}{r} \frac{d}{dr}(r^2 \sigma_r) = 0 \qquad (4.6\text{-}9)$$

The last term is expanded and written as

$$\frac{\nu}{r}\frac{d}{dr}(r^2\sigma_r) = \frac{\nu}{r}\left[r\frac{d(r\sigma_r)}{dr} + r\sigma_r\right] = \nu\frac{d}{dr}(r\sigma_r) + \nu\sigma_r \qquad (4.6\text{-}10)$$

This is substituted for the last term of Eq. (4.6-9), and that equation becomes

$$(1-\nu)\left[\frac{1}{2}\frac{d^2}{dr^2}(r^2\sigma_r) - \sigma_r\right] = 0 \qquad (4.6\text{-}11)$$

In the form of a single differential, omitting the factor $(1-\nu)$, it is expressed as

$$\frac{d}{dr}\left[\frac{1}{r^2}\frac{d}{dr}(r\cdot r^2\sigma_r)\right] = \frac{d}{dr}\left[\frac{1}{r^2}\frac{d}{dr}(r^3\sigma_r)\right] = 0 \qquad (4.6\text{-}12)$$

Integration yields

$$\sigma_r = A - \frac{B}{r^3} \qquad (4.6\text{-}13)$$

and from Eq. (4.6-6) the tangential stress is

$$\sigma_t = A + \frac{B}{2r^3} \qquad (4.6\text{-}14)$$

Note that $\sigma_r + 2\sigma_t = $ constant. The boundary conditions are

$$\sigma_{1r}(r = R_i) = 0 \qquad \sigma_{1r}(r = R_j) = -p$$
$$\sigma_{2r}(r = R_o) = 0 \qquad \sigma_{2r}(r = R_j) = -p \qquad (4.6\text{-}15)$$

with p the radial pressure at the interface of the two hollow spheres. Substitution into Eqs. (4.6-13) and (4.6-14) yields the stresses in terms of the still undetermined contact pressure p, as

$$\sigma_{1r} = -\frac{R_j^3 p}{R_j^3 - R_i^3}\left(1 - \frac{R_i^3}{r^3}\right) \qquad (4.6\text{-}16)$$

$$\sigma_{1t} = -\frac{R_j^3 p}{R_j^3 - R_i^3}\left(1 + \frac{R_i^3}{2r^3}\right) \qquad (4.6\text{-}17)$$

$$\sigma_{2r} = \frac{R_j^3 p}{R_o^3 - R_j^3}\left(1 - \frac{R_o^3}{r^3}\right) \qquad (4.6\text{-}18)$$

$$\sigma_{2t} = \frac{R_j^3 p}{R_o^3 - R_j^3}\left(1 + \frac{R_o^3}{2r^3}\right) \qquad (4.6\text{-}19)$$

The tangential strains in Spheres 1 and 2 are equal at $r = R_j$. From Eq. (4.5-1) we have, setting $\epsilon_{1t} = \epsilon_{2t}$

$$\frac{1}{E_1}\left[(1-\nu_1)\sigma_{1tj} - \nu_1 \sigma_{1rj}\right] + \alpha_1 T_1 = \frac{1}{E_2}\left[(1-\nu_2)\sigma_{2tj} - \nu_2 \sigma_{2rj}\right] + \alpha_2 T_2 \quad (4.6\text{-}20)$$

In Eqs. (4.6-16) through (4.6-19) we set $r = R_j$. The resulting interfacial stresses are substituted into Eq. (4.6-20), which yields

$$\frac{p}{E_1}\left[-(1-\nu_1)\frac{2R_j^3 + R_i^3}{R_j^3 - R_i^3} + \nu_1\right] + \alpha_1 T_1 = \frac{p}{E_2}\left[(1-\nu_2)\frac{2R_j^3 + R_o^3}{R_o^3 - R_j^3} + \nu_2\right] + \alpha_2 T_2$$
$$(4.6\text{-}21)$$

From the preceding equation we obtain the radial pressure, p, as

$$p = \frac{\alpha_1 T_1 - \alpha_2 T_2}{\frac{1}{E_2}\left[\frac{(1-\nu_2)(2R_j^3 + R_o^3)}{R_o^3 - R_j^3} + \nu_2\right] + \frac{1}{E_1}\left[\frac{(1-\nu_1)(2R_j^3 + R_i^3)}{R_j^3 - R_i^3} - \nu_1\right]} \qquad (4.6\text{-}22)$$

Substitution of this value of p into Eqs. (4.6-16) through (4.6-19) yields the radial and tangential stresses in concentric hollow spheres which are at different but uniform temperatures.

Eq. (4.6-22) implies a stress-free state at $T_1 = T_2 = 0$, the datum temperature. In an interference-fit problem the datum will generally be below the ambient or room temperature. The temperatures T_1 and T_2 are then taken as the increase in temperature above that which produces stress-free contact between the hollow spheres.

4.7 RADIAL TEMPERATURE VARIATION IN SPHERES

A solid sphere or a sphere with a central cavity, subjected to a radial temperature variation, will develop stresses and strains which are a function of the radius only. Fig. 4.7-1 shows the hollow and solid spheres.

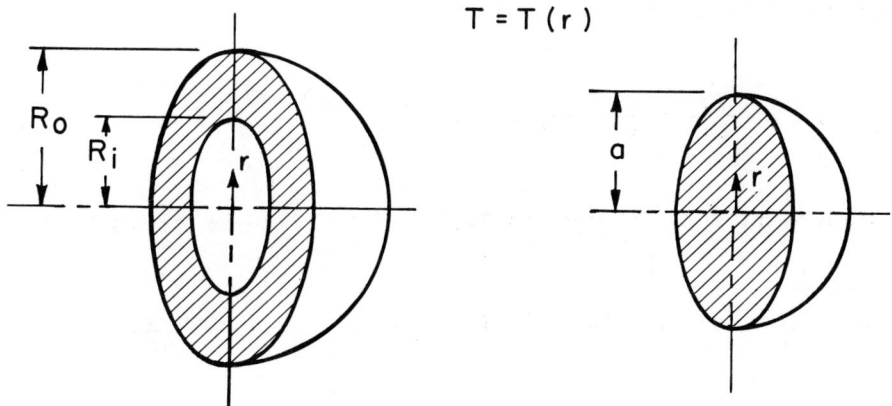

Fig. 4.7-1

Since there is spherical symmetry, the stresses and strains will be equal in any two orthogonal tangential planes, and the stress-strain relations can be written as

$$\epsilon_r = \frac{1}{E}[\sigma_r - 2\nu\sigma_t] + \alpha T \qquad (4.7\text{-}1)$$

$$\epsilon_t = \frac{1}{E}[(1-\nu)\sigma_t - \nu\sigma_r] + \alpha T \qquad (4.7\text{-}2)$$

with $T = T(r)$. Fig. 4.6-2 shows the radial and tangential loads acting on a differential element, and the equation of equilibrium for the spherically symmetrical state of stress, as given by Eq. (4.6-6), is

$$r\sigma_t = \frac{1}{2}\frac{d}{dr}(r^2\sigma_r) \qquad (4.7\text{-}3)$$

Differentiation of the foregoing yields

$$\frac{d}{dr}(r\sigma_t) = \frac{1}{2}\frac{d^2}{dr^2}(r^2\sigma_r) \qquad (4.7\text{-}4)$$

The compatibility equation for spherically symmetrical deformations is, as in the case of cylindrical symmetry,

$$\epsilon_r = \epsilon_t + r\frac{d\epsilon_t}{dr} = \frac{d}{dr}(r\epsilon_t) \qquad (4.7\text{-}5)$$

Substitution of the strains, Eqs. (4.7-1) and (4.7-2), into Eq. (4.7-5) yields

$$(1-\nu)\frac{d}{dr}(r\sigma_t) - \nu\frac{d}{dr}(r\sigma_r) - \sigma_r + 2\nu\sigma_t = -E\alpha r\frac{dT}{dr} \qquad (4.7\text{-}6)$$

Eqs. (4.7-3) and (4.7-4) are used to eliminate σ_t from the foregoing equation. The result, in terms of σ_r alone, is

$$\frac{1-\nu}{2} \frac{d^2}{dr^2}(r^2 \sigma_r) - \nu \frac{d}{dr}(r\sigma_r) - \sigma_r + \frac{\nu}{r}\frac{d}{dr}(r^2 \sigma_r) = -E\alpha r \frac{dT}{dr} \qquad (4.7\text{-}7)$$

The last term on the left-hand side of the equation is expanded and written as

$$\frac{\nu}{r} \frac{d}{dr}(r^2 \sigma_r) = \nu \frac{d}{dr}(r\sigma_r) + \nu \sigma_r$$

Eq. (4.7-7) then becomes

$$\frac{1}{2} \frac{d^2}{dr^2}(r^2 \sigma_r) - \sigma_r = -\frac{E\alpha}{1-\nu} r \frac{dT}{dr} \qquad (4.7\text{-}8)$$

and as a single differential this equation can be written as

$$\frac{r}{2} \frac{d}{dr}\left[\frac{1}{r^2}\frac{d}{dr}(r^3 \sigma_r)\right] = -\frac{E\alpha}{1-\nu} r \frac{dT}{dr} \qquad (4.7\text{-}9)$$

Integration of Eq. (4.7-9) yields the radial stress in terms of undetermined constants A and B. The integration gives

$$\sigma_r = A - \frac{B}{r^3} - \frac{2E\alpha}{(1-\nu)r^3} \int^r T r^2 dr \qquad (4.7\text{-}10)$$

and from Eq. (4.7-3) the tangential stress, in terms of the constants A and B, is

$$\sigma_t = A + \frac{B}{r^3} + \frac{E\alpha}{(1-\nu)r^3} \int^r T r^2 dr - \frac{E\alpha T}{1-\nu} \qquad (4.7\text{-}11)$$

Note that the sum $\sigma_r + 2\sigma_t$ is

$$\sigma_r + 2\sigma_t = 3A - \frac{2E\alpha T}{1-\nu} \qquad (4.7\text{-}12)$$

Hollow Sphere

The boundary conditions for a hollow sphere are

$$\sigma_r(r = R_i) = 0 \qquad \sigma_r(r = R_o) = 0 \qquad (4.7\text{-}13)$$

When set into Eq. (4.7-10) the constants A and B are found to be

$$A = \frac{2E\alpha}{(1-\nu)(R_o^3 - R_i^3)} \int_{R_i}^{R_o} Tr^2 dr = \frac{2}{3} \cdot \frac{E\alpha}{1-\nu} \overline{T} \qquad (4.7\text{-}14)$$

$$B = \frac{2E\alpha}{1-\nu} \left[\frac{R_i^3}{R_o^3 - R_i^3} \int_{R_i}^{R_o} Tr^2 dr - \int_{R_i}^{R_i} Tr^2 dr \right] \qquad (4.7\text{-}15)$$

Setting these into Eqs. (4.7-10) and (4.7-11) the radial and tangential stresses are found to be

$$\sigma_r = \frac{2E\alpha}{(1-\nu)r^3} \left[\frac{r^3 - R_i^3}{R_o^3 - R_i^3} \int_{R_i}^{R_o} Tr^2 dr - \int_{R_i}^{r} Tr^2 dr \right] \qquad (4.7\text{-}16)$$

and

$$\sigma_t = \frac{E\alpha}{(1-\nu)r^3} \left[\frac{2r^3 + R_i^3}{R_o^3 - R_i^3} \int_{R_i}^{R_o} Tr^2 dr + \int_{R_i}^{r} Tr^2 dr - r^3 T \right] \qquad (4.7\text{-}17)$$

The sum of the radial stress and twice the tangential stress, from Eqs. (4.7-12) and (4.7-14), or from the two preceding equations, is

$$\sigma_r + 2\sigma_t = \frac{2E\alpha}{1-\nu}(\overline{T} - T) \qquad (4.7\text{-}18)$$

In the commonly encountered temperature distributions, the inner and outer surfaces of the hollow sphere are the locations where the peak stresses are likely to be found. At these points the radial stress is zero and the tangential stress, from Eq. (4.7-18), is

$$\sigma_{ti} = \frac{E\alpha}{1-\nu}(\overline{T} - T_i) \qquad \sigma_{to} = \frac{E\alpha}{1-\nu}(\overline{T} - T_o) \qquad (4.7-19)$$

The subscripts i and o refer to the inner and outer surfaces.

Solid Sphere

The constant B, in Eq. (4.7-10), must be zero, as otherwise the stress at the center of the sphere would become infinite. Using the boundary condition $\sigma_r(r=a) = 0$, in Eq. (4.7-10), the constant A is found to be

$$A = \frac{2E\alpha}{(1-\nu)} \int_o^a Tr^2 dr = \frac{2}{3}\frac{E\alpha}{(1-\nu)} \overline{T} \qquad (4.7-20)$$

Substitution of A and B into Eq. (4.7-10) gives the radial stress as

$$\sigma_r = \frac{2E\alpha}{1-\nu}\left[\frac{1}{a^3}\int_o^a Tr^2 dr - \frac{1}{r^3}\int_o^r Tr^2 dr\right] \qquad (4.7-21)$$

or

$$\sigma_r = \frac{2E\alpha}{3(1-\nu)}(\overline{T} - \overline{T}_{ri}) \qquad (4.7-22)$$

where $\overline{T}$ is the mean temperature of the sphere, and $\overline{T}_{ri}$ the temperature of the inner portion of the sphere lying between r = o and r. Thus

$$\overline{T} = \frac{3}{a^3}\int_o^a Tr^2 dr \qquad \overline{T}_{ri} = \frac{3}{r^3}\int_o^r Tr^2 dr \qquad (4.7-23)$$

The tangential stress, obtained from Eq. (4.7-11), can be written in integral form or in terms of the mean temperature parameters as follows:

$$\sigma_t = \frac{E\alpha}{1-\nu}\left[\frac{2}{a^3}\int_o^a Tr^2 dr + \frac{1}{r^3}\int_o^r Tr^2 dr - T\right] \qquad (4.7-24)$$

$$\sigma_t = \frac{E\alpha}{1-\nu}(\bar{T} - T) - \frac{\sigma_r}{2} = \frac{E\alpha}{1-\nu}\left(\frac{2}{3}\bar{T} + \frac{1}{3}\bar{T}_{ri} - T\right) \qquad (4.7\text{-}25)$$

To obtain the stress of the center of the sphere from Eq. (4.7-24) the following integration should be noted

$$\frac{1}{r^3}\int_0^r Tr^2 dr = \frac{T_c}{(\Delta r)^3}\int_0^{\Delta r} r^2 dr = \frac{T_c}{3} \qquad (4.7\text{-}26)$$

$$r \to \Delta r \to 0$$

T_c denotes the temperature at the center of the sphere. The stress at the center is

$$\sigma_c = \frac{2}{3}\frac{E\alpha}{1-\nu}(\bar{T} - T_c) \qquad (4.7\text{-}27)$$

This result can be obtained by observing that the sum of the stresses, in accordance with Eq. (4.7-12), is

$$\sigma_r + 2\sigma_t = \frac{2E\alpha}{1-\nu}(\bar{T} - T) \qquad (4.7\text{-}28)$$

By letting $\sigma_r = \sigma_t = \sigma_c$ at $r = 0$, the center stress

$$\sigma_c = \frac{2}{3}\frac{E\alpha}{1-\nu}(\bar{T} - T_c) \qquad \begin{array}{c}(4.7\text{-}27)\\ (\text{rep.})\end{array}$$

is obtained. At the surface of a solid sphere, where $\sigma_r = 0$, it follows from Eq. (4.7-28) that the tangential stress is

$$\sigma_{ta} = \frac{E\alpha}{1-\nu}(\bar{T} - T_a) \qquad (4.7\text{-}29)$$

When a small cavity is present at the center of a sphere (where $\sigma_r = 0$) the central tangential stress, σ_{tc}, is given by Eq. (4.7-19) as

$$\sigma_{tc} = \frac{E\alpha}{1-\nu}(\bar{T} - T_c) \qquad (4.7\text{-}30)$$

$$R_i \to 0$$

SPHERES

and when there is no cavity the central stress is as given by Eq. (4.7-27). From these equations it is seen that the effect of a small central cavity in a sphere with a radial temperature variation, is the introduction of a stress concentration factor of 1.5.

The radial displacement at the surface of the solid sphere is obtained from the expression for tangential strain, Eq. (4.7-2), and from Eq. (4.7-29). Thus

$$u_a = \epsilon_{ta} a = [\alpha(\overline{T} - T_a) + \alpha T_a] a = \alpha \overline{T} a \qquad (4.7-31)$$

At the inner and outer surfaces R_i and R_o of a hollow sphere, the displacements, using Eqs. (4.7-2) and (4.7-19), are

$$u_{r=R_i} = \epsilon_{ti} R_i = [\alpha(\overline{T} - T_i) + \alpha T_i] R_i = \alpha \overline{T} R_i \qquad (4.7-32)$$

$$u_{r=R_o} = \epsilon_{to} R_o = [\alpha(\overline{T} - T_o) + \alpha T_o] R_o = \alpha \overline{T} R_o \qquad (4.7-33)$$

Example (a)   Uniform Heat Generation in a Solid Sphere

A solid sphere of radius, "a", and conductivity, k, is subjected to a uniform volumetric heat generation density, W. The resulting steady state temperature distribution is

$$T = \frac{Wa^2}{6k} \left(1 - \frac{r^2}{a^2}\right) \qquad (1)$$

The mean temperature parameters are

$$\overline{T} = \frac{3}{a^3} \int_0^a T r^2 dr = \frac{Wa^2}{15k} \qquad (2)$$

$$\overline{T}_{ri} = \frac{3}{r^3} \int_0^r T r^2 dr = \frac{Wa^2}{30k} \left(5 - \frac{3r^2}{a^2}\right) \qquad (3)$$

From Eq. (4.7-22) and (4.7-25) we have

$$\sigma_r = \frac{2}{3}\frac{E\alpha}{1-\nu}(\overline{T}-T_{ri}) = -\frac{E\alpha}{1-\nu}\frac{Wa^2}{15k}\left(1-\frac{r^2}{a^2}\right) \tag{4}$$

$$\sigma_t = \frac{E\alpha}{1-\nu}(\overline{T}-T) - \frac{\sigma_r}{2} = \frac{E\alpha}{1-\nu}\frac{Wa^2}{15k}\left(\frac{2r^2}{a^2}-1\right) \tag{5}$$

Example (b) Concentric Hollow Spheres with Radial Temperature Variation

An assembly of two concentric hollow spheres of different materials, similar to that shown in Fig. 4.6-1, is subjected to a radial temperature variation. In Sect. 4.6 the solution was obtained for the case of hollow spheres at uniform but different temperatures. The radial and tangential stresses were then due only to the radial pressure resulting from radial interference. In the present problem there is a radial variation in temperature in each of the hollow spheres. The interference pressure will be as given by Eq.(4.6-22), with $\overline{T}_1$ and $\overline{T}_2$ replacing T_1 and T_2. That is

$$p = \frac{\alpha_1 \overline{T}_1 - \alpha_2 \overline{T}_2}{\frac{1}{E_2}\left[\frac{(1-\nu_2)(2R_j^3+R_0^3)}{R_0^3-R_j^3}+\nu_2\right] + \frac{1}{E_1}\left[\frac{(1-\nu_1)(2R_j^3+R_i^3)}{R_j^3-R_i^3}-\nu_1\right]} \tag{1}$$

This is so since the radial displacements at the inner and outer surfaces of a hollow sphere having a radial temperature variation are as given by Eqs. (4.7-32) and (4.7-3). They are

$$u_i = \alpha \overline{T} R_i \qquad u_0 = \alpha \overline{T} R_0 \tag{2}$$

or the same as the radial displacements of a hollow sphere uniformly heated to a temperature, $\overline{T}$.

The free expansion interference stresses are therefore, as in Eqs. (4.6-16), (4.6-17), (4.6-18) and (4.6-19); namely

$$\sigma_{1rf} = -\frac{R_j^3 p}{R_j^3-R_i^3}\left(1-\frac{R_i^3}{r^3}\right) \tag{3}$$

$$\sigma_{1tf} = -\frac{R_j^3 p}{R_j^3 - R_i^3} \left(1 + \frac{R_i^3}{2r^3}\right) \tag{4}$$

$$\sigma_{2rf} = \frac{R_j^3 p}{R_0^3 - R_j^3} \left(1 - \frac{R_0^3}{r^3}\right) \tag{5}$$

$$\sigma_{2tf} = \frac{R_j^3 p}{R_0^3 - R_j^3} \left(1 + \frac{R_0^3}{2r^3}\right) \tag{6}$$

The distributed stresses due to the temperature variations in the unrestrained hollow spheres are, in accordance with Eqs. (4.7-16) and (4.7-17),

$$\sigma_{1rd} = \frac{2E\alpha}{(1-\nu)r^3} \left[\frac{r^3 - R_i^3}{R_j^3 - R_i^3} \int_{R_i}^{R_j} T_1 r^2 dr - \int_{R_i}^{r} T_1 r^2 dr \right] \tag{7}$$

$$\sigma_{1td} = \frac{E\alpha}{(1-\nu)r^3} \left[\frac{2r^3 + R_i^3}{R_j^3 - R_i^3} \int_{R_i}^{R_j} T_1 r^2 dr + \int_{R_i}^{r} T_1 r^2 dr - r^3 T_1 \right] \tag{8}$$

$$\sigma_{2rd} = \frac{2E\alpha}{(1-\nu)r^3} \left[\frac{r^3 - R_j^3}{R_0^3 - R_j^3} \int_{R_i}^{R_0} T_2 r^2 dr - \int_{R_j}^{r} T_2 r^2 dr \right] \tag{9}$$

$$\sigma_{2td} = \frac{E\alpha}{(1-\nu)r^3} \left[\frac{2r_3 + R_j^3}{R_0^3 - R_j^3} \int_{R_j}^{R_0} T_2 r^2 dr + \int_{R_j}^{r} T_2 r^2 dr - r^3 T_2 \right] \tag{10}$$

The net stresses are the sum of the interference stresses, Eqs. (3), (4), (5) and (6), and the stresses due to the radial temperature distribution, Eqs. (7), (8), (9) and (10). Thus

$$\sigma_{1r} = \sigma_{1rf} + \sigma_{1rd} \qquad \sigma_{1t} = \sigma_{1tf} + \sigma_{1td} \tag{11}$$

$$\sigma_{2r} = \sigma_{2rf} + \sigma_{2rd} \qquad \sigma_{2t} = \sigma_{2tf} + \sigma_{2td} \tag{12}$$

PROBLEMS

4-1 An assembly consists of two radially-thin concentric rings. The inner ring, of thickness h_1, is heated uniformly to a temperature T_1, and the outer ring, of thickness h_2, is heated uniformly to a temperature T_2. Eqs. (2.13-1) show the stresses in such a ring assembly to be

$$\sigma_1 = \frac{E_1 E_2 h_2 (\alpha_2 T_2 - \alpha_1 T_1)}{E_1 h_1 + E_2 h_2}$$

$$\sigma_2 = \frac{E_1 E_2 h_1 (\alpha_1 T_1 - \alpha_2 T_2)}{E_1 h_1 + E_2 h_2}$$

Demonstrate that the expressions for the tangential stresses in uniformly heated concentric thick-walled rings, as given by Eqs. (4.1-29) and (4.1-30) degenerate into the equations for the thin-walled rings when the thicknesses become small. Let $R_j \rightarrow R$, $R_0 \rightarrow R + h_2$, $R_i \rightarrow R - h_1$, and neglect higher order quantities.

4-2 The central area of a homogeneous circular plate is uniformly heated to a temperature T_1 and the outer portion of the plate to a temperature T_2. Using the appropriate expressions for stress distribution, derive the expressions for the stresses in the inner and outer parts of the plate.

4-3 Use the expressions derived in Prob. 4-2 to find the stress distribution in an infinite plate with a circular hot area. Check result with solution of this problem given in Sect. 3.6.

4-4 A steel liner in the form of a circular ring of radius R_i (assumed constant), and cross section area A, is lowered in temperature to $T = -T_r$ below ambient temperature and fitted to the inner rim of an aluminum disc. The thickness of the aluminum disc is h and its inner and outer radii are R_i and R_0 as shown in Fig. 4-4. Assuming the radial thickness of the ring to be small, derive general expressions for the stresses in the ring and disc. Use subscript a for aluminum and s for steel.

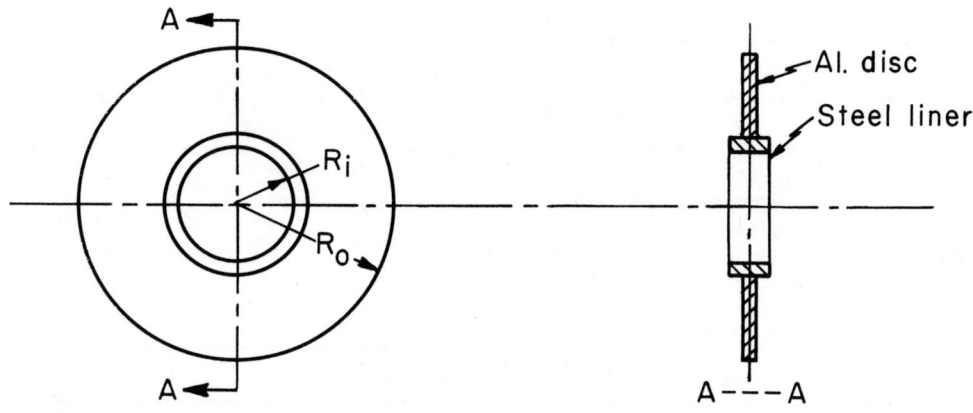

Fig. 4-4

4-5 Derive the expressions for the stresses in an assembly consisting of three concentric circular flat plate rings of different materials and at different but uniform temperatures. Use the notations shown in Fig. 4-5.

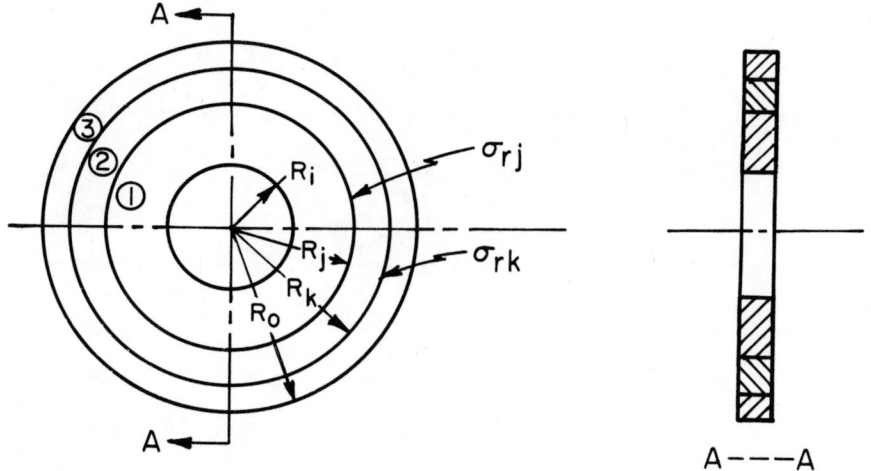

Fig. 4-5

4-6 A no-interference fit exists at ambient temperature between a 1" stainless steel shaft and a copper alloy bearing, $\frac{1}{4}$" thick. Before assembly the contact surfaces of the shaft and bearing are lubricated to permit differential axial expansion. Fig. 4-6 shows the arrangement. Find the stresses in the components when the assembly temperature is increased $300°F$. Use $E_s = 28 \times 10^6$ psi, $\alpha_s = 10 \times 10^{-6}/°F$ for the stainless steel, and for the copper use $E_c = 20 \times 10^6$ psi, $\alpha_c = 5 \times 10^{-6}/°F$; $\nu_s = \nu_c = .3$.

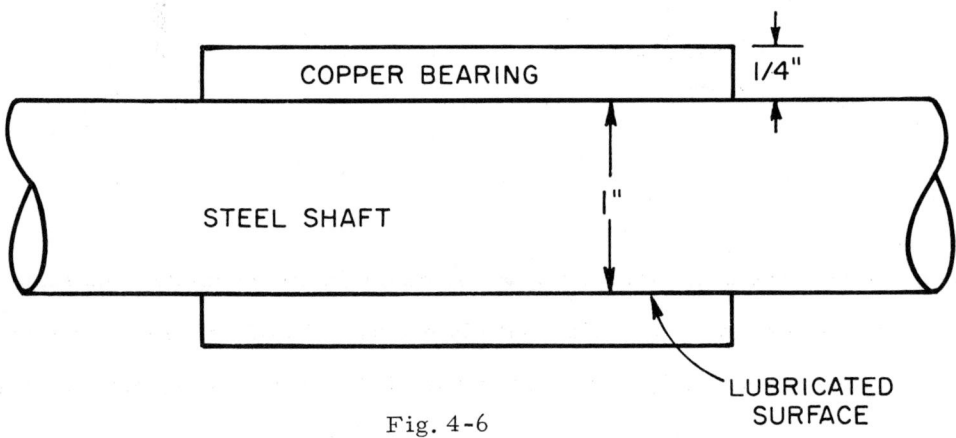

Fig. 4-6

4-7 The central portion of a homogeneous circular plate of radius "b" is subjected to a radial temperature distribution

$$T = T_0(1 - \frac{r}{a})$$

The rest of the plate is at zero temperature as shown in Fig. 4-7. Find the stresses in the central and outer parts of the plate. Verify that the maximum in-plane stress at $r = a$ is always $\tau = E\alpha T_0/6$ regardless of the magnitude of "a" or "b".

PROBLEMS

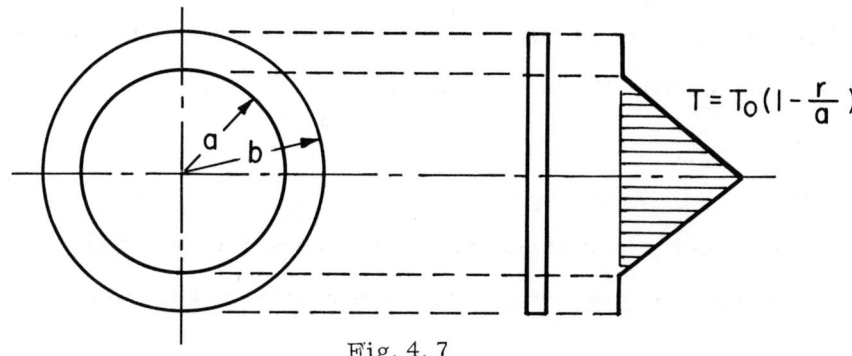

Fig. 4.7

4-8 A uniform heat generation in a circular area in a very large plate produces a temperature distribution expressed as

$$T = T_o \left(1 - \frac{r^2}{a^2}\right)$$

Determine the thermal stresses in the hot area and in the surrounding cold area.

4-9 A circular plate of outer radius "b" is subjected to the same temperature distribution over the inner portion of radius "a", as given in Prob. 4-8. Determine the stresses in both parts of the plate. What can be said of the shear stress at $r = a$? Is this the maximum shear stress in the plate?

4-10 An aluminum radial fin, constructed as shown in Fig. 4.2-4, is subjected to a temperature distribution

$$T = 300 \left(1 - \frac{r}{R_o}\right) \quad °F$$

Using $E = 10 \times 10^6$ psi, $\alpha = 10 \times 10^{-6}/°F$, $\nu = .3$; $R_i = \frac{1}{2}"$, $R_j = 1"$, $R_k = 1\frac{1}{2}"$, $R_o = 2"$, $h_1 = \frac{1}{2}"$, $h_2 = \frac{1}{4}"$, $h_3 = \frac{1}{8}"$, find the stress distribution in all sections of the fin.

4-11 A solid circular fuel plate of thickness h and radius "a" has a variable fissionable material concentration which is equivalent to a steady-state heat source distribution expressed as

$$W = W_o \frac{r}{a}$$

The significant heat conduction is perpendicular to the plane of the plate so that the resulting temperature distribution takes the form

$$T = T_o(1 - \frac{4z^2}{h^2}) \frac{r}{a}$$

Derive the equations for the radial and tangential stresses in the plate.

4-12 Eqs. (4.4-14) through (4.4-17) give the radial and tangential stresses in concentric uniformly heated circular cylinders. These equations are in terms of the interfacial pressure, p, given by Eq. (4.4-24). Assume that the cylinders are relatively thin with h_1 the thickness of the inner cylinder and h_2 the thickness of the outer cylinder. Take the outer radius R_o as $R_o = R_j + h_2$ and the inner radius R_i as $R_i = R_j - h_1$. Show that by neglecting small quantities, the tangential stresses, Eqs. (4.4-15) and (4.4-17), and the axial stresses, Eqs. (4.4-31) and (4.4-32), degenerate into the equations

$$\sigma_{1t} = \sigma_{1z} = \frac{E_1}{1-\nu_1}(\overline{\alpha T}' - \alpha_1 T_1)$$

$$\sigma_{2t} = \sigma_{2z} = \frac{E_2}{1-\nu_2}(\overline{\alpha T}' - \alpha_2 T_2)$$

These equations are derived in Sect. 3. See Eqs. (3.10-18) and (3.10-19).

4-13 In Prob. 4-6 the assumption is made that the shaft and bearing, shown in Fig. 4-6, do not exert any axial restraints on each other so that axial expansion takes place freely. Evaluate the stresses for the case where there is no axial slippage of the shaft within the bearing such that both components are subject to the same axial strain. Use the same dimensions and physical constants given in Prob. 4-6.

4-14 A circular tube, which is free to expand axially, has its inner wall, $r = R_i$, at a temperature T_i and its outer wall, $r = R_o$, at a temperature T_o. Radial heat conduction produces a temperature distribution

$$T - T_o = (T_i - T_o) \frac{\ln(R_o/r)}{\ln(R_o/R_i)}$$

Derive the expressions for σ_r, σ_t and σ_z. Show that for a thin-walled tube the expressions for the stresses become approximately

$$\sigma_t = \sigma_z = \frac{E\alpha}{1-\nu}(\overline{T} - T) \qquad \sigma_r = 0$$

4-15 A double-walled heat exchanger tube consists of a stainless steel inner tube and copper alloy outer tube. The dimensions are as shown in Fig. 4.15.

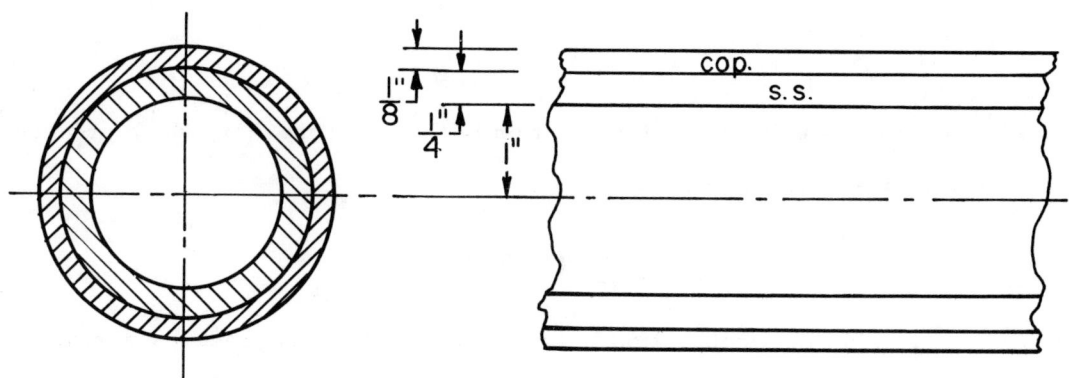

Fig. 4-15

Use the material constants $E_s = 28 \times 10^6$ psi, $\alpha_s = 10 \times 10^{-6}/°F$, $E_c = 20 \times 10^6$ psi, $\alpha_c = 8 \times 10^{-6}/°F$, and $\nu = 0.3$ for both. The temperature at the inner surface $r = 1''$ is $800°F$. At the interface, $r = 1.25''$ it is $1000°F$, and at the outer radius $r = 1.385''$ it is $1200°F$. There is radial heat conduction with the temperature variation as given in Prob. 4-14. Assume that there is no slippage between the tubes and that there is no axial expansion restraints on the tube assembly. Use the results of Example (b) of Sect. 4.5 to obtain the stress distribution in the dual tube assembly.

4-16 A nuclear fuel element is constructed as shown in Fig. 4-16.

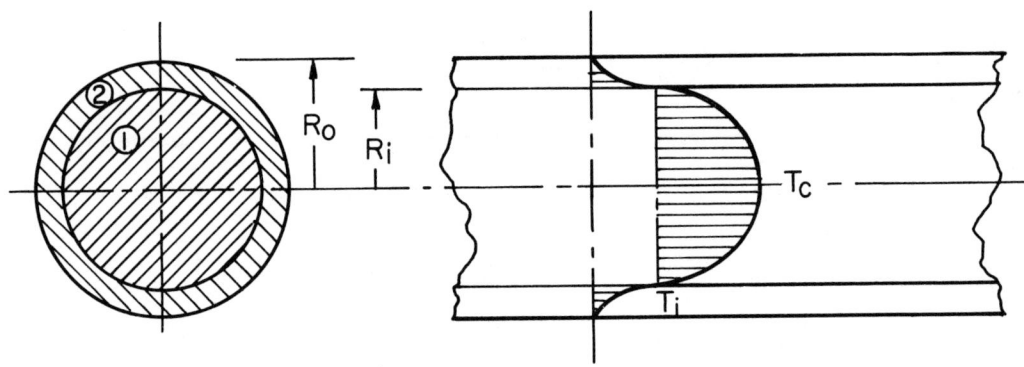

Fig. 4-16

There is a uniform heat generation in the fuel core, (1), which produces a temperature distribution

$$T - T_i = (T_c - T_i)(1 - \frac{r^2}{R_i^2})$$

and radial heat conduction in the cladding (2) which produces a temperature distribution

$$T = T_i \frac{\ln(R_o/r)}{\ln(R_o/R_i)}$$

Use the subscript 1 for the core and the subscript 2 for the cladding. Assume no slippage at the interface and no restraint to axial expansion of the fuel element. Employ the results of Example (b), Sect. 4.5, to derive the general expressions for the stresses in the fuel core and cladding in terms of T_i and T_c.

4-17 Determine the simplified expression for p, given by Eq. (4.6-22), when the concentric hollow spheres become concentric spherical shells of thickness h_1 and h_2. Let $R_o = R_j + h_2$ and $R_i = R_j - h_1$. Show that Eqs. (4.6-16) and (4.6-18) then reduce to $\sigma_{1r} = \sigma_{2r} = 0$ and that Eqs. (4.6-17) and (4.6-19) become

$$\sigma_{1t} = \frac{E_1}{1-\nu_1}(\overline{\alpha T}' - \alpha_1 T_1)$$

$$\sigma_{2t} = \frac{E_2}{1-\nu_2}(\overline{\alpha T}' - \alpha_2 T_2)$$

as derived in Sect. 3.12.

4-18 Derive the expressions for the radial and tangential stresses in an assembly of three concentric hollow spheres. The spheres are of dissimilar materials and different but uniform temperatures, T_1, T_2 and T_3. Use the conventional symbols for the material constants with subscripts 1, 2, and 3 referring to the innermost, the middle, and the outermost hollow sphere, respectively. The inner radius is R_i, the radius at the interface of the first and second hollow spheres is R_j, the radius at the interface of the second and third spheres is R_k, and the outer radius is R_o.

4-19 Show that the radial and tangential stresses in a hollow sphere, given by Eqs. (4.7-16) and (4.7-17) become

$$\sigma_r = 0 \qquad \sigma_t = \frac{E\alpha}{1-\nu}(\bar{T} - T)$$

when the wall thickness of the hollow sphere becomes small. Let $R_o \to R_m + \frac{h}{2}$ and $R_i \to R_m - \frac{h}{2}$.

4-20 A solid sphere of fissionable material is enclosed by an hollow sphere as shown in Fig. 4-20. There is uniform heat generation and radial heat

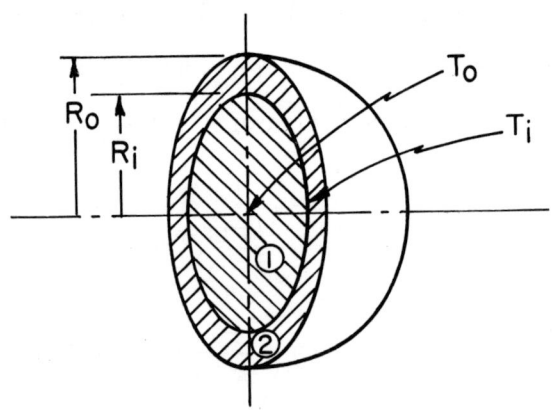

Fig. 4-20

conduction in the solid sphere, and simple radial heat conduction in the outer casing. Determine the temperature distribution in terms of a central temperature T_c an interfacial temperature T_i, and then find the stress distribution. Use the procedure demonstrated in Example (b), Sect. 4.7.

5 | Energy Methods

5.1 ENERGY METHODS OF THERMAL STRESS ANALYSIS

As in isothermal stress analysis, strain energy methods in thermal stress analysis provide useful means of solving problems. Strain energy expressions that have been derived in isothermal analysis are not always directly applicable to thermal stress problems since the presence of the term αT in the classical stress-strain relationships alters the form of some of the classical strain energy relations. However, when equivalent thermal loading is used, wherein thermal body forces and surface tractions are treated as real loads, the classical form of the energy relationships apply. The stresses obtained in solutions employing equivalent thermal loads are displacement stresses, to which equivalent internal pressures must be added.

When the solution of a problem is obtained through the use of the principle of stationary or minimum complementary energy, it is convenient to express the complementary energy in terms of the real stresses rather than the displacement stresses. On the other hand, when solving problems by means of Castigliano's theorem it is preferable to use equivalent thermal loads and the associated displacement stresses. The analytical procedures for obtaining solutions by means of the complementary energy method and by means of Castigliano's theorem are discussed in Sections 5.2 and 5.3.

A third method of solving thermal stress problems, derived from energy principles, is based on the use of the thermoelastic reciprocal theorem. The types of problems that are most suited to this method of analysis are those for which there are readily available solutions of the required auxiliary isothermal problems. This method of analysis is discussed in Section 5.4.

Energy methods may be utilized to obtain approximate solutions to problems. The Rayleigh - Ritz method is one of these methods that permits the approximate determination of displacements or deformations. It is particularly useful in problems having a high degree of redundancy or structural nonuniformity, which make exact methods of analysis difficult to apply. Section 5.5 contains a discussion of this procedure, and a number of illustrative problems.

5.2 COMPLEMENTARY ENERGY

The significance of complementary energy can be demonstrated by referring to the one-dimensional stress-strain equation

$$\epsilon_x = \frac{\sigma_x}{E} + \alpha T \tag{5.2-1}$$

The diagram of Fig. 5.2-1 is a graphical representation of this equation

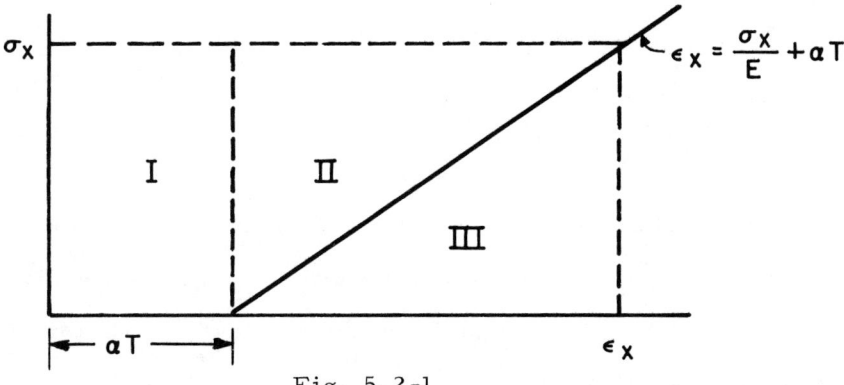

Fig. 5.2-1

The strain energy density which is defined as

$$u_o = \int_{\alpha T}^{\epsilon_x} \sigma_x \, d\epsilon_x = \frac{\sigma_x^2}{2E} \qquad (5.2\text{-}2)$$

is shown as area III in the diagram. The complementary energy density is the difference between total rectangular area, $\sigma_x \epsilon_x$, and the strain energy density, area III. Thus the complementary energy density, u_1, is defined as

$$u_1 = \sigma_x \epsilon_x - \sigma_x^2/2E = \int_0^{\sigma_x} \epsilon_x \, d\sigma_x \qquad (5.2\text{-}3)$$

Since u_1 is equal to the sum of areas I and II, with area I equal to $\sigma_x \alpha T$ and areas II and III each equal to $\sigma_x^2/2E$, the complementary energy density can be written as

$$u_1 = I + II = \sigma_x \alpha T + \frac{\sigma_x^2}{2E} \qquad (5.2\text{-}4)$$

The total complementary energy in a body in which there is present only a normal stress σ_x, is therefore

$$U_1 = \int_V \left(\frac{\sigma_x^2}{2E} + \sigma_x \alpha T \right) dV \tag{5.2-5}$$

In a body at equilibrium the complementary energy is a minimum. If the complementary energy is expressed in terms of either statically indeterminate stresses, statically indeterminate internal loads, or in terms of a series of undetermined coefficients of functions which describe a state of stress, then the derivative of the complementary energy with respect to each of the undetermined stress or load related quantities, X_i, will be zero. Thus, by setting

$$\frac{\partial U_1}{\partial X_i} = 0 \tag{5.2-6}$$

we can obtain the statically indeterminate loads, stresses, or stress function coefficients. Eq. (5.2-6) applies to one-dimensional, two-dimensional, and three-dimensional stress systems.

One-Dimensional Complementary Energy

In the complementary energy analysis of one-dimensional elements which undergo extensional deformation, it is sometimes convenient to express the strain energy in terms of a statically indeterminate mean stress. The distributed stress then follows directly from the mean stress by means of the relationships derived below.

In extensional deformations, the axial strain over a cross section is constant. Thus with T and σ_x functions of the cross section coordinates y and z we have

$$\epsilon_x = \frac{1}{A} \int_A \epsilon_x dA = \frac{1}{A} \int (\frac{\sigma_x}{E} + \alpha T) dA = \frac{\overline{\sigma}_x}{E} + \alpha \overline{T}(x) \qquad (5.2-7)$$

and as $\epsilon_x = \sigma_x/E + \alpha T$, we obtain the following relationship between the mean and distributed values of stress and temperature:

$$\sigma_x(y, z) - \overline{\sigma}_x = E\alpha(\overline{T}(x) - T(x, y, z)) \qquad (5.2-8)$$

The mean quantities represent average values over the cross section A. Equation (5.2-8) holds for one-dimensional parallel and series assemblies with any specified axial boundary restraint. By means of it the axial distributed stresses can be expressed in terms of the mean axial cross section stresses.

When two parallel elements undergoing extensional deformation are of different materials we obtain from Eq. (5.2-7)

$$\epsilon = \frac{\overline{\sigma}_{1x}}{E_1} + \alpha_1 \overline{T}_1(x) = \frac{\overline{\sigma}_{2x}}{E_2} + \alpha_2 \overline{T}_2(x) = \frac{\sigma_{1x}}{E_1} + \alpha_1 T_1(x) = \frac{\sigma_{2x}}{E_2} + \alpha_2 T_2(x) \qquad (5.2-9)$$

The complementary energy in parallel elements of dissimilar materials, can, with the use of the foregoing equations, be written in terms of the mean stresses $\overline{\sigma}_1$ and $\overline{\sigma}_2$. Equilibrium requirements permit $\overline{\sigma}_2$ to be expressed in terms of $\overline{\sigma}_1$, and $\overline{\sigma}_1$ then determined by minimizing the complementary energy.

Problems of beam bending are also one-dimensional stress problems which may be solved by means of complementary energy analysis. Beam bending is treated in Chapter 6, and the application of complementary energy methods to the solution of beam problems is discussed in Section 6.9. It is shown there that statically indeterminate loads in thermal beams may be obtained by means of the following equation, derived from complementary energy principles:

$$\int_0^L M \frac{\partial M}{\partial X_i} dx = -E\alpha \int_0^L M_t \frac{\partial M}{\partial X_i} dx \tag{5.2-10}$$

In this equation M is expressed in terms of the indeterminate beam reactions, and the quantities X_i represent statically indeterminate loads. M_t is the thermal moment over the cross section.

Plane Complementary Energy

In the case of plane stress, with $\sigma_z = \tau_{xz} = \tau_{yz} = 0$, the complementary energy is written as

$$U_1 = \frac{h}{2E} \int_A \left[\sigma_x^2 + \sigma_y^2 - 2\nu\sigma_x\sigma_y + 2(1+\nu)\tau_{xy} + \alpha T(\sigma_x + \sigma_y) \right] dA \tag{5.2-11}$$

with h the plate thickness. In a simply connected body, the formulation as well as the solution of the plane stress problem does not contain Poisson's ratio. It follows that an arbitrary value can be assigned to ν without affecting the results. If we let $\nu = -1$, Eq. (5.2-11) takes the form

$$U_1 = \frac{h}{2E} \int_A \left[(\sigma_x + \sigma_y)^2 + \alpha T(\sigma_x + \sigma_y) \right] dA \tag{5.2-12}$$

On the other hand if we let $\nu = 0$, Eq. (5.2-11) becomes

$$U_1 = \frac{h}{2E} \int_A \left[\sigma_x^2 + \sigma_y^2 + 2\tau_{xy}^2 + \alpha T(\sigma_x + \sigma_y) \right] dA \tag{5.2-13}$$

The stress function, φ, is commonly used in the solution of plane stress problems. With the normal and shear stresses defined as

$$\sigma_x = \frac{\partial^2 \varphi}{\partial y^2} \qquad \sigma_y = \frac{\partial^2 \varphi}{\partial x^2} \qquad \tau_{xy} = -\frac{\partial^2 \varphi}{\partial x \partial y} \tag{5.2-14}$$

the complementary energy is expressed in terms of φ as follows.

$$U_1 = \frac{h}{2E} \int_A \left[\left(\frac{\partial^2 \varphi}{\partial x^2}\right)^2 + 2\left(\frac{\partial^2 \varphi}{\partial x \partial y}\right)^2 + \left(\frac{\partial^2 \varphi}{\partial y^2}\right)^2 + \alpha T \left(\frac{\partial^2 \varphi}{\partial x^2} + \frac{\partial^2 \varphi}{\partial y^2}\right) \right] dA \qquad (5.2\text{-}15)$$

This form of the plane complementary energy is often used to obtain solutions of thermal stress problems in rectangular plates having temperature distribution which are functions of the plane coordinates. A solution can be assumed in the form

$$\varphi = C_0 \varphi_0 + C_1 \varphi_1 + C_2 \varphi_2 + \cdots \qquad (5.2\text{-}16)$$

with the functions φ_i and the undetermined coefficients C_i taken such that the boundary conditions are satisfied. The constants, C_i, are determined by minimizing the complementary energy. That is, it is required that

$$\frac{\partial U_1}{\partial C_i} = 0 \qquad (5.2\text{-}17)$$

Three-Dimensional Complementary Energy

In a three-dimensional stress system the strain energy U_0 is

$$U_0 = \frac{1}{2E} \int_V \left[(\sigma_x^2 + \sigma_y^2 + \sigma_z^2) - 2\nu(\sigma_x \sigma_y + \sigma_x \sigma_z + \sigma_y \sigma_z) \right.$$
$$\left. + 2(1+\nu)(\tau_{xy}^2 + \tau_{xz}^2 + \tau_{yz}^2) \right] dV \qquad (5.2\text{-}18)$$

and the complementary energy, U_1, is

$$U_1 = U_0 + \int_V \alpha T (\sigma_x + \sigma_y + \sigma_z) \, dV \qquad (5.2\text{-}19)$$

Example (a) Bar Held Between Rigid Walls

We shall assume, at first, that the temperature is a function of the axial coordinate, x, as in Fig. 1.8-4. The complementary energy, is given by Eq. (5.2-5) as

$$U_1 = \int_0^L \left(\frac{\sigma_x^2}{2E} + \sigma_x \alpha T \right) A \, dx \tag{1}$$

The constant axial stress, σ_x is statically indeterminate. The derivative of U_1 with respect to σ_x is set equal to zero, in accordance with Eq. (5.2-6). Thus

$$\frac{dU_1}{d\sigma_x} = \int_0^L \left(\frac{\sigma_x}{E} + \alpha T \right) A \, dx = 0 \tag{2}$$

or

$$\sigma_x = -\frac{E\alpha}{L} \int_0^L T \, dx = -E\alpha \overline{T} \tag{3}$$

We now assume that the temperature is a symmetrical function of y and z, that is $T(x, y, z) = T(x, -y, -z)$, and set the stress, from Eq. (5.2-8), into Eq. (5.2-5). This yields

$$U_1 = \int_V \left\{ \frac{[\overline{\sigma}_x + E\alpha(\overline{T}(x) - T)]^2}{2E} + [\overline{\sigma}_x + E\alpha(\overline{T}(x) - T)]\alpha T \right\} dV \tag{4}$$

Minimizing the complementary energy with respect to $\overline{\sigma}_x$ we have

$$\frac{\partial U_1}{\partial \overline{\sigma}_x} = 0 = \int_V [\overline{\sigma}_x + E\alpha(\overline{T}(x) - T) + E\alpha T] \, dV \tag{5}$$

or

$$\int_V [\bar{\sigma}_x + E\alpha \bar{T}(x)] \, dV = 0 \tag{6}$$

Since $\bar{\sigma}_x$ is constant, we obtain from Eq. (6)

$$\bar{\sigma}_x = -E\alpha \bar{T} \tag{7}$$

and from Eq. (5.2-8) we obtain the modified (See Sect. 2.4) distributed stress as

$$\sigma_x = E^*\alpha (\bar{T}(x) - \bar{T} - T) \tag{8}$$

Example (b)   Concentric Tube Assembly

In the concentric tube assembly, shown in Fig. 5.3-3, let σ_1 be the indeterminate axial stress in the inner tube. The stress in the outer tube is therefore

$$\sigma_2 = -\sigma_1 A_1/A_2 \tag{1}$$

The complementary energy is

$$U_1 = \left[\frac{\sigma_1^2 A_1}{2E_1} + \frac{\sigma_2^2 A_2}{2E_2} + \sigma_1 \alpha_1 T_1 A_1 + \sigma_2 \alpha_2 T_2 A_2 \right] L \tag{2}$$

Substitution of σ_2 from Eq. (1) yields

$$U_1 = \left[\frac{\sigma_1^2 A_1}{2E_1} + \left(\frac{\sigma_1 A_1}{A_2}\right)^2 \frac{A_2}{2E_2} + \sigma_1 \alpha_1 T_1 A_1 - \left(\frac{\sigma_1 A_1}{A_2}\right) \sigma_2 T_2 A_2 \right] L \tag{3}$$

For stationary complementary energy we have

$$\frac{dU_1}{d\sigma_1} = 0 = \left[\frac{\sigma_1 A_1}{E_1} + \frac{\sigma_1 A_1^2}{A_2 E_2} + \alpha_1 T_1 A_1 - \alpha_2 T_2 A_1 \right] L \tag{4}$$

from which

$$\sigma_1 = \frac{E_1 E_2 A_2 (\alpha_2 T_2 - \alpha_1 T_1)}{E_1 A_1 + E_2 A_2} \tag{5}$$

and

$$\sigma_2 = \frac{E_1 E_2 A_1 (\alpha_1 T_1 - \alpha_2 T_2)}{E_1 A_1 + E_2 A_2} \tag{6}$$

Example (c) Redundant Truss

Consider the simply supported redundant truss shown in Fig. 5.2-2.

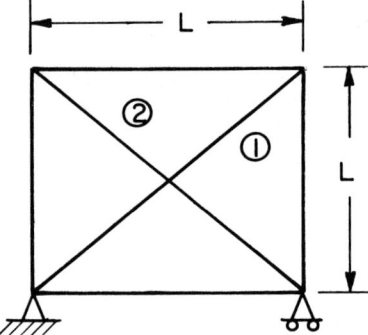

Fig. 5.2-2

The cross section areas, A, and moduli of elasticity, E, are assumed the same for all members. Diagonal 1 is heated to a temperature T_1 and diagonal 2 to a temperature T_2.

Let σ_1 be the indeterminate stress in diagonal 1. The stress σ_s, in each of the vertical and horizontal members is then

$$\sigma_s = -\frac{\sigma_1}{\sqrt{2}} \tag{1}$$

and the stress in diagonal 2, σ_2, is the same as in diagonal 1. The complementary energy is therefore

$$U_1 = \frac{\sigma_1^2}{2E} AL\sqrt{2} + \frac{\sigma_2^2}{2E} AL\sqrt{2} + \frac{4\sigma_s^2}{2E} AL \qquad (2)$$

$$+ \sigma_1 \alpha_1 T_1 AL\sqrt{2} + \sigma_2 \alpha_2 T_2 AL\sqrt{2}$$

and with the use of Eq. (1) we express U_1 as

$$U_1 = [\frac{\sigma_1^2}{E}(\sqrt{2}+1) + \sigma_1 \sqrt{2}(\alpha_1 T_1 + \alpha_2 T_2)] AL \qquad (3)$$

The derivative of the complementary energy, U_1 with respect to σ_1 is set equal to zero. This yields the stresses in the members as

$$\sigma_1 = -\frac{E}{2+\sqrt{2}}(\alpha_1 T_1 + \alpha_2 T_2) \qquad (4)$$

$$\sigma_2 = -\frac{E}{2+\sqrt{2}}(\alpha_1 T_1 + \alpha_2 T_2) \qquad (5)$$

$$\sigma_s = \frac{E}{2+2\sqrt{2}}(\alpha_1 T_1 + \alpha_2 T_2) \qquad (6)$$

5.3 CASTIGLIANO'S THEOREM

Castigliano's theorem is commonly used to find deflections or statically indeterminate loads in structures such as trusses, frames and beams. It is a useful tool in thermal stress analysis when used in conjunction with equivalent thermal loads. The structure can be loaded with the equivalent body forces and surface tractions and Castigliano's theorem applied, treating the thermal loads as real loads. One obtains from such an analysis the true displacements, and the displacement loads and displacement stresses.

CASTIGLIANO'S THEOREM

Castigliano's theorem states that the displacement at a point, at which there is a real or dummy load, is given by the derivative of the strain energy with respect to the load. The displacement strain energy U_o', is defined as the strain energy that is due to the combination of real loads, thermal body forces, and surface tractions. In a three-dimensional problem the displacement strain energy U_o' is

$$U_o' = \frac{1}{2E} \int_V [(\sigma_x')^2 + (\sigma_y')^2 + (\sigma_z')^2 - 2\nu(\sigma_x'\sigma_y' + \sigma_x'\sigma_z' + \sigma_y'\sigma_z') + 2(1+\nu)((\tau_{xy}')^2 + (\tau_{xz}')^2 + (\tau_{yz}')^2)] \, dV \quad (5.3\text{-}1)$$

and in a one-dimensional problem the displacement strain energy is

$$U_o' = \frac{1}{2E} \int_V (\sigma')^2 \, dV \quad (5.3\text{-}2)$$

The deflection, δ_i, in the direction of the load X_i is the derivative of the displacement strain energy. Thus

$$\delta_i = \frac{dU_o'}{dX_i} \quad (5.3\text{-}3)$$

The procedures that are commonly used in applying Castigliano's theorem to isothermal problems apply in every respect. The displacements due to equivalent thermal loading are the true displacements. Eq. (5.3-3) will therefore give the true displacements. Castigliano's theorem is sometimes used to obtain statically indeterminate loads or stresses by setting δ_i equal to zero at those points in a structure at which there can be no displacement (as in Example (b) at the end of this section). In such a problem the indeterminate loads and stresses that are obtained are the displacement loads and stresses, and these are then converted to the true stresses by the addition of the equivalent internal pressures.

Statically Indeterminate Structures

To obtain displacements in a statically indeterminate structure, one can proceed in one of two ways. The statically indeterminate loads and the dummy load at the point of interest can represent several unknown quantities in the displacement strain energy expression. The strain energy is first differentiated with respect to each of the indeterminate loads and set equal to zero, and then differentiated with respect to the dummy load and set equal to the deflection. In this manner the indeterminate loads and the deflection are obtained concurrently.

Another means of determining the deflection at a specified point in an indeterminate structure, is to place a dummy load P_d' at the point and compute the deflection due to P_d'. The dummy load P_d' is set equal to zero after the minimization of the strain energy with respect to P_d'. In performing this operation through the straightforward application of Castigliano's theorem, we would first compute the stress distribution in the indeterminate structure due to the equivalent thermal and external loading P, and then compute the stress distribution due to the dummy load P_d' by performing an analysis of the stress distribution due to P_d' in the <u>indeterminate</u> structure. This procedure, although correct, can be shown to be unnecessarily lengthy, as the computations would be found to contain sets of terms whose sum is identically zero. It is shown in the analysis that follows that a correct solution is obtained if the stresses due to the dummy load P_d' are computed with one or more restraints arbitrarily removed so as to make the structure statically determinate. That is, the stresses due to P_d' can (and should) be computed on the structure which is arbitrarily altered so that it is statically <u>determinate</u>.

CASTIGLIANO'S THEOREM

Consider a structure such as a truss or a beam with a number of redundant members or indeterminate reactions as shown in Figs. 5.3-1(a) and (b). Let X_{ip} represent the statically indeterminate loads, such as

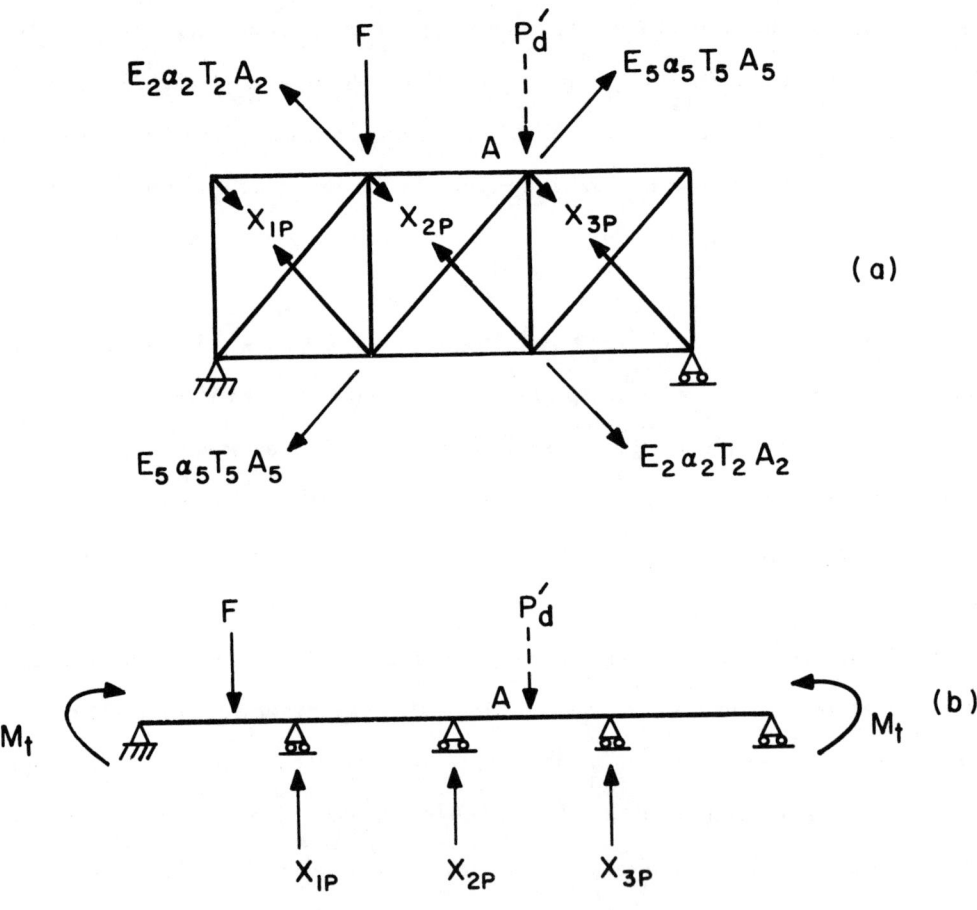

Fig. 5.3-1

X_{1p}, X_{2p} and X_{3p}, in Fig. 5.3-1(a) and (b). The loading P includes the real loads, F, and the thermal loads $E_2 \alpha_2 T_2 A_2$ and $E_5 \alpha_5 T_5$ in the case of the truss, and M_t in the case of the beam. The strain energy in a structure, U_{op}, is a function of the applied loading, P, and the undetermined loads, X_{ip}. That is

$$U_{op} = U_{op}(P, X_{1p}, X_{2p}, X_{3p}) \qquad (5.3-4)$$

The stress σ_{op} is the sum of the stresses due to the external loads and the indeterminate loads, which themselves are a function of the external loading. All of these loads acting on the statically determinate structure produce a stress

$$\sigma_{op} = \sigma_p + \sigma_{x1p} + \sigma_{x2p} + \sigma_{x3p} \qquad (5.3-5)$$

The strain energy U_{op} is

$$U_{op} = \frac{1}{2E} \int_V \sigma_{op}^2 \, dV \qquad (5.3-6)$$

and minimization of the strain energy with respect to the statically interminate loads X_{ip}, yields

$$\int_V \sigma_{op} \frac{\partial \sigma_{x1p}}{\partial X_{1p}} dV = \int_V \sigma_{op} \frac{\partial \sigma_{x2p}}{\partial X_{2p}} dV = \int_V \sigma_{op} \frac{\partial \sigma_{x3p}}{\partial X_{3p}} dV = 0 \qquad (5.3-7)$$

These three equations permit the computation of the three statically indeterminate loads.

The partial derivatives in Eq. (5.3-7) may be written as

$$\frac{\partial \sigma_{x1p}}{\partial X_{1p}} = \sigma(X_1 = 1) \quad \frac{\partial \sigma_{x2p}}{\partial X_{2p}} = \sigma(X_2 = 1) \quad \frac{\partial \sigma_{x3p}}{\partial X_{3p}} = \sigma(X_3 = 1) \qquad (5.3-8)$$

They represent the stress distributions in the statically determinate structure due to a unit load acting at each of the points where an indeterminate load is present. We then have

$$\int_V \sigma_{op} \cdot \sigma(X_1 = 1) dV = \int_V \sigma_{op} \cdot \sigma(X_2 = 1) dV = \int_V \sigma_{op} \cdot \sigma(X_3 = 1) dV = 0 \qquad (5.3\text{-}9)$$

We wish to determine the vertical deflection at point A in each of the structures. A dummy load P_d' is applied at point A, as shown in Fig. 5.3-1, and the deflection is computed by minimizing the total strain energy with respect to P_d'. The strain energy is that which is due to both the applied loading and the dummy load, P_d'. The statically indeterminate loads, X_{ip}, which are produced by the thermal and real loads, are determined by the simultaneous solution of Eqs. (5.3-7). We compute next the indeterminate loads, X_{id}, due the dummy load P_d'. Following the same procedure as in the computation of the X_{ip}, we determine the stress distribution due to the dummy load P_d' and the indeterminate loads X_{1d}, X_{2d} and X_{3d}, acting on the statically determinate truss. The strain energy in the structure produced by the dummy load P_d' and the indeterminate loads X_{1d}, is

$$U_{od} = \frac{1}{2E} \int_V \sigma_{od} \, dV \qquad (5.3\text{-}10)$$

The stress σ_{od} consists of the stresses σ_d due to the dummy load and σ_{1xd}, σ_{2xd} and σ_{3xd} which are the stresses developed by indeterminate loads -- all computed on the statically determinate structure. We have then

$$\sigma_{od} = \sigma_d + \sigma_{1xd} + \sigma_{2xd} + \sigma_{3xd} \qquad (5.3\text{-}11)$$

Substitution of Eq. (5.3-11) into Eq. (5.3-10) and minimizing the strain energy with respect to each of the indeterminate loads, yields

$$\int_V \sigma_{od} \frac{\partial \sigma_{x1d}}{\partial X_{1d}} dV = \int_V \sigma_{od} \frac{\partial \sigma_{x2d}}{\partial X_{2d}} dV = \int_V \sigma_{od} \frac{\partial \sigma_{x3d}}{\partial X_{3d}} dV = 0 \quad (5.3-12)$$

As in the case of the applied load computations, the partial derivatives represent the stress distribution in the structure due to unit loads applied in place of the indeterminate loads.

We are now ready to compute the deflections at the points A in Fig. 5.3-1. In accordance with Castigliano's theorem, the deflection is equal to the derivative of the strain energy with respect to P'_d, with P'_d subsequently set equal to zero. The pertinent stresses, σ_o, consist of the stresses due to both the applied real and thermal loads, and the dummy load, P'_d. The addition of Eqs. (5.3-5) and (5.3-11) yields

$$\sigma_o = \sigma_{op} + \sigma_{od} = \sigma_p + \sigma_{x1p} + \sigma_{x2p} + \sigma_{x3p} + \sigma_d + \sigma_{x1d} + \sigma_{x2d} + \sigma_{x3d} \quad (5.3-13)$$

The total strain energy is

$$U_o = \frac{1}{2E} \int_V \sigma_o^2 \, dV \quad (5.3-14)$$

and the deflection is

$$\delta = \frac{\partial U_o}{\partial P'_d} = \frac{1}{E} \int_V \sigma_{op} \left(\frac{\partial \sigma_d}{\partial P'_d} + \frac{\partial \sigma_{x1d}}{\partial P'_d} + \frac{\partial \sigma_{x2d}}{\partial P'_d} + \frac{\partial \sigma_{x3d}}{\partial P'_d} \right) dV$$

$$P'_d \to 0 \quad (5.3-15)$$

The partial derivatives of the stresses σ_{1xd}, σ_{2xd} and σ_{3xd}, with respect to P'_d may be written as

$$\frac{\partial \sigma_{x1d}}{\partial P'_d} = \frac{\partial \sigma_{x1d}}{\partial X_{1d}} \cdot \frac{\partial X_{1d}}{\partial P'_d}$$

$$\frac{\partial \sigma_{x2d}}{\partial P'_d} = \frac{\partial \sigma_{x2d}}{\partial X_{2d}} \cdot \frac{\partial X_{2d}}{\partial P'_d} \qquad (5.3-16)$$

$$\frac{\partial \sigma_{x3d}}{\partial P'_d} = \frac{\partial \sigma_{x3d}}{\partial X_{3d}} \cdot \frac{\partial X_{3d}}{\partial P'_d}$$

The quantities $\frac{\partial X_{1d}}{\partial P'_d}$, $\frac{\partial X_{2d}}{\partial P'_d}$ and $\frac{\partial X_{3d}}{\partial P'_d}$ represent the loads in the indeterminate members when a unit dummy load is acting, and are therefore constant when integrating over the volume of the structure.

Substitution of Eqs. (5.3-16) into Eq. (5.3-15) yields

$$\delta = \frac{1}{E}\left[\int_V \sigma_{op} \frac{\partial \sigma_d}{\partial P'_d} dV + \frac{\partial X_{1d}}{\partial P'_d}\int_V \sigma_{op}\frac{\partial \sigma_{x1d}}{\partial X_{1d}} dV \right.$$
$$\left. + \frac{\partial X_{2d}}{\partial P'_d}\int_V \sigma_{op}\frac{\partial \sigma_{x2d}}{\partial X_{2d}} dV + \frac{\partial X_{3d}}{\partial P'_d}\int_V \sigma_{op}\frac{\partial \sigma_{x3d}}{\partial X_{3d}} dV\right] \qquad (5.3-17)$$

The stress distribution due to a unit indeterminate load is

$$\frac{\partial \sigma_{x1d}}{\partial X_{1d}} = \frac{\partial \sigma_{x1p}}{\partial X_{1p}} = \sigma(X_1 = 1)$$

$$\frac{\partial \sigma_{x2d}}{\partial X_{2d}} = \frac{\partial \sigma_{x2p}}{\partial X_{2p}} = \sigma(X_2 = 1) \qquad (5.3-18)$$

$$\frac{\partial \sigma_{x3d}}{\partial X_{3d}} = \frac{\partial \sigma_{x3p}}{\partial X_{3p}} = \sigma(X_3 = 1)$$

Substitution of the foregoing into Eqs. (5.3-7) shows that the last three terms in Eq. (5.3-17) are zero. The deflection can therefore be written as

$$\delta = \frac{1}{E}\int_V \sigma_{op} \cdot \frac{\partial \sigma_d}{\partial P'_d} \cdot dV \qquad (5.3-19)$$

Note the significance of each of the quantities in the foregoing equation. The quantity σ_{op} represents the stress distribution in the indeterminate structure due to all applied loads. These could be real loads and thermal loads. The quantity $\dfrac{\partial \sigma_d}{\partial P'_d}$ represents, in accordance with our definition, the distributed stresses in the __determinate__ structure due to a unit dummy load acting at the point whose deflection is desired. As the quantity σ_{op} remains the same regardless which of the redundancies or restraints are taken to be the indeterminate ones, the choice as to the statically determinate structure to which the unit dummy load is applied, is completely arbitrary. In the structure shown in Fig. 5.3-1 (a), for example, the dummy load P'_d could be applied to the statically determinate structure formed when all of the upper truss chord members are removed, or when all of the lower truss chord members are removed. In the beam of Fig. 5.3-1(b) the structure could be made determinate by removing any three of the five supports.

Rather than use a deflection formula in the form of Eq. (5.3-19) it is sometimes more convenient to express the strain energy in terms of the combined stresses due to the thermal and real loads acting on the statically indeterminate structure, and the stresses due to P'_d acting on the arbitrarily chosen statically determinate structure. These combined stresses, σ_{oc}, are

$$\sigma_{oc} = \sigma_p + \sigma_{x1p} + \sigma_{x2p} + \sigma_{x3p} + \sigma_d \qquad (5.3\text{-}20)$$

The strain energy U_{oc} is then

$$U_{oc} = \frac{1}{2E} \int_V (\sigma_p + \sigma_{x1p} + \sigma_{x2p} + \sigma_{x3p} + \sigma_d)^2 \, dV \qquad (5.3\text{-}21)$$

and we obtain the deflection as

$$\delta = \frac{\partial U_{oc}}{\partial P'_d} = \int_V \underbrace{(\sigma_p + \sigma_{x1p} + \sigma_{x2p} + \sigma_{x3p})}_{\text{stat. ind.}} \cdot \underbrace{\sigma_d (P'_d = 1)}_{\text{stat. det.}} \cdot dV \qquad (5.3\text{-}22)$$

which is another form of Eq. (5.3-19).

When thermal loads are part of the loading, it is clear that the computed stresses in the one-dimensional thermally loaded structure are displacement stresses, σ', with the real stresses equal to $\sigma = \sigma' - E\alpha T$.

Beams

Castigliano's theorem is used extensively in the computation of isothermal beam deflections and statically indeterminate reactions in beams. It is equally useful in computing the statically indeterminate reactions and the deflections of thermal beams. Section 6.10 contains a discussion of beam analysis by means of Castigliano's theorem. There we find that the application of Castigliano's theorem for the purpose of obtaining statically indeterminate reactions, leads to the relationship

$$\int_0^L M \frac{\partial M}{\partial X_i} dx = -E\alpha \int_0^L M_t \frac{\partial M}{\partial X_i} dx \qquad (5.3\text{-}23)$$

This is the same expression as that obtained by means of complementary energy analysis. Both methods represent basically the application of the least work principle.

The deflection and rotation at a point in a beam is obtained in Chapter 6 by means of Castigliano's theorem. They are given by Eq. (6.10-7) as

$$\delta_i = \int_0^L \left(\frac{M}{EI} + \frac{\alpha}{I} M_t\right) \frac{\partial M}{\partial P_i} dx \qquad \theta_i = \int_0^L \left(\frac{M}{EI} + \frac{\alpha}{I} M_t\right) \frac{\partial M}{\partial M_i} \qquad (5.3\text{-}24)$$

M is the moment due to real loads and statically determinate reactions. In the case of a statically determinate beam on which no real loads are acting, M is equal to zero.

Example (a) Bar Extension

The bar shown in Fig. 5.3-2 has temperature distribution which is a function of the linear coordinate. We wish to find the lengthening of the bar.

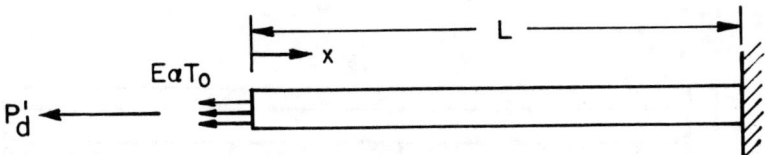

Fig. 5.3-2

A dummy load, P'_d, is applied to the end of the bar. The axial displacement stress in the bar due to the surface traction, body forces, and dummy load P'_d is

$$\sigma' = \frac{P'_d}{A} + E\alpha T_o - \int_0^x - E\alpha \frac{dT}{dx} \, dx \tag{1}$$

or

$$\sigma' = \frac{P'_d}{A} + E\alpha T$$

The displacement strain energy is

$$U'_o = \int \frac{(\sigma')^2}{2E} \, dV = \frac{A}{2E} \int_0^L \left(\frac{P'_d}{A} + E\alpha T\right)^2 dx \tag{2}$$

and the axial extension of the bar, in accordance with Castigliano's theorem, is

$$\delta = \frac{dU'_o}{dP'} = \frac{1}{E} \int_0^L \left(\frac{P'_d}{A} + E\alpha T\right) dx \tag{3}$$

We let P'_d go to zero and obtain

$$\delta = \alpha \overline{T} L \tag{4}$$

Example (b)   Stresses in Concentric Tube Assembly

The concentric tube assembly shown in Fig. 5.3-3 consists of tubes of dissimilar materials, heated to temperatures T_1 and T_2. Fig. 5.3-3 shows the thermal loads acting on the assembly.

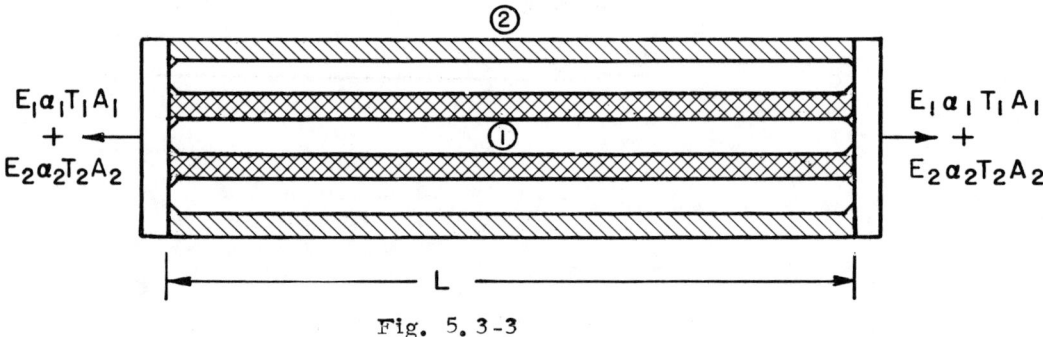

Fig. 5.3-3

Let $P'_1 = \sigma'_1 A_1$ be the axial indeterminate equivalent load in tube 1. The equivalent load in tube 2, $P'_2 = \sigma'_2 A_2$, is then

$$P'_2 = E_1 \alpha_1 T_1 A_1 + E_2 \alpha_2 T_2 A_2 - P'_1 \tag{1}$$

The displacement strain energy is

$$U'_o = \frac{(\sigma'_1)^2}{2E_1} A_1 L + \frac{(\sigma'_2)^2}{2E_2} A_2 L \tag{2}$$

or

$$U'_o = \frac{(P'_1)^2 L}{2A_1 E_1} + \left[\frac{E_1 \alpha_1 T_1 A_1 + E_2 \alpha_2 T_2 A_2 - P'_1}{A_2}\right]^2 \frac{A_2 L}{2E_2} \tag{3}$$

As tube 1 is not discontinuous, $\dfrac{dU'_o}{dP'_1} = 0$, and we have

$$\dfrac{P'_1}{A_1 E_1} - \dfrac{E_1 \alpha_1 T_1 A_1 + E_2 \alpha_2 T_2 A_2 - P'_1}{E_2 A_2} = 0 \tag{4}$$

The displacement stresses are

$$\sigma'_1 = \dfrac{P'_1}{A_1} = E_1 \dfrac{E_1 \alpha_1 T_1 A_1 + E_2 \alpha_2 T_2 A_2}{E_1 A_1 + E_2 A_2} \tag{5}$$

and

$$\sigma'_2 = \dfrac{P'_2}{A_2} = E_2 \dfrac{E_1 \alpha_1 T_1 A_1 + E_2 \alpha_2 T_2 A_2}{E_1 A_1 + E_2 A_2} \tag{6}$$

and the thermal stresses are

$$\sigma_1 = \sigma'_1 - E_1 \alpha_1 T_1 = \dfrac{E_1 E_2 A_2 (\alpha_2 T_2 - \alpha_1 T_1)}{E_1 A_1 + E_2 A_2} \tag{7}$$

and

$$\sigma_2 = \sigma'_2 - E_2 \alpha_2 T_2 = \dfrac{E_1 E_2 A_1 (\alpha_1 T_1 - \alpha_2 T_2)}{E_1 A_1 + E_2 A_2} . \tag{8}$$

Example (c)   Deflection of a Statically Determinate Truss

Castigilano's theorem affords a convenient means of determining displacements of points on a heated truss. The statically determinate truss shown in Fig. 5.3-4 has its diagonals heated to temperatures T_1 and T_2, the lower chord members to temperatures T_3 and T_4 and the central vertical to a temperature T_5. Each member has a cross section area of one sq. inch. It is desired to find the deflection at the midpoint, at the bottom of the truss.

It was shown in Section 3.3 that no stresses are produced in a statically determinate truss whose members are heated to different temperatures. The displacement stresses due to the thermal loads shown in Fig. 5.3-4 are equilibrated by the internal pressures, and as a result no thermal stresses are developed. In the present problem we compute the displacement strain

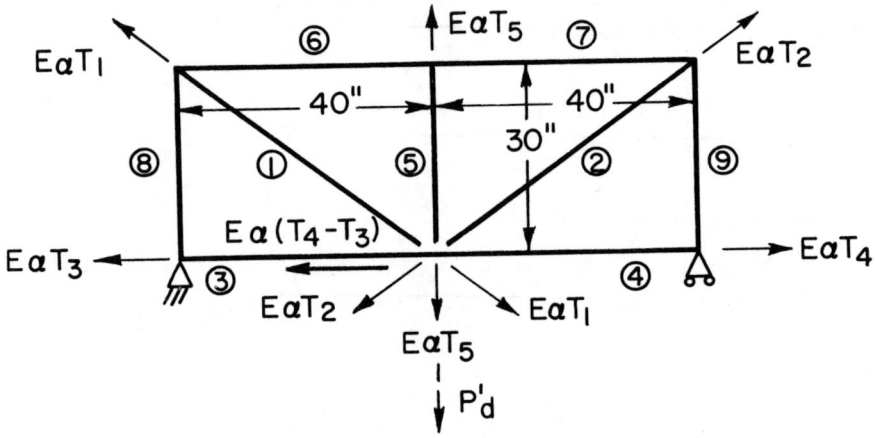

Fig. 5.3-4

energy due to the thermal tractions and the dummy load P'_d placed at the midpoint of the lower chord. The displacement stresses in each of the members are

$$\sigma'_1 = E\alpha T_1 + \frac{5P'_d}{6} \qquad \sigma'_2 = E\alpha T_2 + \frac{5P'_d}{6} \qquad \sigma' = E\alpha T_3$$

$$\sigma'_4 = E\alpha T_4 \qquad \sigma'_5 = E\alpha T_5 \qquad \sigma'_6 = -\frac{2P'_d}{3} \qquad (1)$$

$$\sigma'_7 = -\frac{2P'_d}{3} \qquad \sigma'_8 = -\frac{P'_d}{2} \qquad \sigma'_9 = -\frac{P'_d}{2}$$

The deflection at the point of application of P'_d is, from Eq. (5.3-19)

$$\delta = \frac{dU'_o}{dP'_d} = \frac{1}{E}\int_V \sigma'_i \frac{d\sigma'_i}{dP'_d} A\, dL = \frac{A}{E}\sum_1^9 \sigma'_i \frac{d\sigma'_i}{dP'_d} L_i \qquad (2)$$

and when we let $P'_d \to 0$ we obtain

$$\delta = \frac{1}{E}\left[E\alpha T_1 \left(\frac{5}{6}\right)(50) + E\alpha T_2 \left(\frac{5}{6}\right)(50)\right] = \frac{125\alpha(T_1+T_2)}{3} \qquad (3)$$

Only σ'_1 and σ'_2 were contributors to the deflection since only these displacement stresses were dependent on both the temperature of the member and the dummy load P_d'.

Example (d)   Statically Indeterminate Truss

The redundant structural truss shown in Fig. 5.3-5 is analyzed by means of Castigliano's theorem.

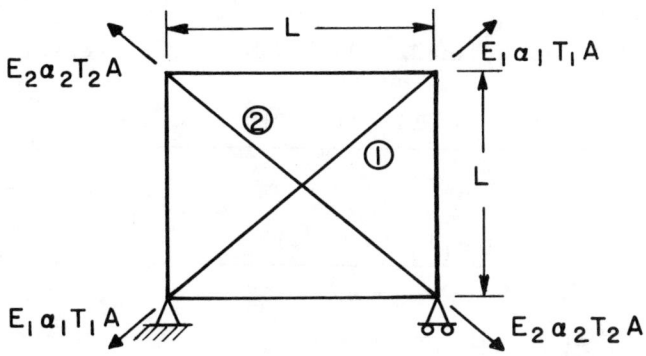

Fig. 5.3-5

The members are of uniform modulus, E, and cross section A, and the diagonals are at temperatures T_1 and T_2. Assume P_1' to be the unknown displacement load in diagonal 1. The displacement loads in the verticals and horizontals, P_s', are then

$$P_s' = (E\alpha_1 T_1 A - P_1') \frac{\sqrt{2}}{2} \tag{1}$$

and the load in diagonal 2 is

$$P_2' = E\alpha_2 T_2 A - E\alpha_1 T_1 A + P_1' \tag{2}$$

The displacement strain energy is, from Eq.(5.3-2),

$$U'_o = \left[\frac{L_1}{2E}\left(\frac{P'_1}{A}\right)^2 + \frac{4L}{2E}\left(\frac{P'_s}{A}\right)^2 + \frac{L_2}{2E}\left(\frac{P'_2}{A}\right)^2\right] A \qquad (3)$$

or

$$U'_o = \frac{L}{2AE}\left[\sqrt{2}\,(P'_1)^2 + 2(E\alpha_1 T_1 A - P'_1)^2 + \sqrt{2}\,(E\alpha_2 T_2 A - E\alpha_1 T_1 A + P'_1)^2\right] \qquad (4)$$

Structural continuity requires that $\dfrac{dU'_o}{dP_1} = 0$. This yields

$$\frac{\sqrt{2}}{2} P'_1 - (E\alpha_1 T_1 A - P'_1) + \frac{\sqrt{2}}{2}(E\alpha_2 T_2 A - E\alpha_1 TA + P'_1) = 0 \qquad (5)$$

from which P'_1 is obtained. The displacement stresses are

$$\sigma'_1 = \frac{P'_1}{A} = \frac{-E\alpha_2 T_2 + (1+\sqrt{2})E\alpha_1 T_1}{2+\sqrt{2}} \qquad (6)$$

$$\sigma'_2 = \frac{P'_2}{A} = \frac{-E\alpha_1 T_1 + (1+\sqrt{2})E\alpha_2 T_2}{2+\sqrt{2}} \qquad (7)$$

$$\sigma'_s = \frac{P'_s}{A} = \frac{E\alpha_1 T_1 + E\alpha_2 T_2}{2+2\sqrt{2}} \qquad (8)$$

The stresses in each of the truss members are $\sigma_i = \sigma'_i - E\alpha_i T_i$, or

$$\sigma_1 = -\frac{E}{2+\sqrt{2}}(\alpha_1 T_1 + \alpha_2 T_2) \qquad (9)$$

$$\sigma_2 = -\frac{E}{2+\sqrt{2}}(\alpha_1 T_1 + \alpha_2 T_2) \qquad (10)$$

$$\sigma_s = \frac{E}{2+2\sqrt{2}}(\alpha_1 T_1 + \alpha_2 T_2) \qquad (11)$$

These are the same values for the stresses as found in Section 5.2 Example (c) where the truss was analyzed by means of complementary energy.

We compute next the deflection of the upper right corner of the

CASTIGLIANO'S THEOREM   305

redundant truss in the direction of the applied dummy load P'_d. The redundant truss is made statically determinate by eliminating diagonal 1, in Fig. 5.3-6. The thermal loads act as shown in Fig. 5.3-6.

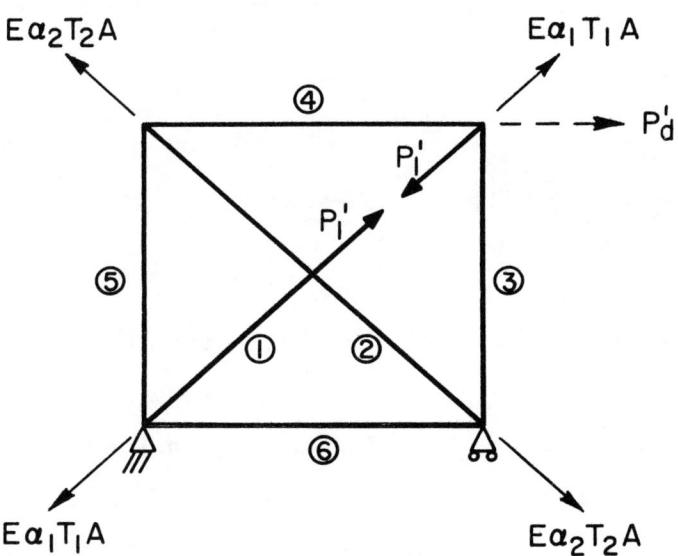

Fig. 5.3-6

The dummy load P'_d produces the following displacement stresses in the statically determinate truss (diagonal 1 absent).

$$\sigma'_{1d} = 0 \qquad \sigma'_{2d} = -\sqrt{2}\, P'_d \qquad \sigma'_{3d} = 0$$

$$\sigma'_{4d} = P'_d \qquad \sigma'_{5d} = P'_d \qquad \sigma'_{6d} = P'_d \tag{12}$$

Eqs. (6), (7) and (8) give the displacement thermal stresses in each of the members. In order to contribute to the displacement strain energy we observe, from Eq. (2) of Example (c), that both σ'_i and $\dfrac{d\sigma'_i}{dP_d}$ must be non-zero. Accordingly only members ②, ④, ⑤ and ⑥ contribute to the strain energy.

From Eq.(2) of Example (c), the deflection is expressed as

$$\delta = \frac{A}{E} \sum_{2,4,5,6} \sigma'_i \frac{d\sigma'_i}{dP_d} L_i = \frac{AL}{E} \left[\frac{-E\alpha_1 T_1 + (1+\sqrt{2})E\alpha_2 T_2}{2+\sqrt{2}} (-\sqrt{2})(\sqrt{2}) \right.$$

$$\left. + 3 \left(\frac{E\alpha_1 T_1 + E\alpha_2 T_2}{2 + 2\sqrt{2}} \right)(1)(1) \right] \quad (13)$$

which yields

$$\delta = \frac{AL}{1+\sqrt{2}} \left[\left(\frac{3}{2}+\sqrt{2}\right) \alpha_1 T_2 - \left(\frac{1}{2}+\sqrt{2}\right) \alpha_2 T_2 \right] \quad (14)$$

5.4 THERMOELASTIC RECIPROCAL THEOREM

The thermoelastic reciprocal theorem is applicable to the analysis of elastic bodies subjected to distributed temperatures. It is useful in determining displacements or rotations at particular points; it can be used for the determination of generalized displacements and stresses which are functions of the body coordinates, and it can be employed to determine volume changes. The reciprocal theorem requires the use of an auxiliary load system. If we use an auxiliary load system in the form of a concentrated load at a point on a beam, the reciprocal theorem will give us the deflection of the point on the beam due to a specified temperature distribution. If we apply an auxiliary moment at a point, the reciprocal theorem will give us the beam rotation at that point, due to the specified temperature distribution.

The reciprocal theorem states that the volume integral of the product of the auxiliary body loads and auxiliary surface loads, and the displacements produced at the auxiliary load locations by a specified temperature distribution, is equal to the volume integral of the free expansion, αT, multiplied by the sum of the principal stresses generated by the auxiliary loads. In the text which follows the starred quantities represent auxiliary loads and also the stresses and displacements produced by the auxiliary loads.

To derive the thermoelastic reciprocal theorem, we start with Betti's reciprocal theorem which states that the volume integral of the body forces X', Y', Z' and surface tractions $\overline{X}', \overline{Y}', \overline{Z}'$, multiplied by displacements u^*, v^*, w^* produced by an auxiliary load system X^*, Y^*, Z^* and $\overline{X}^*, \overline{Y}^*, \overline{Z}^*$, is equal to the volume integral of the product of the auxiliary loads and the displacements u, v, w that are due to the real loads. This statement is written as

$$\int_V (X'u^* + Y'v^* + Z'w^*) dV + \int_S (\overline{X}'u^* + \overline{Y}'v^* + \overline{Z}'w^*) dS$$
$$= \int_V (X^*u + Y^*v + Z^*w) dV + \int_S (\overline{X}^*u + \overline{Y}^*v + \overline{Z}^*w) dS \qquad (5.4\text{-}1)$$

We use the equivalent load analysis, and replace the real body loads and surface tractions, $X', Y', Z', \overline{X}', \overline{Y}', \overline{Z}'$, by equivalent thermal loads. Eqs. (1.2-7) and (1.2-9) are employed, and Eq. (5.4-1) becomes

$$-\frac{E\alpha}{1-2\nu} \int_V (u^* \frac{\partial T}{\partial x} + v^* \frac{\partial T}{\partial y} + w^* \frac{\partial T}{\partial z}) dV$$
$$+ \frac{E\alpha}{1-2\nu} \int_S (u^*\ell + v^*m + w^*n) T \, dS \qquad (5.4\text{-}2)$$
$$= \int_V (X^*u + Y^*v + Z^*w) dV + \int_S (\overline{X}^*u + \overline{Y}^*v + \overline{Z}^*w) dS$$

Gauss' divergence theorem states that

$$\int_V (\underline{\nabla} \cdot (\underline{r}^*T)) dV = \int_S (\underline{r}^*T) \cdot \underline{dS} \qquad (5.4\text{-}3)$$

where the underlined quantities are vectors defined as

$$\underline{r}^* = iu^* + jv^* + kw^*$$

$$\underline{dS} = i\,dS_x + j\,dS_y + k\,dS_z = i\ell\,dS + jm\,dS + kn\,dS$$

and as

$$\nabla \cdot (\underline{r}^*T) = \frac{\partial}{\partial x}(u^*T) + \frac{\partial}{\partial y}(v^*T) + \frac{\partial}{\partial z}(w^*T)$$

$$= u^*\frac{\partial T}{\partial x} + v^*\frac{\partial T}{\partial y} + w^*\frac{\partial T}{\partial z} + T\frac{\partial u^*}{\partial x} + T\frac{\partial v^*}{\partial y} + T\frac{\partial w^*}{\partial z} \quad (5.4\text{-}4)$$

$$= u^*\frac{\partial T}{\partial x} + v^*\frac{\partial T}{\partial y} + w^*\frac{\partial T}{\partial z} + Te^*$$

and

$$(\underline{r}^*T)\cdot \underline{dS} = (u^*\ell + v^*m + w^*n)\,T\,dS \quad (5.4\text{-}5)$$

Eq. (5.4-3) becomes

$$\int_V \left(u^*\frac{\partial T}{\partial x} + v^*\frac{\partial T}{\partial y} + w^*\frac{\partial T}{\partial z}\right) dV + \int_V Te^*\,dV$$

$$= \int_S (u^*\ell + v^*m + w^*n)\,T\,dS \quad (5.4\text{-}6)$$

From the foregoing we observe that the left hand side of Eq. (5.4-2) is equal to $\frac{E\alpha}{1-2\nu}\int_V Te^*\,dV$. Eq. (5.4-2) therefore becomes

$$\int_V (X^*u + Y^*v + Z^*w)\,dV + \int_S (\overline{X}^*u + \overline{Y}^*v + \overline{Z}^*w)\,dS = \frac{E\alpha}{1-2\nu}\int_V e^*T\,dV \quad (5.4\text{-}7)$$

Since

$$\frac{E}{1-2\nu}(\epsilon_x^* + \epsilon_y^* + \epsilon_z^*) = \sigma_x^* + \sigma_y^* + \sigma_z^* \quad (5.4\text{-}8)$$

or, in terms of e^* and θ^*

$$\frac{E}{1-2\nu}e^* = \theta^*$$

we can write Eq. (5.4-7) as

$$\int_V (X^*u + Y^*v + Z^*w)\,dV + \int_S (\overline{X}^*u + \overline{Y}^*v + \overline{Z}^*w)\,dS = \alpha \int_V \theta^*T\,dV \quad (5.4\text{-}9)$$

Eq. (5.4-9) is a statement of the thermoelastic theorem. When the coefficient of expansion is not the same throughout the body, Eq.(5.4-9) is written as

$$\int_V (X^* u + Y^* v + Z^* w) \, dV + \int_S (\overline{X}^* u + \overline{Y}^* v + \overline{Z}^* w) \, dS = \int_V \alpha \theta^* T \, dV \tag{5.4-10}$$

If we wish to find the deflection or rotation at a point in a heated body, we apply an auxiliary load or moment at the point, calculate the auxiliary stresses, $\theta^* = \sigma_x^* + \sigma_y^* + \sigma_z^*$, and evaluate the integral on the right hand side of Eq.(5.4-10). The volume integral on the left hand side is zero, and the surface integral takes the form

$$\int_S \overline{N}^* \cdot \delta \, dS = P^* \delta = \int_V \alpha \theta^* T \, dV \qquad \int_S (\overline{N}^* h \, dS)(\frac{\delta}{h}) = M^* \varphi = \int_V \alpha \theta^* T \, dV \tag{5.4-11}$$

or

$$\delta = \int_V \alpha \theta_1^* T \, dV \qquad \varphi = \int_V \alpha \theta_1^* T \, dV \tag{5.4-12}$$

P^* and M^* represent an auxiliary load or an auxiliary moment, respectively, and δ and φ are the corresponding deflections or rotations. θ_1^* represents the sum of the principal stresses due to a unit auxiliary load or unit auxiliary moment.

Volume changes are readily calculated by using a uniform surface auxiliary load p^*, which causes a uniform internal auxiliary stress $-p^*$. Noting that $(u + v + w) \, ds = -\Delta V$, Eq. (5.4-10) becomes

$$\int_S p^* (u + v + w) \, ds = -3 \int_V \alpha p^* T \, dV$$

or

$$\Delta V = 3 \int_V \alpha T \, dV \tag{5.4-13}$$

It can be shown that Eq.(5.4-13) is valid for either simply connected or multiply connected bodies.

Thermal displacements can be computed with the use of either Castigliano's theorem or the thermoelastic reciprocal theorem. The reciprocal theorem can be used to advantage when solutions of auxiliary indeterminate problems are easily computed or are known. Solutions by Castigliano's theorem have the advantage of being simpler to compute when the solution of the statically indeterminate thermal loading problem is known beforehand or easily computed. Castigliano's theorem appears to have a greater versatility in that indeterminate internal loads and external reactions are determined, as well as displacements.

An important difference in the application of the reciprocal theorem and Castigliano's theorem should also be noted. Dummy loads, used in Castigliano's theorem to obtain desired deflections or rotations, are set equal to zero after the derivative of the strain energy has been computed. Auxiliary loads, which are always used when applying the reciprocal theorem, are not set equal to zero, but automatically disappear since they are present as factors on both sides of the reciprocal theorem equation.

Example (a)   Unrestrained Rod with Axially Varying Temperature

The rod shown in Fig. 5.4-1 is subjected to a temperature variation $T = T(x)$. We wish to find the increase in the length of the bar.

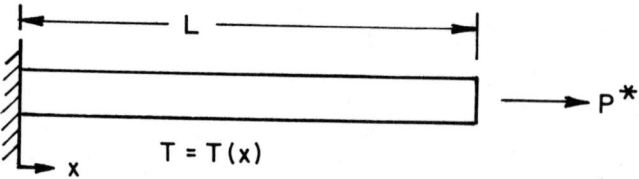

Fig. 5.4-1

In order to find u_L, the thermal elongation of the bar, an auxiliary load P^* is placed at the end of the bar. From Eq. (5.4-11)

THERMOELASTIC RECIPROCAL THEOREM

$$P^* u_L = \alpha \int_0^L \sigma_x^* \, T \, dV \tag{1}$$

The auxiliary load stress is $\sigma_x^* = P^*/A$. Eq. (1) then gives the displacement u_L as

$$u_L = \alpha \overline{T} L \tag{2}$$

Example (b) Concentric Tube Assembly

The concentric tube assembly has an axial auxiliary load P^* applied at each end, as shown in Fig. 5.4-2, since we wish to obtain the elongation of the assembly. The tubes are of different materials and at temperatures T_1 and T_2.

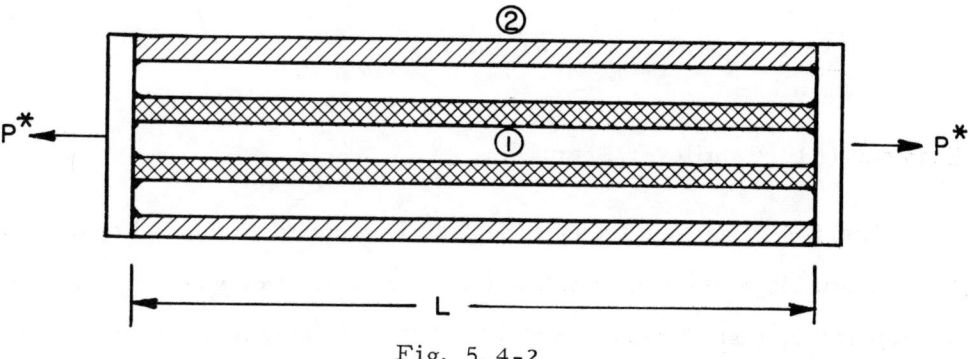

Fig. 5.4-2

The load P^* is distributed in each of the tubes in proportion to their rigidities $E_1 A_1$ and $E_2 A_2$. Thus

$$P_1^* = P^* \frac{E_1 A_1}{E_1 A_1 + E_2 A_2} \qquad P_2^* = P^* \frac{E_2 A_2}{E_1 A_1 + E_2 A_2} \tag{1}$$

The auxiliary stresses in the tubes are therefore

$$\sigma_1^* = \frac{P_1^*}{A_1} = P^* \frac{E_1}{E_1 A_1 + E_2 A_2} \qquad \sigma_2^* = \frac{P_2^*}{A_2} = P^* \frac{E_2}{E_1 A_1 + E_2 A_2} \, . \tag{2}$$

From the reciprocal theorem

$$P^* u_L = \int_{V_1} \frac{P^* E_1}{E_1 A_1 + E_2 A_2} (\alpha_1 T_1 \cdot dA_1 dx) + \int_{V_2} \frac{P^* E_2}{E_1 A_1 + E_2 A_2} (\alpha_2 T_2 \cdot dA_2 dx) \tag{3}$$

This gives the displacement u_L as

$$u_L = \left(\frac{E_1 \alpha_1 T_1 A_1}{E_1 A_1 + E_2 A_2} + \frac{E_2 \alpha_2 T_2 A_2}{E_1 A_1 + E_2 A_2} \right) L \tag{4}$$

The strain ϵ is uniform and equal to u_L / L, or

$$\epsilon = \frac{u_L}{L} = \frac{E_1 \alpha_1 T_1 A_1 + E_2 \alpha_2 T_2 A_2}{E_1 A_1 + E_2 A_2} = \overline{\alpha T} \tag{5}$$

The knowledge of the strains permits us to obtain the stresses. They are

$$\begin{aligned} \sigma_1 &= E_1(\epsilon - \alpha_1 T_1) = E_1(\overline{\alpha T} - \alpha_1 T_1) \\ \sigma_2 &= E_2(\epsilon - \alpha T_2) = E_2(\overline{\alpha T} - \alpha_2 T_2) \end{aligned} \tag{6}$$

Example (c) Cantilever Beam

Fig. 5.4-3 shows a transverse load P^* applied at the end of the cantilever beam. It is desired to obtain the beam deflection when it is subjected to a temperature distribution $T = T_o y/h$. The auxiliary moment at any point is $-P^* x$ and the bending stress is

$$\sigma_x^* = \frac{M^* y}{I} = -\frac{P^* xy}{I} \tag{1}$$

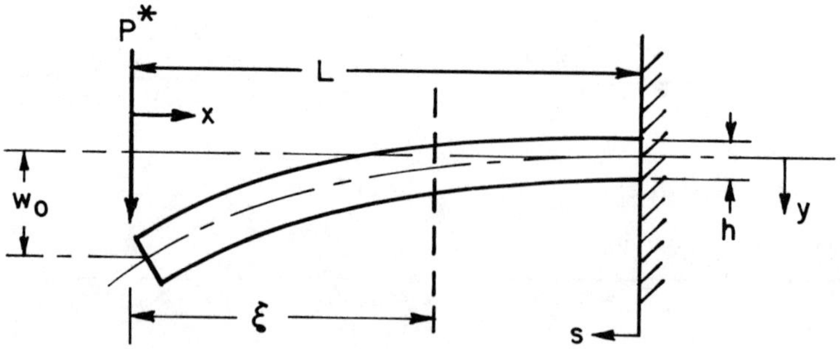

Fig. 5.4-3

From the reciprocal theorem

$$P^* w_o = \alpha \int_V \sigma_x^* T \, dV = -\alpha \int_{-\frac{h}{2}}^{\frac{h}{2}} \int_0^L \frac{P^*_{xy}}{I} \cdot \frac{T_o y}{h} \, dx \, dA \qquad (2)$$

We note that

$$I = \int_{-\frac{h}{2}}^{\frac{h}{2}} y^2 \, dA$$

Eq. (2) then gives w_o as

$$w_o = -\frac{\alpha T_o}{h} \left[\frac{x^2}{2} \right]_0^L = -\frac{\alpha T_o L^2}{2h} \qquad (3)$$

The minus sign denotes that the thermal displacement is upwards. To obtain the displacement at any point, $x - \xi$, a distance s from the wall measured as shown in Fig. 5.4-3, the load P^* is placed at point s, and from Eq. (5.4-11) we have

$$w = -\frac{\alpha T_o}{h} \left[\frac{(x-\xi)^2}{2} \right]_\xi^L = -\frac{\alpha T_o}{2h}(L-\xi)^2 = -\frac{\alpha T_o s^2}{2h} \qquad (4)$$

The beam curvature is

$$K = -\frac{1}{R} = -\frac{d^2 w}{ds^2} = \frac{\alpha T_o}{h} \qquad (5)$$

and the strain is

$$\epsilon = \frac{y}{R} = \frac{\alpha T_o y}{h} \qquad (6)$$

The stress is $\sigma = E(\epsilon - \alpha T)$. Using ϵ from Eq. (6) we have

$$\sigma = E\left(\frac{\alpha T_o y}{h} - \frac{\alpha T_o y}{h}\right) = 0 \tag{7}$$

The stress is obviously zero since the temperature distribution is linear and the beam is statically determinate.

Example (d)   Deflection of a Truss

The reciprocal theorem is an expedient means of obtaining the deflections of points in a redundant truss. Before applying this method to a redundant truss we apply it first to the statically determinate truss shown in Fig. 5.3-4. Instead of the load P'_d we apply a load P^* at the midpoint of the lower chord. The stresses in the members due to P^* are

$$\sigma_1^* = \sigma_2^* = \frac{5P^*}{6} \qquad \sigma_3^* = \sigma_4^* = \sigma_5^* = 0 \qquad \sigma_6^* = \sigma_7^* = -\frac{2P^*}{3} \qquad \sigma_8^* = \sigma_9^* = \frac{P^*}{2} \tag{1}$$

and the temperatures in members ① through ⑤ are, T_1, T_2, T_3, T_4, T_5 respectively; the others are at zero temperature. From the reciprocal theorem we have (with $A = 1$)

$$P^* \delta = \alpha \int_L \sigma_i^* T_i \, dL_i \tag{2}$$

or

$$\delta = \alpha \sum_1^9 \frac{\sigma_i^*}{P^*} T_i L_i \tag{3}$$

The only members in which either the auxiliary stress and temperature are not zero, are members ① and ② . Eq. (3) therefore yields

$$\delta = \alpha \left[T_1 \left(\frac{5}{6}\right)(50) + T_2 \left(\frac{5}{6}\right)(50) \right] = \frac{125\alpha(T_1 + T_2)}{3} \tag{4}$$

This is the same result obtained earlier in Example (c) of Section 5.3.

We now proceed to calculate the deflection of the upper corner of the redundant truss shown in Figs. 5.3-5 and 5.3-6. In the application of the reciprocal theorem, we replace the dummy load P'_d with an auxiliary load P^*, and compute the stresses in each of the six members of the truss due to P^*. The difference in the present procedure from that used in the Castigliano's theorem analysis should be recognized. In the application of Castigliano's theorem, the redundant member is disconnected and the stresses due to the dummy load are determined in the resulting statically determinate truss. In the present reciprocal theorem analysis the stresses due to the auxiliary load P^* are computed in the indeterminate truss.

The load in a single redundant member of a truss is given as

$$P_1^* = -\frac{\sum_{1}^{n} p_i^* x_i^* L_i / A_i E_i}{\sum_{1}^{n} (x_i^*)^2 L_i / A_i E_i} \qquad (5)$$

where P_1^* is the load in member ① due to the auxiliary load P^*. The quantity p_i^* represents the loads in the members due to P^* when member ① is disconnected. The quantities x_i^* represents the loads in the members due to a unit load acting in place of the loads P'_1 shown in Fig. 5.3-5.

The loads due to P^* are

$$p_1^* = 0 \qquad p_2^* = -\sqrt{2}\, P^* \qquad p_3^* = 0 \qquad p_4^* = p_5^* = p_6^* = P^* \qquad (6)$$

and the loads x_i due to a unit load in member ① are

$$x_1 = 1 \qquad x_2 = 1 \qquad x_3 = x_4 = x_5 = x_6 = -\frac{\sqrt{2}}{2} \qquad (7)$$

The summations in Eq. (5), with $A_i = A$, are

$$\frac{L}{AE} \sum_{1}^{6} p_i^* x_i^* \frac{L_i}{L} = \frac{LP^*}{AE} (0 - (\sqrt{2})(\sqrt{2}) + 0 - \frac{3\sqrt{2}}{2}) = \frac{LP^*}{AE}(-2 - \frac{3\sqrt{2}}{2}) \qquad (8)$$

$$\frac{L}{AE} \sum_1^6 (x_i^*)^2 \frac{L_i}{L} = \frac{L}{AE} (2(1)^2(\sqrt{2}) + 4(-\frac{\sqrt{2}}{2})^2(1)) = \frac{L}{AE} (2\sqrt{2} + 2) \qquad (9)$$

Therefore from Eq. (5) the auxiliary loads in the members are

$$P_1^* = \frac{1+\frac{3\sqrt{2}}{4}}{1+\sqrt{2}} P^* \qquad P_2^* = -\frac{1+\frac{\sqrt{2}}{4}}{1+\sqrt{2}} P^* \qquad P_3^* = -\frac{1+\frac{3\sqrt{2}}{4}}{2+\sqrt{2}} P^*$$

$$P_4^* = \frac{1+\frac{\sqrt{2}}{4}}{2+\sqrt{2}} P^* \qquad P_5^* = \frac{1+\frac{\sqrt{2}}{4}}{2+\sqrt{2}} P^* \qquad P_6^* = \frac{1+\frac{\sqrt{2}}{4}}{2+\sqrt{2}} P^* \qquad (10)$$

Applying the reciprocal theorem we have

$$\delta = \alpha_1 \frac{P_1^*}{P^*} T_1 L_1 A + \alpha_2 \frac{P_2^*}{P^*} T_2 L_2 A \qquad (11)$$

With $L_1 = L_2 = L\sqrt{2}$ we obtain the deflection as

$$\delta = \frac{AL}{1+\sqrt{2}} \left[(\frac{3}{2} + \sqrt{2}) \alpha_1 T_1 - (\frac{1}{2} + \sqrt{2}) \alpha_2 T_2 \right] \qquad (12)$$

This is the same result as obtained by means of Castigliano's theorem in Example (d) of Section 5.3.

Most of the labor in the foregoing computation is in the determination of the loads due to P^*, in the members of the redundant truss. This is the auxiliary load problem whose solution, in other types of problems, is known beforehand. In that case computations are very much simplified.

Example (e) Radial Displacement of a Hollow Cylinder.

The hollow cylinder is shown in Fig. 5.4-4. We take p^* as the auxiliary internal pressure in order to determine the radial displacement, u_a of the inside wall of the cylinder. The well-known solution of the problem of a thick-walled cylinder with an internal pressure p^* gives the sum of the

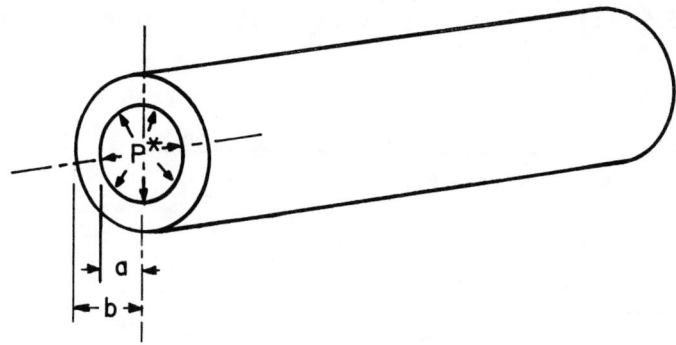

Fig. 5.4-4

radial and tangential stresses as

$$\sigma_r^* + \sigma_t^* = \frac{2p^* a^2}{b^2 - a^2} \tag{1}$$

with $\sigma_z = 0$.

From the reciprocal theorem

$$p^* u_a (2\pi a) = \alpha \int_V \theta^* T dV = \alpha \int_a^b \frac{2p^* a^2}{b^2 - a^2} \cdot 2\pi r dr \tag{2}$$

and

$$u_a = \frac{\alpha a}{\pi(b^2 - a^2)} \int_a^b T \cdot 2\pi r\, dr = \alpha \overline{T}\, a \tag{3}$$

Example (f) Volume Change of a Hollow Sphere

We apply Eq.(5.4-13) which is valid for both simply and multiply connected bodies. For the hollow sphere the volume change is

$$\Delta V = 3\alpha \int_V T dV = (3\alpha)[\tfrac{4}{3}\pi (b^3 - a^3) \overline{T}] = 4\pi (b^3 - a^3) \alpha \overline{T} \tag{1}$$

We can verify this result by finding separately the radial displacements of the internal and external surfaces of the sphere. Let p^* be the internal auxiliary pressure. From the well-known classical solution of this problem we have the result that

$$\theta^* = \sigma_r^* + 2\sigma_t^* = -\frac{p_a^{*} a^3(b^3-r^3)}{r^3(b^3-a^3)} + \frac{p_a^{*} a^3(2r^3+b^3)}{r^3(b^3-a^3)} \qquad (2)$$

or

$$\theta^* = \frac{3a^3 p^*}{b^3 - a^3}$$

From the reciprocal theorem

$$p^* u_a (4\pi a^2) = \alpha \int_V \theta^* T dV = \frac{3\alpha a^3 p^*}{b^3 - a^3} \int_V T dV \qquad (3)$$

from which

$$u_a = \alpha \overline{T} a \qquad (4)$$

In a similar manner the outer radial displacement is found to be

$$u_b = \alpha \overline{T} b \qquad (5)$$

The volume change ΔV is therefore

$$\Delta V = 4\pi b^2 u_b - 4\pi a^2 u_a = 4\pi(b^3 - a^3)\alpha \overline{T} \qquad (6)$$

which is the same result as found by the more direct application of volume change reciprocal theorem.

5.5 RAYLEIGH-RITZ METHOD

The Rayleigh-Ritz method of analysis is used to determine approximate displacements of points in a body. Solutions are generally obtained as series of functions which describe the overall deformation of a body. It is employed where exact methods are difficult to use because of either irregular body geometry, or the presence of a large number of restraints.

The Rayleigh-Ritz procedure requires, to begin with, the assumption of a deflection curve or deformation pattern which satisfies the boundary conditions. The deformation is expressed in terms of one or more functions with arbitrary coefficients, selected to conform as closely as possible to the estimated form of the deformation, and also to satisfy the prescribed boundary conditions. The potential energy is then expressed in terms of the selected displacement functions, and minimized with respect to each of the arbitrary coefficients. In most thermal stress problems the potential energy and the strain energy are the same since the body under consideration will generally be either free of surface loads, or have rigid boundary restraints. In either case there will be no external work.

We will consider here the application of the Rayleigh-Ritz method to one-dimensional problems only. Eq.(5.2-18) gives the three-dimensional form of the strain energy. If we assume that only the stresses σ_x are of consequence, the one-dimensional form of the strain energy U_o, or the potential energy, V, which it equals, becomes

$$V = U_o = \frac{1}{2E} \int_A \int_0^L \sigma_x^2 \, dx \, dA \qquad (5.5-1)$$

The foregoing potential energy can also be expressed in terms of the one-dimensional strain ϵ_x. With the substitution $\sigma_x = E\epsilon_x - E\alpha T$, Eq. (5.5-1) becomes

$$V = \frac{E}{2} \int_A \int_0^L (\epsilon_x - \alpha T)^2 \, dx \, dA \qquad (5.5-2)$$

Problems of extensional deformation of bars and the bending of beams are typical one-dimensional problems. In the case of extensional deformations, that could be due, for example, to a combined axial temperature distribution and symmetrical transverse temperature distribution, the axial displacement u could be selected in the form of a polynomial such as

$$u = a_0 + a_1 x + a_2 x^2 + a_3 x^3 + \ldots \qquad (5.5-3)$$

and

$$\epsilon_x = a_1 + 2a_2 x + 3a_3 x^2 + \ldots \qquad (5.5-4)$$

One or more of the coefficients, a_i, will be predetermined by the boundary conditions. For a free bar it can be specified that $u(o) = 0$, and it follows that $a_0 = 0$. The remaining coefficients a_i are determined by minimizing the potential energy, Eq.(5.5-2) with respect to each of the undetermined coefficients. Thus

$$\frac{\partial V}{\partial a_1} = 0 \qquad \frac{\partial V}{\partial a_2} = 0 \ldots \qquad \frac{\partial V}{\partial a_n} = 0 \qquad (5.5-5)$$

and performing this operation on Eq. (5.5-2) we obtain

$$\int_A \int_0^L (\epsilon_x - \alpha T) \frac{\partial \epsilon_x}{\partial a_i} \, dx \, dA = 0 \qquad (5.5-6)$$

In the case of a beam, the stress is given by Eq. (6.2-9) in terms of the second derivative of the beam deflection, v, as

$$\sigma_x = E\alpha \left(\overline{T}(x) - \frac{y}{\alpha} \frac{d^2 v}{dx^2} - T(x, y)\right) \qquad (5.5-7)$$

We can then select a polynomial to represent the beam deflection. Such a polynomial could be

$$v = b_o + b_1 x + b_2 x^2 + b_3 x^3 + b_4 x^4 + \ldots \qquad (5.5-8)$$

with

$$\frac{d^2 v}{dx^2} = 2 b_2 + 6 b_3 x + 24 b_4 x^2 + \ldots \qquad (5.5-9)$$

At least two of the constants b_i will be determined by the boundary conditions. For a simply supported beam, for example, we have $v(o) = v(L) = 0$. The potential energy Eq. (5.5-1) is minimized with respect to the remaining coefficients. Setting Eq. (5.5-7) into Eq. (5.5-1) and minimizing with respect to b_i, we have

$$\int_A \int_0^L (\alpha \overline{T}(x) - y \frac{d^2 v}{dx^2} - \alpha T(x, y)) \frac{\partial}{\partial b_i} \left(\frac{\partial^2 v}{\partial x^2}\right) y \, dA \, dx = 0 \qquad (5.5-10)$$

Example (a)   Restrained Rod with Axial Temperature Variation

In Example (d) of Sect. 1.8 the expression for the strain in a restrained rod is found to be

$$\epsilon_x = -\alpha(\overline{T} - T) \qquad (1)$$

We now treat the problem by means of the Rayleigh-Ritz method. Assume that the temperature distribution, with the coordinate x measured from the center of the rod, is

$$T = T_o \cos \frac{\pi x}{L} \qquad (2)$$

as shown in Fig. 5.5-1. We take the displacement as an odd polynomial in order that the strain be symmetrical about the center of the rod. The displacement is taken as

$$u = a_0 + a_1 x + a_3 x^3 \qquad (3)$$

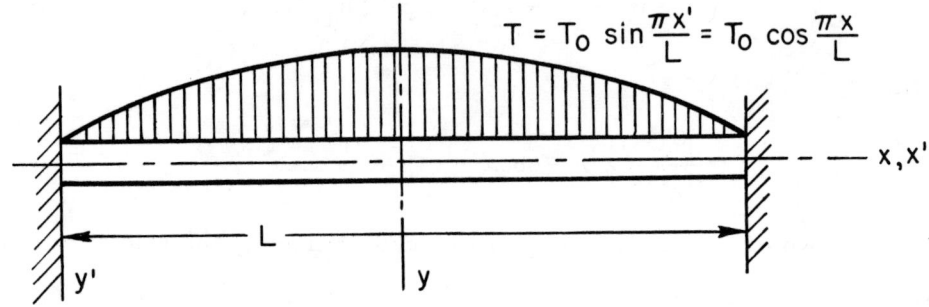

Fig. 5.5-1

Since $u(-\frac{L}{2}) = 0$ and $u(\frac{L}{2}) = 0$ in order to satisfy the boundary conditions, a_0 must be zero and $a_1 = -a_3 \frac{L^2}{4}$. The displacement is then written in terms of a single arbitrary coefficient, a_3, as

$$u = a_3 (x^3 - \frac{L^2 x}{4}) \qquad (4)$$

and $\epsilon_x = \frac{du}{dx}$ is

$$\epsilon_x = a_3 (3x^2 - \frac{L^2}{4}) \qquad (5)$$

The potential energy is

$$V = \frac{AE}{2} \int_{-L/2}^{L/2} (\epsilon_x - \alpha T)^2 \, dx \qquad (6)$$

or

$$V = \frac{AE}{2} \int_{-L/2}^{L/2} \left[a_3 (3x^2 - \frac{L^2}{4}) - \alpha T_0 \cos \frac{\pi x}{L} \right]^2 dx \qquad (7)$$

Minimization of the potential energy yields

$$\frac{\partial U_o}{\partial a_3} = \int_{-L/2}^{L/2} \left[a_3(3x^2 - \frac{L^2}{4}) - \alpha T_o \cos\frac{\pi x}{L} \right] (3x^2 - \frac{L^2}{4}) \, dx = 0 \qquad (8)$$

The integration yields

$$a_3 = \frac{20 \alpha T_o}{L^2} \left(\frac{1}{\pi} - \frac{12}{\pi^3} \right) \qquad (9)$$

and the strain distribution, Eq. (5) becomes

$$\epsilon_x = \frac{20 \alpha T_o}{L^2} \left(\frac{1}{\pi} - \frac{12}{\pi^3} \right)(3x^2 - \frac{L^2}{4}) \qquad (10)$$

The true strain distribution, which is obtained from Eqs. (1) and (2) is

$$\epsilon_x = \alpha T_o \left(\cos\frac{\pi x}{L} - \frac{2}{\pi} \right) \qquad (11)$$

A comparison of Eqs. (10) and (11) shows close agreement between the approximate solution and the exact solution.

Good judgment in the selection of the form of the displacement function leads to accurate results. We could, for example, select some form of a sinusoidal variation of strain to conform to the sinusoidal temperature variation. This temperature variation is now expressed as $T = T_o \sin\frac{\pi x'}{L}$ with x' measured from the left end of the rod. Let us assume that the displacement can be expressed as

$$u = a_o + a_1 x' + a_2 \cos\frac{\pi x'}{L} \qquad (12)$$

so that

$$\epsilon_x = \frac{du}{dx} = a_1 - \frac{a_2 \pi}{L} \sin\frac{\pi x'}{L} \qquad (13)$$

The conditions $u(o) = u(L) = 0$ gives

$$a_2 = -a_o \qquad\qquad a_1 = -\frac{2a_o}{L} \qquad (14)$$

Eqs. (13) and (14) then become

$$u = a_o \left(1 - \frac{2x'}{L} - \cos\frac{\pi x'}{L}\right) \tag{15}$$

$$\epsilon_x = \frac{2a_o}{L}\left(\frac{\pi}{2}\sin\frac{\pi x'}{L} - 1\right) \tag{16}$$

The strain ϵ_x is set into Eq. (5.5-2), and the minimization $\partial V/\partial a_o = 0$ yields

$$\int_0^L \left(-\frac{2a_o}{L} + \frac{a_o \pi}{L}\sin\frac{\pi x'}{L} - \alpha T_o \sin\frac{\pi x'}{L}\right)\left(\frac{\pi}{2}\sin\frac{\pi x'}{L} - 1\right) dx = 0 \tag{17}$$

from which we obtain

$$a_o = \frac{\alpha T_o L}{\pi} \tag{18}$$

The displacement and strain are

$$u = \frac{\alpha T_o L}{\pi}\left(1 - \frac{2x'}{L} - \cos\frac{\pi x'}{L}\right) \tag{19}$$

$$\epsilon_x = \alpha T_o \left(\sin\frac{\pi x'}{L} - \frac{2}{\pi}\right) \tag{20}$$

This solution of the problem, which was obtained through the "judicious" selection of the form of the displacement function, is actually the exact solution of the problem, as can be ascertained by means of Eq. (1).

Example (b) Deflection of a Beam

The Rayleigh-Ritz method is now applied to obtain the deflection curve of a simply supported rectangular beam of unit width having a transverse and axial temperature distribution. The coordinate system is shown in Fig. 5.5-2. The temperature distribution is taken as the product of two functions, with the

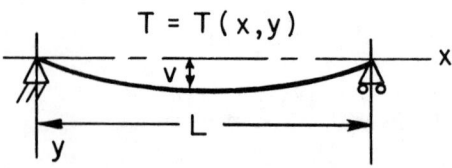

Fig. 5.5-2

axial distribution expressed as a sine series satisfying the end conditions, $T(o, y) = T(L, y) = 0$. That is

$$T = T(y)T(x) = T(y) \sum_{n=1}^{\infty} t_n \sin \frac{n\pi x}{L} \qquad (1)$$

with t_n the modal constants of the conventional Fourier series, obtained when the axial temperature variation is expressed as a sine series. The deflection curve is taken as

$$v = \sum_{1}^{\infty} a_n \sin \frac{n\pi x}{L} \qquad (2)$$

This satisfies the end conditions, $v(o) = v(L) = 0$. The stress is given by Eq. (5.5-7) as

$$\sigma_x = E\alpha(\overline{T}(x) - \frac{y}{\alpha} \frac{d^2 v}{dx^2} - T(x, y)) \qquad (3)$$

which, substituted into Eq. (5.5-1), gives

$$V = \frac{1}{2E} \int_{-\frac{h}{2}}^{\frac{h}{2}} \int_0^L E\alpha(\overline{T}(x) - \frac{y}{\alpha} \frac{d^2 v}{dx^2} - T(x, y))^2 \, dx \, dy \qquad (4)$$

and minimization of V with respect to any coefficient a_m, yields

$$\frac{\partial V}{\partial a_m} = \int_{-\frac{h}{2}}^{\frac{h}{2}} \int_0^L (\alpha \overline{T}(x) - y \frac{d^2 v}{dx^2} - \alpha T(x, y)) \frac{\partial}{\partial a_m} (\frac{d^2 v}{dx^2}) \, dx \cdot y \, dy = 0 \qquad (5)$$

We carry out the integration over the cross section area A, observing that

$$\int_A \overline{T}(x) \, y \, dA = 0 \qquad (6)$$

since the indefinite integral is an even function of y. We then obtain for Eq. (5)

$$\int_0^L (I\frac{d^2v}{dx^2} + \alpha M_t)\frac{\partial}{\partial a_m}(\frac{d^2v}{dx^2})\,dx = 0 \tag{7}$$

with I the cross section moment of inertia and M_t defined as

$$M_t = \int_{-\frac{h}{2}}^{\frac{h}{2}} Ty\,dy = M_t(x) \tag{8}$$

Substitution of T and v from Eqs. (1) and (2) yields

$$\int_0^L (-\frac{I\pi^2}{L^2}\sum_1^\infty a_n n^2 \sin\frac{n\pi x}{L} + \alpha M_{ty}\sum_1^\infty t_n \sin\frac{n\pi x}{L})(\sin\frac{m\pi x}{L})\,dx = 0 \tag{9}$$

with

$$M_{ty} = \int_{-\frac{h}{2}}^{\frac{h}{2}} T(y)\cdot y\,dy = \text{const.} \tag{10}$$

All harmonics except those containing m disappear and Eq. (9) becomes

$$a_m(\frac{m^2\pi^2}{L^2})\int_0^L \sin^2(\frac{m\pi x}{L})\,dx = \frac{\alpha M_{ty} t_m}{I}\int_0^L \sin^2(\frac{m\pi x}{L})\,dx \tag{11}$$

from which

$$a_m = \frac{L^2 \alpha M_{ty}}{\pi^2 I}(\frac{t_m}{m^2}) \tag{12}$$

The deflection of the beam is therefore given by the series

$$v = \frac{L^2}{\pi^2 I}\alpha M_{ty}\sum_1^\infty t_n \frac{\sin\frac{n\pi x}{L}}{n^2} \tag{13}$$

In the case of a simple sinusoidal temperature variation (n = 1) we have

$$T = T(y)\sin\frac{\pi x}{L} \tag{14}$$

Eq. (13) gives the deflection curve as

$$v = \frac{L^2 \alpha M_{ty}}{\pi^2 I}\sin\frac{\pi x}{L} \tag{15}$$

PROBLEMS

5-1 Using complementary energy derive the expression for the stress in an axially restrained bar of varying cross section. The temperature is a symmetrical function of the transverse coordinate and a slowly varying function of the axial coordinate. Make use of the fact that equilibrium requires that $\bar{\sigma}(x)A(x) = $ const.

5-2 A bimetallic sandwich bar with core physical constants E_1, α_1, A_1 and outer layer constants E_2, α_2, A_2, is subjected to symmetrical temperature distributions $T_1(x,y)$ and $T_2(x,y)$ which change slowly in the axial, x, direction. The coordinate system is as shown in Fig. 2.7-1. Using complementary energy analysis find the axial stress and strain distribution, for the case of the bar rigidly held at its extremities.

5-3 Perform a complementary energy analysis of a concentric ring assembly consisting of two concentric rings of dissimilar materials, such as those shown in Fig. 2.14-1. The assembly is subjected to a temperature distribution $T = T(r)$. Use the same symbols as in Fig. 2.14-1.

5-4 Derive the expressions for the thermal stresses in the disc and ring assembly shown in Fig. 3.13-1 by means of complementary energy. Assume, in the present problem, that the disc is at a uniform temperature T_1 and that the ring has a radially varying temperature $T = T(r)$. Take as the indeterminate stress, the uniform stress in the disc, σ_1.

5-5 The truss shown in Fig. 2.18-1 consists of four carbon steel bars, each having a modulus of elasticity, $E = 30 \times 10^6$ psi, a coefficient of expansion $\alpha = 6.5 \times 10^{-6}/°F$, and a cross section area of one square inch. The horizontal member, (1), is 10' long, and members (2), (3), and (4) make angles of $60°$, $45°$, and $30°$, respectively, with the vertical.

Bars (1), (2), (3) and (4) are heated to temperatures $100°F$, $200°F$, $300°F$, and $400°F$ respectively. Assuming members (2) and (3) to be the redundant members, obtain the thermal stresses in the members by means of a complementary energy analysis.

5-6 Find the stresses in each of the bars of the truss shown in Fig. 2-29 by means of complementary energy analysis. Use the constants given in Prob. 2-29.

5-7 Using complementary energy analysis, derive the expression for the stresses in the bars of Fig. 2-24. Let $\bar{\sigma}_1$ be the statically indeterminate stress. Hint: Employ Eq. (5.2-9) to relate the mean stresses in bars (1) and (2), and again to relate the mean stresses in bars (3) and (4); also the fact that $\bar{\sigma}_1 A_1 + \bar{\sigma}_2 A_2 = \bar{\sigma}_3 A_3 + \bar{\sigma}_4 A_4$.

5-8 Derive the expression for the stresses in the bars shown in Fig. 2.16-4, by means of complementary energy. Use the conditions stated in Example (c) of Sect. 2.16. Take σ_o as the statically indeterminate load. Hint: Make use of Eq. (5.2-8) to relate σ_1 and $\bar{\sigma}_1$.

5-9 By means of Castigliano's theorem, compute concurrently the stresses in the rectangular bars (1) and (2) and the axial displacement of the junction point B in Fig. 5-9. The temperature distributions are

$$T_1 = T_{1,0} \cos \frac{\pi y_1}{h_1} \qquad T_2 = T_{2,0} \cos \frac{\pi y_2}{h_2}$$

and the dimensions are as shown.

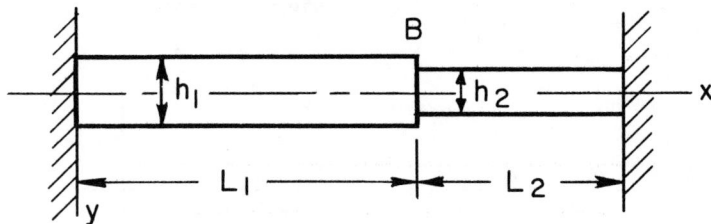

Fig. 5-9

5-10 Employing Castigliano's theorem derive the expression for the vertical deflection of point B on the truss shown in Fig. 5-10. Assume a uniform cross section area A for all rods, a length L for all vertical and horizontal members and $\sqrt{2}\,L$ for all diagonal members. The temperature are T_1, T_2, T_3, T_4, T_5, and T_6 corresponding to the numbers assigned to the members.

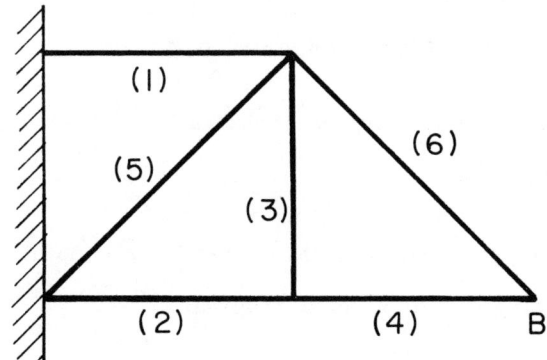

Fig. 5-10

5-11 The rectangular cantilever beam shown in Fig. 5-11 has a temperature distribution

$$T = \frac{T_o}{2}\left(1 + \frac{2y}{h}\right)$$

Employ Castigliano's theorem to obtain the vertical deflection, v, of the free end of the beam.

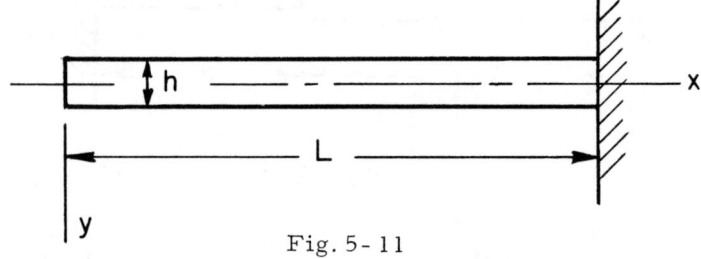

Fig. 5-11

Hint: Use the first of Eqs.(5.3-24) with $M = 0$ and also that

$$M_t = \int_{-\frac{h}{2}}^{\frac{h}{2}} T y \, dy = \frac{T_o}{2} \int_{-\frac{h}{2}}^{\frac{h}{2}} (1 + \frac{2y}{h}) y \, dy$$

5-12 In Prob. 2-29 and Prob. 5-6 the stresses in the redundant truss, shown in Fig. 2-29, were determined. Use Castigliano's theorem to obtain the horizontal and vertical displacements of point B.

5-13 Compute the axial displacement of junction B of Prob. 5-9 by means of the reciprocal theorem.

5-14 Compute the horizontal and vertical deflections of point B in the truss of Prob. 2-29, using the reciprocal theorem.

5-15 Compute the vertical and horizontal deflections of point B on the truss of Fig. 5-10 by means of the reciprocal theorem.

5-16 Using the reciprocal theorem, obtain the deflection of the center, $x = \frac{L}{2}$, of the clamped beam shown in Fig. 5-16. The temperature distribution is

$$T = T_o \frac{xy}{Lh}$$

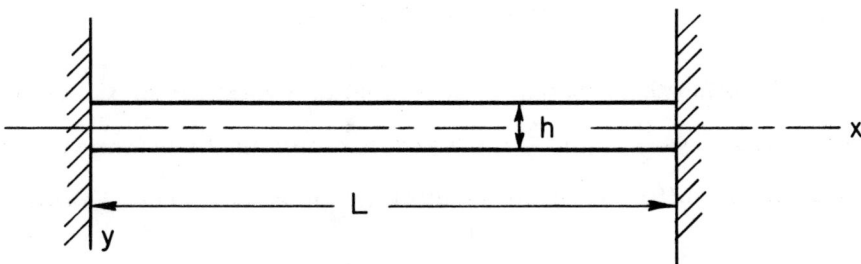

Fig. 5-16

5-17 Find the volume change when the rectangular body shown in Fig. 5-17 is subjected to temperature distribution

$$T = T_o \frac{x}{L} \cdot \frac{y^2}{h^2} \sin \frac{\pi z}{w}$$

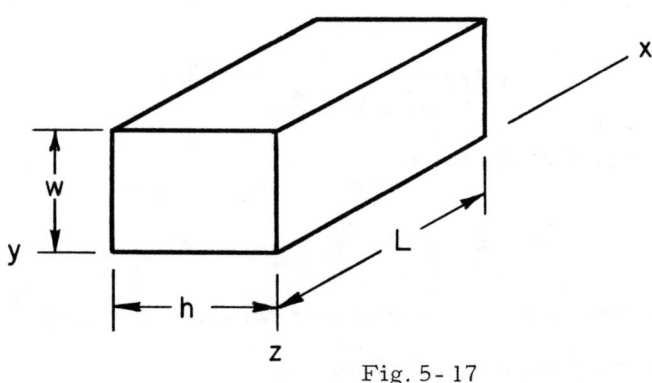

Fig. 5-17

5-18 Example (a) of Sect. 5.5 contains the Rayleigh-Ritz solution for the distributed axial displacments of a restrained bar with an axial temperature variation

$$T = T_o \cos \frac{\pi x}{L}$$

Take as an approximate solution

$$u = a_1 x + a_3 x^3 + a_5 x^5$$

and determine the constants a_1, a_3 and a_5 by satisfying the boundary conditions $u(-\frac{L}{2}) = u(\frac{L}{2}) = 0$ and minimizing the potential energy. Compare result to accurate solution given by Eq.(11) of Example (a).

5-19 Obtain, by means of the Rayleigh-Ritz procedure the approximate deflection curve of a simply supported beam of depth h and length L with a temperature distribution

$$T = \frac{T_o}{4} \left(\frac{2y}{h}+1\right)^2 \frac{x}{L}\left(1-\frac{x}{L}\right)$$

The coordinate x is measured from the left end of the beam and y is measured downward from the beam center line. Put the temperature distribution in the form

$$T = T(y) \sum_{n=1,3,5\ldots}^{\infty} t_n \sin \frac{n\pi x}{L}$$

and take the deflection in the form

$$v = \sum_{1}^{\infty} a_n \sin \frac{n\pi x}{L}$$

Use the first three terms of the series and compare this approximate solution with the exact solution which is

$$v = \frac{\alpha T_o L^2}{12h} \left(\frac{x^4}{L^4} - \frac{2x^3}{L^3} + \frac{x}{L}\right)$$

6 | Beams

6.1 BEAM THERMAL STRESS ANALYSIS

Isothermal beam stress analysis lends itself readily to strength of materials methods since fairly accurate results are obtainable by these methods. The errors in the computation of bending stresses and deflections are of the order of a few percent. The inaccuracies that are present in the computation of the thermal stresses in beams by means of strength of materials methods, can however be considerably larger. A primary source of error arises from the neglect of lateral and transverse stresses in cross section planes. In wide beams this can introduce a non-conservative error of as much as thirty percent. In narrow beams with a transverse temperature distribution, strength of materials analysis gives, more or less, the same degree of accuracy as that obtained in isothermal beam analysis. Also, it has been demonstrated in Sect. 3.3, (Class 4: Plane Harmonic Cross Section Temperature Distributions) that one-dimensional beam analysis is accurate when the cross section Laplacian $\nabla^2 T$ is zero. Such a plane harmonic temperature distribution will be produced when there is lateral and (or) transverse heat condition, without internal heat generation.

In general, the error that it is possible to introduce becomes larger as the lateral dimension (perpendicular to the plane of bending) is increased. In the case of a rectangular beam, shown in Fig. 6.1-1 with a transverse temperature distribution, $T = T(y)$, the error will increase as the width w in-

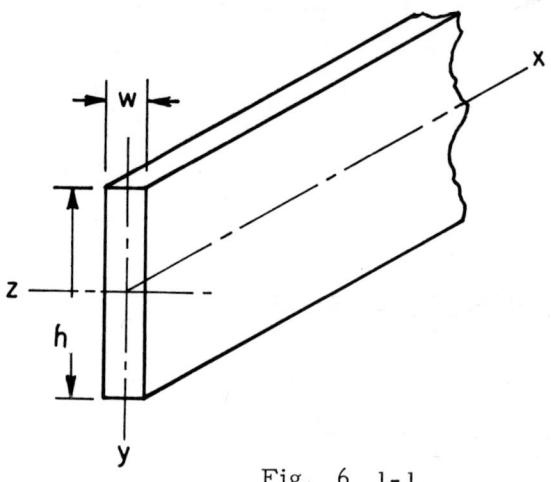

Fig. 6.1-1

creases. When the width is much greater than the depth, the beam becomes a plate and the stress is increased by a factor $1/(1-\nu)$. A conservative estimate, based on the solution of plane stress problems in rectangular plates with varying dimensions, (Sect. 3.5), is that the axial stress should be increased by a factor $1/(1-\nu)$ when the width is equal to or greater than the depth. It should be recognized that this is an approximation.

In order to improve the accuracy of stress computations in <u>unrestrained</u> thermal beams the following rule of thumb may be used: When the width of the beam is less than the depth, the increase in stress may be obtained by using a modified modulus of elasticity E^*, defined as

$$E^* = \frac{E}{1-\nu \frac{w}{h}} \qquad \frac{w}{h} \leq 1 \qquad (6.1-1)$$

and when the beam width is greater than the depth the increase in stress may be obtained by using a modified modulus of elasticity defined as

$$E^* = \frac{E}{1-\nu} \qquad \frac{w}{h} \geq 1 \qquad (6.1\text{-}2)$$

This rule is applicable also in bars having a transverse temperature distribution, in which there is no bending, as discussed in Sect. 2.4. In applying this rule to a restrained (statically indeterminate) beam, it is necessary to remove restraints so that the beam becomes statically determinate. The modification given by Eqs. (6-1-1) or (6.1-2) is applied to the thermal stress computed for the statically determinate beam. The stresses due to the statically indeterminate reactions are then added to the modified thermal stresses. It is necessary to follow this procedure since the modification does not apply to the isothermal stresses arising from the indeterminate beam reactions.

A second source of error is due to relatively rapid changes of temperature along the beam length. One-dimensional beam analysis does not permit rapid temperature changes along the beam length, since in that case it becomes a two-dimensional plane stress problem. The error that is introduced by permitting slow axial temperature variation arises from the transverse expansion incompatibility that an axial temperature variation introduces. Eq. (3.3-23) gives the order of magnitude of error that temperature variations in the plane of a plate introduce. In a beam the maximum transverse stress, σ_y, that an axial temperature variation produces is approximately

$$\sigma_y \approx \frac{E\alpha h}{2} \left(\frac{\Delta T}{\Delta x}\right)_{max} \qquad \Delta x \geq h \qquad (6.1\text{-}3)$$

A third source of error that is often present in thermal beam analysis is the neglect of the axial load that is generated when a beam on immovable supports is heated. Large axial loads can be developed as a result of the restrained axial expansion, and they can have a significant effect on the beam deflection. This source of axial load differs from the self-generated axial loads that are produced when isothermal beams are on immovable supports. In isothermal beams the modification in beam deflection can be disregarded when the deflection is small in comparison with the depth of the beam, as small deflections result in small axial loads. In heated beams, on the other hand, large axial loads can be developed even when the beam deflection is small.

In heated beams, as well as in isothermal beams, the ratio of axial load to buckling load, P/P_{cr}, determines the degree to which the beam deflection is increased. If δ_o is the beam deflection calculated without consideration of axial thrust, then the modification due to the axial thrust P, generated by restrained thermal expansion as well as deflection, is approximately

$$\frac{\delta}{\delta_o} = \frac{1}{1-P/P_{cr}} \qquad (6.1-4)$$

with δ the modified beam deflection. An accurate analysis requires that the beam be treated as a beam-column. Eq. (6.1-4) can however be used directly when δ is considerably less than the beam depth h. Then $P \approx E\alpha\bar{T}A$.

A fourth source of error lies in the changes that occur in the modulus of elasticity, coefficient of expansion, and yield strength of a material as the temperature is increased. The errors due to these changes are, of course, not restricted to beams alone, but represent a source of error in all thermal stress analysis problems.

In spite of these sources of error, which are a part of thermal beam analysis, strength of materials methods afford a means of rapid stress and deformation computations and are used extensively in engineering analysis. The inaccuracies are mitigated in part by the self-limiting and self-relieving character of thermal stresses. Large thermal bending stresses in a beam will generally not cause rupture, and the structural integrity will generally not be impaired. In cyclic or repeated loading the possibility of thermal stress fatigue failure exists. Beams which are subjected to cyclic thermal stresses, which are so large that they produce yielding, will undergo favorable self-prestressing. The effect of the self-prestressing or self-springing as it is sometimes called, is to reduce the peak stresses that are obtained during thermal cycling.

6.2 THERMAL BEAM RELATIONSHIPS

As in isothermal beam analysis, the transverse and lateral normal stresses σ_y and σ_z are assumed to be zero. The axial normal stress, σ, is related to the axial strain, ϵ, by the equation

$$\epsilon = \frac{\sigma}{E} + \alpha T \qquad (6.2\text{-}1)$$

with $T = T(x, y)$ and the temperature restricted to a slow rate of change in the axial direction. The strain consists of two parts, a uniform extensional strain ϵ_e and a bending strain ϵ_b which varies linearly over the depth of the beam. The bending strain is related to the beam curvature $\frac{d^2 v}{dx^2} = \frac{1}{R}$ by the expression

$$\epsilon_b = \frac{y}{R} = -y \frac{d^2 v}{dx^2}$$

with y the distance from the bending axis, located at the centroid of the beam section. In contrast to isothermal beams, the stress and strain at the bending axis need not be zero in a heated beam. The total strain is

$$\epsilon = \epsilon_e + \epsilon_b = \epsilon_e + \frac{y}{R} \qquad (6.2\text{-}2)$$

Eq. (6.2-2) is substituted for ϵ in Eq. (6.2-1), which becomes

$$\epsilon_e + \frac{y}{R} = \frac{\sigma}{E} + \alpha T \qquad (6.2\text{-}3)$$

Both sides of this equation are multiplied by dA, an element of cross section area, and integrated over the cross section. We note that

$$\int_A y\,dA = 0$$

about the centroidal or bending axis, and also that

$$\int_A \sigma\,dA = 0$$

since there are no axial loads on the beam. The integration of Eq. (6.2-3) gives

$$\epsilon_e = \alpha \overline{T}(x) \qquad (6.2\text{-}4)$$

Eq. (6.2-3) is now multiplied by ydA and integrated over the cross section area. Since ϵ_e is constant over any cross section, we have

$$\int_A \epsilon_e y\,dA = 0$$

and Eq. (6.2-3) becomes

$$\frac{E}{R} \int_A y^2\,dA = \int_A \sigma y\,dA + E\alpha \int_A Ty\,dA \qquad (6.2\text{-}5)$$

We define a thermal moment $M_t = M_t(x)$ as

$$M_t = \int_A Ty\,dA \qquad (6.2\text{-}6)$$

and observing that the integral on the left hand side of Eq. (6.2-5) is the rectangular moment of inertia, I, we can write this equation as

$$\frac{1}{R} = -\frac{d^2v}{dx^2} = \frac{M}{EI} + \frac{\alpha}{I} M_t \tag{6.2-7}$$

or

$$M = -EI \frac{d^2v}{dx^2} - E\alpha M_t \tag{6.2-8}$$

The foregoing is one form of the thermal beam deformation equation. It can be integrated to obtain thermal beam deflections.

Substitution of $1/R$ from Eq. (6.2-7), and ϵ_e from Eq. (6.2-4), into Eq. (6.2-3) yields the thermal beam stress equation

$$\sigma = E\alpha \left(\overline{T} - \frac{y}{\alpha} \frac{d^2v}{dx^2} - T \right) \tag{6.2-9}$$

With the use of Eq. (6.2-7) we can also write this equation in the form

$$\sigma = E\alpha \left(\overline{T} + \frac{M_t y}{I} - T \right) + \frac{My}{I} \tag{6.2-10}$$

When a thermal beam is statically determinate, i.e, when there are no external restraints to beam deformation, $M = 0$ and Eq. (6.2-10) becomes

$$\sigma = E\alpha \left(\overline{T} + \frac{M_t y}{I} - T \right) \tag{6.2-11}$$

The moment M in Eq. (6.2-10) or the curvature $\frac{d^2v}{dx^2}$ in Eq. (6.2-9) is found from the integration of the thermal beam deformation equation, Eq. (6.2-8). It should be recognized that $\overline{T} = \overline{T}(x)$ is the mean temperature of the beam at a given coordinate location x, rather than the mean temperature of the entire beam. As stated earlier, the temperature T is a function of x and y, with the restriction that the axial temperature variation takes place slowly. This restriction is required in order to minimize transverse free expansion incompatibilities.

6.3 STRESS AND DEFORMATION EQUATIONS

An element of beam, shown in Fig. 6.3-1, of length dx and of unit width is acted upon by an external distributed load q and moments and shears at

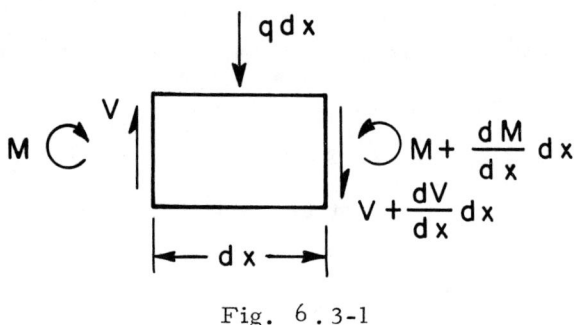

Fig. 6.3-1

the ends of the element. The required equilibrium of the loads and moments yields the following equations.

$$-\frac{dV}{dx} = q \qquad (6.3\text{-}1)$$

$$V = \frac{dM}{dx} \qquad (6.3\text{-}2)$$

Substitution of the second equation into the first gives

$$-\frac{d^2M}{dx^2} = q \qquad (6.3\text{-}3)$$

The expression for M, from Eq. (6.2-8) is set into the foregoing equation. The result is

$$EI\frac{d^4v}{dx^4} + E\alpha\frac{d^2M_t}{dx^2} = q \qquad (6.3\text{-}4)$$

STRESS AND DEFORMATION EQUATIONS

This is the most general form of the thermal beam deformation equation. It can be solved with the use of prescribed boundary conditions of deflection, slope, moment, and shear. From Eq. (6.2-8)

$$\frac{d^2 v}{dx^2} = -\frac{M}{EI} - \frac{\alpha}{I} M_t \qquad (6.3\text{-}5)$$

$$\frac{d^3 v}{dx^3} = -\frac{V}{EI} - \frac{\alpha}{I} \frac{dM_t}{dx} \qquad (6.3\text{-}6)$$

At a point in the beam where the moment is zero, as at a simple support or a free end, Eq. (6.3-5) gives the boundary condition

$$\frac{d^2 v}{dx^2} = -\frac{\alpha}{I} M_t \qquad (M = 0) \qquad (6.3\text{-}7)$$

and at a point in a beam where the shear is zero, as at a free end, Eq. (6.3-6) gives the boundary condition

$$\frac{d^3 v}{dx^3} = -\frac{\alpha}{I} \frac{dM_t}{dx} \qquad (V = 0) \qquad (6.3\text{-}8)$$

When there is no external distributed load, q = 0, then Eq. (6.3-4) becomes

$$\frac{d^4 v}{dx^4} = -\frac{\alpha}{I} \frac{d^2 M_t}{dx^2} \qquad (6.3\text{-}9)$$

which, upon integration, yields

$$\frac{d^2 v}{dx^2} = -\frac{\alpha}{I} M_t + Ax + B \qquad (6.3\text{-}10)$$

A more explicit form of this equation is obtained by designating V_o and M_o as the shear and moment in the beam at x = 0. The moment is then M = $M_o + V_o x$ and Eq. (6.3-5) takes the form

$$\frac{d^2v}{dx^2} = -\frac{\alpha}{I} M_t - \frac{V_o x}{EI} - \frac{M_o}{EI} \qquad (6.3-11)$$

In a statically determinate beam there are no bending restraints and no reactions at the supports. The end moment and shear M_o and V_o as well as the moment and shear M and V throughout the beam are zero. Eq. (6.3-11) is then written as

$$\frac{d^2v}{dx^2} = -\frac{\alpha}{I} M_t \qquad (6.3-12)$$

The deflection curve of a statically determinate beam is obtained by integrating this equation.

6.4 SHEAR STRESSES

As in isothermal beams, shear stresses will be present in thermal beams, only when the axial normal stress, σ, is a function of the axial coordinate. There will be shear stresses only when $\frac{\partial \sigma}{\partial x} \neq 0$. In a statically determinate beam having a transverse temperature distribution $T = T(y)$, there is no change in the normal stress along the beam length, and consequently no shear stresses are developed. Near the extreme ends of the beam, however, local shear stresses will be developed as the normal axial stresses fall off to zero. This will be the case in all beams which develop thermal stresses, whether statically determinate or indeterminate. Section 3.6 contains a discussion of the end shear stress problem. At the present time we are concerned with shear stresses distributed throughout the beam.

It is clear that in statically indeterminate thermal beams in which $T = T(y)$ shear stresses will be generated, since in such beams σ will be a function of x even though T is not a function of x. When $T = T(x, y)$, then $\sigma = \sigma(x, y)$ and shear stresses will be generated in statically determinate and indeterminate beams.

Fig. 6.4-1 shows an element of beam of length dx and variable width

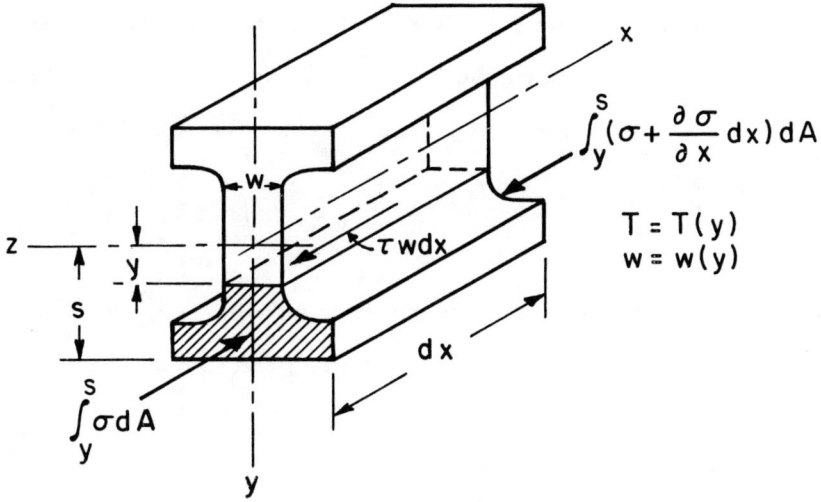

Fig. 6.4-1

$w = w(y)$. The temperature $T = T(x, y)$ is a function of the axial and transverse coordinates so that bending takes place in the x-y plane. The shaded portion of the beam element is in equilibrium under the normal and shear loads at the surfaces. Axial load equilibrium is expressed as

$$\tau w \, dx + \int_y^s \frac{\partial \sigma}{\partial x} dx dA = 0 \qquad (6.4\text{-}1)$$

which yields the shear stress τ as

$$\tau = -\frac{1}{w} \int_y^s \frac{\partial \sigma}{\partial x} dA \qquad (6.4\text{-}2)$$

Substitution of σ from Eqs. (6.2-9) and (6.2-10) yields respectively

$$\tau = -\frac{E\alpha}{w} \frac{\partial}{\partial x} \left[\overline{T} \int_y^s dA - \frac{1}{\alpha} \frac{d^2 v}{dx^2} \int_y^s y dA - \int_y^s T dA \right] \qquad (6.4\text{-}3)$$

and

$$\tau = -\frac{E\alpha}{w}\frac{\partial}{\partial x}\left[\overline{T}\int_y^s dA + \frac{M_t}{I}\int_y^s y\,dA - \int_y^s T\,dA\right] + \frac{1}{wI}\frac{dM}{dx}\int_y^s y\,dA \qquad (6.4\text{-}4)$$

We define A_y as the area lying between the variable coordinate y and the surface s. The static moment Q_y is the moment of A_y about the centroidal plane, and the mean temperature of A_y is $\overline{T}_y$. Thus

$$A_y = \int_y^s dA \qquad Q_y = \int_y^s y\,dA \qquad A_y\overline{T}_y = \int_y^s T\,dA \qquad (6.4\text{-}5)$$

The shear V is the derivative of the moment with respect to x, and we define a thermal shear V_t as the derivative of the temperature moment with respect to x. That is

$$V = \frac{dM}{dx} \qquad V_t = \frac{dM_t}{dx} \qquad (6.4\text{-}6)$$

Using the defined quantities of Eqs. (6.4-5) and (6.4-6), we can rewrite Eqs. (6.4-3) and (6.4-4) as follows.

$$\tau = -\frac{E\alpha}{w}\left[A_y\frac{d\overline{T}}{dx} - \frac{d}{dx}(A_y\overline{T}_y) - \frac{Q_y}{\alpha}\frac{d^3v}{dx^3}\right] \qquad (6.4\text{-}7)$$

and

$$\tau = -\frac{E\alpha}{w}\left[A_y\frac{d\overline{T}}{dx} - \frac{d}{dx}(A_y\overline{T}_y) + \frac{V_t Q_y}{I}\right] - \frac{VQ_y}{wI} \qquad (6.4\text{-}8)$$

Note, in the foregoing equation, that when $T = 0$ the shear stress is

$$\tau = -\frac{VQ_y}{wI} \qquad (6.4\text{-}9)$$

which is the well-known expression for the shear stress in an isothermal beam.

SHEAR STRESSES

As found in the analysis of shear stresses in bars (Sect. 3.7), the shear stresses tend to become large when the axial temperature changes relatively rapidly, as expressed for example by a high order polynomial. (Beam analysis is however valid only when σ_y as given by Eq. (6.1-3) is not large.) The shear stress was also shown to increase as the ratio of beam depth to length became large. The same criteria are found to apply to beams. A beam with a high rate of axial temperature change tends to develop high shear stresses, and a deep beam will develop larger shear stresses than a shallow beam.

In rectangular beams of uniform width, the shear stresses are smallest near the upper and lower surfaces of the beams as a result of A_y being small, whereas the normal stresses are a maximum at the extreme upper and lower fibres. Since the position of the peak shear stresses and normal stresses do not coincide, the combined stress, or stress intensity, in rectangular beams will generally not be larger than the normal stress alone. Beams of other shapes such as I-beams and wide-flange beams can develop large thermal shear stresses near the upper and lower surfaces of the beam. A critical point in such beams is the junction of web and flange. At that point the static moment Q_y is large and the shear stress is correspondingly large. Since the normal bending stress and extensional stress are generally also large at this point, the combined shear and normal stress will often exceed the maximum normal extreme fibre stress.

Example (a) Shear Stress Distribution in a Beam

An unrestrained rectangular beam is subjected to a temperature variation expressed as

$$T = \frac{T_o}{4}\left(\frac{2y}{h}+1\right)^2 \frac{x^2}{L^2} \tag{1}$$

The change in transverse temperature distribution along the length of the beam is shown in Fig. 6.4-2.

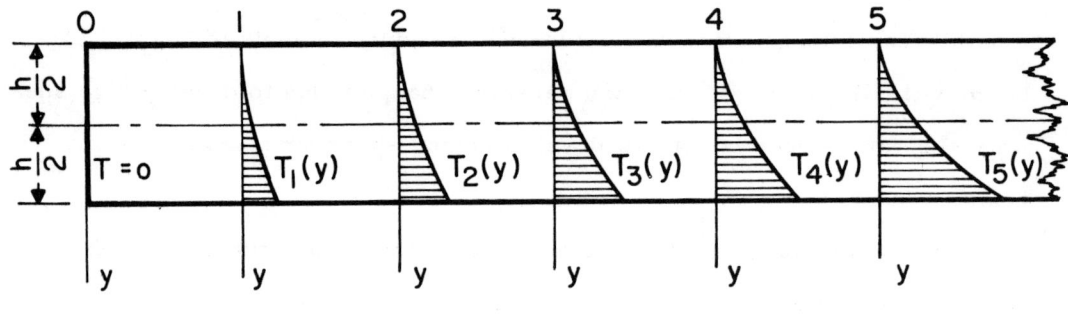

Fig. 6.4-2

The mean temperature and thermal moment are

$$\overline{T} = \frac{T_o}{3}\frac{x^2}{L^2} \qquad M_t = \frac{T_o I}{3}\frac{x^2}{L^2} \tag{2}$$

and the shear parameters A_y, Q_y, and $A_y\overline{T}_y$ are

$$A_y = w\left(\frac{h}{2}-y\right) \qquad Q_y = \frac{w}{2}\left(\frac{h^2}{4}-y^2\right)$$

$$A_y\overline{T}_y = \frac{whT_o}{3}\left[1-\frac{1}{8}\left(\frac{2y}{h}+1\right)^3\right]\frac{x^2}{L^2} \tag{3}$$

The expression for shear stress, as given by Eq. (6.4-7) becomes

$$\tau = \frac{E\alpha T_o xh}{3L^2}\left[1+\frac{2y}{h}-\frac{1}{4}\left(\frac{2y}{h}+1\right)^3\right] + \frac{Eh^2}{2}\left(\frac{1}{4}-\frac{y^2}{h^2}\right)\frac{d^3v}{dx^3} \tag{4}$$

In this form the shear stress is composed of two parts, the first term is due to the extensional deformation, and the second is due to flexure.

6.5 SOLUTIONS OF THERMAL BEAM EQUATIONS

Thermal beam problems differ from isothermal beam problems in several respects. One important respect is that strength of materials analysis of thermal beams contains sources of errors not present in isothermal beam analysis. Sect. 6.1 indicates how these can be taken into account when using the conventional strength of materials methods in thermal beam analysis.

It is also necessary to recognize some important differences in the application of boundary conditions when integrating the thermal beam deformation equation, Eq. (6.3-4). The boundary conditions relating to slope and deflection are the same for both thermal and isothermal problems. The boundary conditions which specify moment and shear are somewhat different. In isothermal problems, the boundary condition at a point, say $x = a$, where the moment is zero, is $\left(\dfrac{d^2 v}{dx^2}\right)_a = 0$. In a thermal beam however, the boundary condition at a point, $x = a$, where the moment is zero, is $\left(\dfrac{d^2 v}{dx^2}\right)_a = -\dfrac{\alpha}{I}(M_t)_a$. Secondly, the boundary condition in an isothermal beam at a location, $x = a$, where the shear is zero is $\left(\dfrac{d^3 v}{dx^3}\right)_a = 0$, while in a thermal beam the boundary condition at point a, where the shear is zero, is $\left(\dfrac{d^3 v}{dx^3}\right)_a = -\dfrac{\alpha}{I}\left(\dfrac{dM_t}{dx}\right)_a$. These differences should be kept in mind when integrating thermal beam equations.

Example (a) Simply Supported Beam

A simply supported rectangular beam of width w and depth h has a transverse temperature distribution

$$T = \frac{T_o}{4}\left(\frac{2y}{h} + 1\right)^2 \tag{1}$$

Fig. 6.5-1 shows the beam coordinate system and the temperature

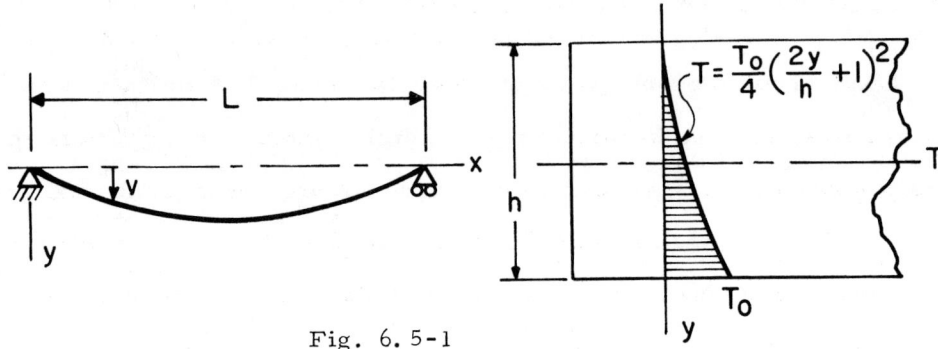

Fig. 6.5-1

distribution. The mean temperature and thermal moment are

$$\overline{T} = \frac{1}{h}\int_{-\frac{h}{2}}^{\frac{h}{2}} T\,dy = \frac{T_o}{3} \tag{2}$$

$$M_t = \int_{-\frac{h}{2}}^{\frac{h}{2}} Ty\,dA = \frac{T_o w h^2}{12} = \frac{T_o I}{h} \tag{3}$$

As the beam is statically determinate the flexure equation is

$$\frac{d^2 v}{dx^2} = -\frac{\alpha}{I} M_t = -\frac{\alpha T_o}{h} \tag{4}$$

The boundary conditions are $w(o) = w(L) = 0$.

Integration of Eq. (4) yields

$$v = \frac{\alpha T_o L^2}{2h}\left(\frac{x}{L} - \frac{x^2}{L^2}\right) \tag{5}$$

SOLUTIONS OF THERMAL BEAM EQUATIONS

and the normal stress is

$$\sigma = E\alpha\left(\overline{T} + \frac{M_t y}{I} - T\right) = E\alpha T_o \left[\frac{1}{3} + \frac{y}{h} - \frac{1}{4}\left(\frac{2y}{h}+1\right)^2\right] \qquad (6)$$

There are no shear stresses as the beam is statically determinate and the temperature is a function of the y coordinate only.

Example (b)   Cantilever Beam

A rectangular cantilever beam of depth h has a transverse temperature distribution

$$T = \frac{T_o}{4}\left(\frac{2y}{h}+1\right)^2 \qquad (1)$$

As in Example (a), the mean temperature and thermal moment are

$$\overline{T} = \frac{T_o}{3} \qquad M_t = \frac{T_o I}{h} \qquad (2)$$

Fig. 6.5-2 shows the beam coordinate system. The beam is statically

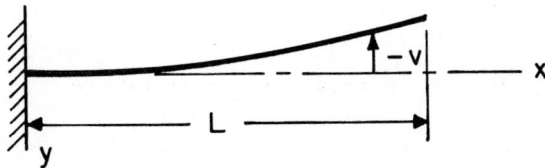

Fig. 6.5-2

determinate so that the flexure equation is

$$\frac{d^2 v}{dx^2} = -\frac{\alpha}{I} M_t = \frac{\alpha T_o}{h} \qquad (3)$$

The boundary conditions are $w(o) = w'(o) = 0$. Integration of the foregoing equation yields

$$v = -\frac{\alpha T_o L^2}{2h}\left(\frac{x^2}{L^2}\right) \tag{4}$$

The normal stress, as in Example (a), is

$$\sigma = E\alpha T_o\left[\frac{1}{3} + \frac{y}{h} - \frac{1}{4}\left(\frac{2y}{h} - 1\right)^2\right] \tag{5}$$

and there are no shear stresses as the beam is statically determinate and $T = T(y)$.

Example (c) Fixed-Pinned Beam

A fixed-pinned rectangular beam of depth h has a transverse temperature distribution

$$T = \frac{T_o}{4}\left(\frac{2y}{h} + 1\right)^2 \tag{1}$$

As in the preceding examples the mean temperature and thermal moment are

$$\overline{T} = \frac{T_o}{3} \qquad M_t = \frac{T_o I}{h} \tag{2}$$

Fig. 6.5-3 shows the beam coordinate system. The statically indeterminate

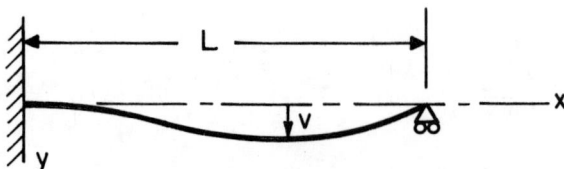

Fig. 6.5-3

SOLUTIONS OF THERMAL BEAM EQUATIONS 351

form of the thermal beam equation is employed, i.e.,

$$\frac{d^2 v}{dx^2} = -\frac{\alpha}{I} M_t + Ax + B \tag{3}$$

subject to the boundary conditions

$$v(o) = v(L) = v'(o) = 0 \qquad v''(L) = -\frac{\alpha}{I} M_t \tag{4}$$

Integration of Eq. (3) yields

$$v = \frac{\alpha T_o L^2}{4h} \left(\frac{x^2}{L^2} - \frac{x^3}{L^3} \right) \tag{5}$$

$$v'' = \frac{\alpha T_o}{4h} \left(2 - \frac{6x}{L} \right) \tag{6}$$

The normal stress, from Eq. (6.2-9) is

$$\sigma = E\alpha T_o \left[\frac{1}{3} - \frac{y}{4h} \left(2 - \frac{6x}{L} \right) - \frac{1}{4} \left(\frac{2y}{h} + 1 \right)^2 \right] \tag{7}$$

and the shear stress, obtained from Eq. (6.4-2) is

$$\tau = \frac{1}{w} \int_y^{\frac{h}{2}} \frac{\partial \sigma}{\partial x} w \, dy = \frac{3 E \alpha T_o}{2hL} \int_y^{\frac{h}{2}} y \, dy = \frac{3}{4} \frac{E\alpha T_o h}{L} \left(\frac{1}{4} - \frac{y^2}{h^2} \right) \tag{8}$$

Example (d) Fixed-Fixed Beam

A fixed-fixed rectangular beam of depth h has a transverse temperature distribution

$$T = \frac{T_o}{4} \left(\frac{2y}{h} + 1 \right)^2 \tag{1}$$

The mean temperature and thermal moment are

$$\bar{T} = \frac{T_o}{3} \qquad M_t = \frac{T_o I}{h} \tag{2}$$

Fig. 6.5-4 shows the beam coordinates and support. The right end of

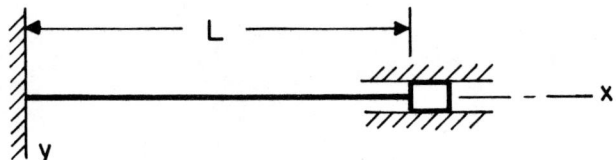

Fig. 6.5-4

the beam is attached to a piston which gives the beam freedom for axial expansion but prevents end rotation. The statically indeterminate thermal beam equation

$$\frac{d^2 v}{dx^2} = -\frac{\alpha}{I} M_t + A x + B \tag{3}$$

is used, with the boundary conditions

$$v(o) + v(L) = v'(o) = v'(L) = 0 \tag{4}$$

Integration of Eq. (3) with boundary conditions (4) yields

$$v = 0 \tag{5}$$

Since $\frac{d^2 v}{dx^2}$ is zero, the normal stress is

$$\sigma = E\alpha (\bar{T} - T) = E\alpha T_o \left[\frac{1}{3} - \frac{1}{4}\left(\frac{2y}{h} + 1\right)^2 \right] \tag{6}$$

Eq. (6) shows the stress to be independent of x. The shear stress is therefore zero.

If axial expansion is prevented as shown in Fig. 6.5-5, and the

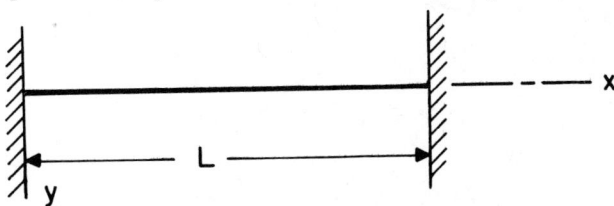

Fig. 6.5-5

axial load, $E\alpha \overline{T} A$, is less than the buckling load, the stress, using Eq. (6.2-3) with ϵ_e and $\frac{y}{R}$ both zero, is

$$\sigma = -E\alpha T = -\frac{E\alpha T_o}{4}\left(\frac{2y}{h}+1\right)^2 \tag{7}$$

Example (e) Pinned-Pinned Beam with Axial Temperature Variation

Pertinent parameters for this example are given in Example (a) of Sect. 6.4. The transverse and axial temperature variation in the rectangular beam is expressed as

$$T = \frac{T_o}{4}\left(\frac{2y}{h}+1\right)^2\left(\frac{x^2}{L^2}\right) \tag{1}$$

The deflection, obtained by integrating Eq. (6.3-12) is

$$v = \frac{\alpha T_o L^2}{12h}\left(\frac{x}{L}-\frac{x^4}{L^4}\right) \tag{2}$$

with the first, second, and third derivatives of the deflection as listed below.

$$\frac{dv}{dx} = \frac{\alpha T_o L}{12h}\left(1-\frac{4x^3}{L^3}\right) \tag{3}$$

$$\frac{d^2v}{dx^2} = -\frac{\alpha T_o}{h}\left(\frac{x^2}{L^2}\right) \qquad (4)$$

$$\frac{d^3v}{dx^3} = -\frac{2\alpha T_o}{hL}\left(\frac{x}{L}\right) \qquad (5)$$

The axial normal stress is, from Eq. (6.2-9)

$$\sigma = E\alpha T_o \left[\frac{1}{3} + \frac{y}{h} - \frac{1}{4}\left(\frac{2y}{h}+1\right)^2\right]\left(\frac{x^2}{L^2}\right) \qquad (6)$$

and the shear stress, using Eq. (4) of Example (a), Sect. 6.4, is

$$\tau = E\alpha T_o \left[\frac{1}{6} - \frac{2}{3}\left(\frac{y}{h}\right)^2\right]\frac{xy}{L^2} \qquad (7)$$

Example (f) Cantilever Beam with Axial Temperature Variation

Fig. 6.5-2 shows the coordinate system that is used. The transverse and axial temperature variation in the rectangular beam is

$$T = \frac{T_o}{4}\left(\frac{2y}{h}+1\right)^2 \frac{x^2}{L^2} \qquad (1)$$

Integration of the statically determinate thermal beam deformation equation, Eq. (6.3-12), yields the deflection and its derivatives as listed below.

$$v = \frac{\alpha T_o L^2}{12h}\left(\frac{x^4}{L^4}\right) \qquad (2)$$

$$\frac{dv}{dx} = -\frac{\alpha T_o L}{3h}\left(\frac{x^3}{L^3}\right) \qquad (3)$$

$$\frac{d^2 v}{dx^2} = -\frac{\alpha T_o}{h}\left(\frac{x^2}{L^2}\right) \tag{4}$$

$$\frac{d^3 v}{dx^3} = -\frac{2\alpha T_o}{hL}\left(\frac{x}{L}\right) \tag{5}$$

From Eq. (6.2-9) the normal stress is found to be

$$\sigma = E\alpha T_o\left[\frac{1}{3} + \frac{y}{h} - \frac{1}{4}\left(\frac{2y}{h} + 1\right)^2\right]\left(\frac{x^2}{L^2}\right) \tag{6}$$

and from Eq. (4) of Example (a), Sect. 6.4, the shear stress is

$$\tau = E\alpha T_o\left[\frac{1}{6} - \frac{2}{3}\left(\frac{y}{h}\right)^2\right]\frac{xy}{L^2} \tag{7}$$

Example (g) Fixed-Pinned Beam with Axial Temperature Variation

Fig. 6.5-3 shows the coordinate system that is used. The transverse and axial temperature variation in the rectangular beam is given as

$$T = \frac{T_o}{4}\left(\frac{2y}{h} + 1\right)^2\left(\frac{x^2}{L^2}\right) \tag{1}$$

Integration of the thermal beam deformation equation, Eq. (6.3-10), subject to the same boundary conditions as in Example (c), yields the deflection and its derivatives as listed below

$$v = \frac{\alpha T_o L^2}{h}\left(\frac{x^2}{8L^2} - \frac{x^3}{24L^3} - \frac{x^4}{12L^4}\right) \tag{2}$$

$$\frac{dv}{dx} = \frac{\alpha T_o L}{h}\left(\frac{x}{4L} - \frac{x^2}{8L^2} - \frac{x^3}{3L^3}\right) \tag{3}$$

$$\frac{d^2v}{dx^2} = \frac{\alpha T_o}{h}\left(\frac{1}{4} - \frac{x}{4L} - \frac{x^2}{L^2}\right) \tag{4}$$

$$\frac{d^3v}{dx^3} = -\frac{2\alpha T_o}{hL}\left(\frac{1}{8} + \frac{x}{L}\right) \tag{5}$$

The axial stress, from Eq. (6.2-9) is

$$\sigma = E\alpha T_o\left[\frac{x^2}{3L^2} + \frac{y}{h}\left(\frac{x^2}{L^2} + \frac{x}{4L} - \frac{1}{4}\right) - \frac{1}{4}\left(\frac{2y}{h} + 1\right)^2\left(\frac{x^2}{L^2}\right)\right] \tag{6}$$

and from Eq. (4) of Example (a), Sect. 6.4, we obtain the shear stress

$$\tau = \frac{E\alpha T_o}{3}\left[1 + \frac{2y}{h} - \frac{1}{4}\left(\frac{2y}{h} + 1\right)^3 - 3\left(\frac{1}{4} - \frac{y^2}{h^2}\right)\left(\frac{L}{8x} + 1\right)\right]\frac{xh}{L^2} \tag{7}$$

Example (h)   Fixed-Fixed Beam with Axial Temperature Variation

Fig. 6.5-7 shows the manner in which the beam is supported.

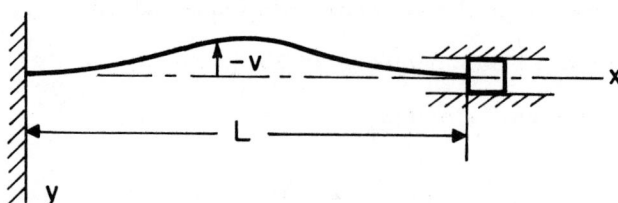

Fig. 6.5-7

SOLUTIONS OF THERMAL BEAM EQUATIONS

The temperature distribution is

$$T = \frac{T_o}{4}\left(\frac{2y}{h} + 1\right)^2\left(\frac{x^2}{L^2}\right) \tag{1}$$

The deflection and its derivatives obtained by integrating Eq. (6.3-10) subject to the boundary conditions $v(o) = v(L) = v'(o) = v'(L) = 0$, are listed below.

$$v = -\frac{\alpha T_o x^2}{12h}\left(1 - \frac{2x}{L} + \frac{x^2}{L^2}\right) \tag{2}$$

$$\frac{dv}{dx} = -\frac{\alpha T_o x}{6h}\left(1 - \frac{3x}{L} + \frac{2x^2}{L^2}\right) \tag{3}$$

$$\frac{d^2 v}{dx^2} = -\frac{\alpha T_o}{h}\left(\frac{x^2}{L^2} - \frac{x}{L} + \frac{1}{6}\right) \tag{4}$$

$$\frac{d^3 v}{dx^3} = -\frac{-2\alpha T_o}{hL}\left(\frac{x}{L} - \frac{1}{2}\right) \tag{5}$$

From Eq. (6.2-9) we obtain the axial stress as

$$\sigma = E\alpha T_o\left[\left(\frac{x^2}{L^2} - \frac{x}{L} + \frac{1}{6}\right)\frac{y}{h} - \frac{1}{4}\left(\frac{2y}{h} + 1\right)^2 \frac{x^2}{L^2} + \frac{x^2}{3L^2}\right] \tag{6}$$

and from Eq. (4) of Example (a), Sect. 6.4, we obtain the shear stress as

$$\tau = \frac{E\alpha T_o}{3}\left[1 + \frac{2y}{h} - \frac{1}{4}\left(\frac{2y}{h} + 1\right)^3 - 3\left(\frac{1}{4} - \frac{y^2}{h^2}\right)\left(1 - \frac{2L}{x}\right)\right]\frac{xh}{L^2} \tag{7}$$

In Example (b) of Sect. 2.3 it was shown that a clamped beam with a transverse temperature distribution does not deflect. In Example (d) of this section it is verified that a clamped beam with a transverse parabolic temperature distribution does not deflect. We find, however, in Example (h) of this section that a clamped beam which has both a transverse and axial temperature variation, does deflect. The reason for this behavior is clear.

Consider first the case of the clamped beam which has a transverse temperature variation only. We first remove the end restraints and permit the beam to deform freely. The resulting curvature will be uniform and equal to $-\frac{\alpha}{I} M_t$ throughout the beam. We then apply restoring end moments of sufficient magnitude to eliminate the end rotations. These restoring end moments produce a uniform curvature in the beam which is exactly equal and opposite to the curvature, $-\frac{\alpha}{I} M_t$, produced by the transverse temperature distribution. The restoring end moments therefore have the effect of restoring the beam to its original undeflected form. On the other hand, when a beam has an axial temperature variation superposed on the transverse variation, the curvature resulting from unrestrained thermal deformation is not uniform over the length of the beam, and the end moments required to restore the horizontal tangencies at the beam ends do not produce curvatures that cancel the free thermal curvatures at all points in the beam.

In Example (h) of this section, we note that the deflection of the clamped beam is upward even though the bottom of the beam is at a higher temperature that the top at all points along the length of the beam. This may at first appear to be odd, since a simply supported beam having a temperature distribution in which the bottom of the beam is hotter than the top, will deflect downward. The direction in which a clamped beam, with a combined transverse and axial temperature variation deflects, depends upon the axial variation. Consider the case of a clamped beam with a symmetrical axial temperature variation superposed on a transverse temperature variation in which the temperature increases monotonically from top to bottom. We find that when the mean transverse temperature gradient of the central area of the clamped beam is higher than the mean gradient near the ends, the beam will deflect downward; and when the mean temperature gradient near the ends of the beam is higher than the mean gradient of the central portion, the clamped beam will deflect upward. This deflection response of a clamped beam to combined transverse and axial temperature variations becomes clear when the effect of end restoring moments, superposed on the unrestrained thermal deflection curve, is considered. When the unrestrained beam curvature is greater at the center than at the ends, the restoring moments at the ends will not alter the direction of the free deflection. When the unrestrained beam curvature is greater at the ends, the restoring moments at the ends will reverse the direction of the free deflection.

Example (i)   Beam with High Mid-Span Temperature Gradient

A fixed-fixed rectangular beam has a transverse and axial temperature distribution expressed as

$$T = \frac{T_o y}{h} \sin \frac{\pi x}{L} \tag{1}$$

It is seen that the temperature increases from top to bottom, and that the temperature gradient at the center is greater than the temperature gradient at the ends.

Fig. 6.5-8 shows the beam coordinate system and the final

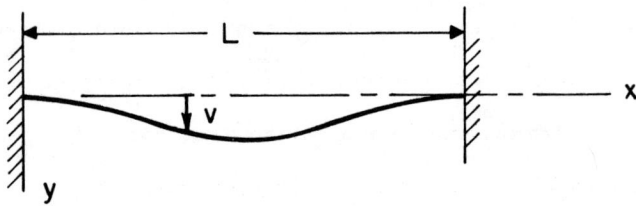

Fig. 6.5-8

deflected shape of the beam. The mean temperature and thermal moment are

$$\overline{T} = 0 \qquad M_t = \frac{T_o I}{h} \sin \frac{\pi x}{L} \qquad (2)$$

The statically indeterminate form of the thermal beam equation is

$$\frac{d^2 v}{dx^2} = -\frac{\alpha}{I} M_t + A x + B \qquad (3)$$

and the boundary conditions are

$$v(o) = v(L) = v'(o) = v'(L) = 0 \qquad (4)$$

Integration yields the deflection as

$$v = \frac{\alpha T_o L^2}{\pi h} \left(\frac{1}{\pi} \sin \frac{\pi y}{L} + \frac{x^2}{L^2} - \frac{x}{L} \right) \qquad (5)$$

and with the use of Eq. (6.2-9) we obtain the stress distribution

$$\sigma = -\frac{2}{\pi} E \alpha T_o \left(\frac{y}{h} \right) \qquad (6)$$

At the center of the beam, $x = \frac{L}{2}$, the deflection is

$$v_{L/2} = \frac{\alpha T_o L^2}{\pi h} \left(\frac{1}{\pi} - \frac{1}{4} \right) \qquad (7)$$

which is positive, signifying that the beam deflects downward. This is consistent with the observation that when the unrestrained curvature at the center is larger than at the ends, the beam will deflect in the same direction as the free deflection. The free deflection is obviously downward, as the bottom is everywhere hotter than the top.

Example (j) Beam with High End Temperature Gradient

A fixed-fixed rectangular beam has a combined transverse and axial temperature distribution

$$T = \frac{T_o y}{h}\left(1 - \sin\frac{\pi x}{L}\right) \tag{1}$$

which shows that the bottom of the beam is everywhere hotter than the top. At the ends of the beam the temperature gradients are greater than at the center. The temperature distribution thus produces a greater free beam curvature at the extremities. We express the temperature as the sum of two parts. Thus

$$T = T_1 + T_2 \qquad T_1 = \frac{T_o y}{h} \qquad T_2 = -\frac{T_o y}{h}\sin\frac{\pi x}{L} \tag{2}$$

The temperature distribution T_1 produces no deflection in the clamped beam since it is a transverse temperature variation. The stress due to T_1 is therefore

$$\sigma_1 = -E\alpha T_1 = -\frac{E\alpha T_o y}{h} \tag{3}$$

The deflection and stress due to T_2 are of the same magnitude as in the preceding Example (i), but of opposite sign. Thus

$$v_2 = -\frac{\alpha T_o L^2}{\pi h}\left(\frac{1}{\pi}\sin\frac{\pi x}{L} + \frac{x^2}{L^2} - \frac{x}{L}\right) \tag{4}$$

and the stress is

$$\sigma_2 = \frac{2}{\pi} E\alpha T_o \left(\frac{y}{h}\right) \tag{5}$$

The net deflection and stress are therfore

$$v = v_1 + v_2 = -\frac{\alpha T_o L^2}{\pi h}\left(\frac{1}{\pi}\sin\frac{\pi x}{L} + \frac{x^2}{L^2} - \frac{x}{L}\right) \tag{6}$$

$$\sigma = \sigma_1 + \sigma_2 = \left(\frac{2}{\pi} - 1\right) E\alpha T_o \left(\frac{y}{h}\right) \tag{7}$$

Fig. 6.6-9 shows the beam deflection curve. It is seen that the

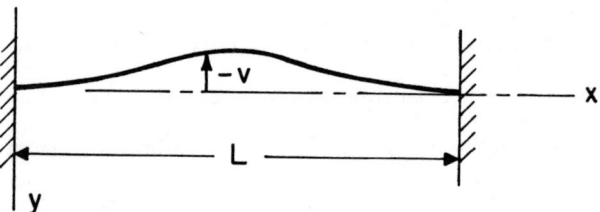

Fig. 6.6-9

clamped beam now deflects in the negative (upward) direction, even though the lower part of the beam is everywhere at a higher temperature than the upper part. The restoring end moments, in this case, have to counteract the large slopes produced by the large end curvatures. The large end moments required to equilibrate the large end slopes cause the center of the beam to be pushed up beyond the straight position, so that it deflects in the direction opposite to the free deflection.

6.6 TWO-PLANE BENDING

A temperature distribution which is a function of the transverse, lateral, and axial coordinates, y, z and x, will produce bending about an axis which goes through the centroid and is inclined to the principal axes of inertia. The principal axes of inertia (those relative to which the products of inertia are zero) are taken as the y and z axes, as, for example, in Fig. 6.4-1. The temperature is specified as being a function of the y and z coordinates but is also permitted to vary slowly with the axial coordinate, x. As is commonly done in extensional and bending analysis, the assumption is made that plane sections remain plane, the rotation of the planes taking place about the y and z axes. In lieu of Eq. (6.2-3), which applies to bending about the z axis only, the stress-strain relationship for biaxial bending is

$$\epsilon_e + \frac{y}{R_z} + \frac{z}{R_y} = \frac{\sigma}{E} + \alpha T \qquad (6.6\text{-}1)$$

We multiply this equation by an element of cross section area, dA, and integrate over the section area, noting that about the centroidal axes,

$$\int_A y\,dA = 0 \qquad \int_A z\,dA = 0$$

and that in the absence of axial loads,

$$\int_A \sigma\,dA = 0$$

The integration gives the uniform extensional strain ϵ_e as

$$\epsilon_e = \alpha \overline{T} \qquad (6.6\text{-}2)$$

where $\overline{T} = \overline{T}(x)$ represents the mean cross section temperature.

Eq. (6.6-1) is multiplied by ydA and integrated, noting that a principal axis of inertia is a centroidal axis, so that

$$\int_A \epsilon_e y dA = 0$$

and as the products of inertia about the principal axes is zero, we have

$$\frac{1}{R_y} \int_A yz dA = 0$$

The integration gives

$$\frac{1}{R_z} = -\frac{d^2 v}{dx^2} = \frac{M_z}{EI_z} + \frac{\alpha}{I_z} M_{tz} \qquad (6.6-3)$$

with the subscript z denoting the axis about which bending is taking place and about which the various moments are computed. In a similar manner the equation for bending about the y axis is obtained. It is

$$\frac{1}{R_y} = -\frac{d^2 w}{dx^2} = \frac{M_y}{EI_y} + \frac{\alpha}{I_y} M_{ty} \qquad (6.6-4)$$

The expressions for the radii of curvature, Eqs. (6.6-3) and (6.6-4) and the extensional strain, Eq. (6.6-2) are substituted into Eq. (6.6-1). This gives the stress distribution as

$$\sigma = E\alpha \left(\overline{T} - \frac{y}{\alpha} \frac{d^2 v}{dx^2} - \frac{z}{\alpha} \frac{d^2 w}{dx^2} - T \right) \qquad (6.6-5)$$

or in terms of the moments it is

$$\sigma = E\alpha \left(\overline{T} + \frac{M_{tz} y}{I_z} + \frac{M_{ty} z}{I_y} - T \right) + \frac{M_z y}{I_z} + \frac{M_y z}{I_y} \qquad (6.6-6)$$

In a statically determinate beam M_z and M_y are zero, and the foregoing equation becomes

$$\sigma = E\alpha \left(\overline{T} + \frac{M_{tz} y}{I_z} + \frac{M_{ty} z}{I_y} - T \right) \tag{6.6-7}$$

In practice, the solution of a two-plane bending problem is carried out by solving the two separate bending problems formulated by Eqs. (6.6-3) and (6.6-4), each with its own deflection, slope, moment, and shear boundary conditions. The curvatures in the x-z and x-y planes are determined and set into Eq. (6.6-5) to obtain the stress.

6.7 BOUNDARY LOAD SUPERPOSITION

A commonly used procedure in thermal stress analysis, when dealing with statically indeterminate problems, is to remove sufficient boundary restraints to transform the problem into one which is statically determinate. This will permit the body to undergo free thermal expansion. Boundary loads, which are of sufficient mganitude to eliminate the boundary displacements produced by the unrestrained thermal expansion, are then applied at the points where the redundant boundary restraints are removed. The boundary conditions are thereby satisfied; and the stresses produced in the body are the sum of the stresses caused by the unrestrained expansion, and those caused by the applied boundary loads.

The treatment of a problem by a two-step procedure, obtaining first the thermal stresses in an unrestrained body, and then adding the stresses due to the restraints, is a procedure that must be followed in applying the corrections due to the presence of a biaxial state of stress. These corrections, given in Sects. 2.4 and 6.1, apply only to the thermal stresses in the unrestrained body.

The stresses due to the reactions or restraints are not modified. The superposition method is therefore essential for statically indeterminate beams in which a correction is called for.

When boundary restraints are removed to transform a statically indeterminate beam into one that is statically determinate, the thermal deformation is unrestrained, and the curvature of the beam at every point is given by the equation

$$\frac{d^2 v}{dx^2} = -\frac{\alpha}{I} M_t \tag{6.7-1}$$

The slopes and deflections are readily determined at all points in the beam by direct integration. The undetermined boundary loads are applied and the slopes and deflections produced by these loads are determined. Solutions of the latter problems are often well-known, and may be found in handbooks. The slopes and deflections obtained from the free deformation problem and those obtained from the undetermined boundary loads are superposed and the boundary loads determined by satisfying the boundary conditions. The deflection of the beam v, is the sum of the unrestrained thermal deflection v_1, and the deflection v_2 produced by the boundary loads. The stress in accordance with Eqs. (6.2-9) and (6.2-11) then follows as $\sigma = \sigma_1 + \sigma_2$ with

$$\sigma_1 = E^*\alpha(\overline{T} + \frac{M_t y}{I} - T) \qquad \sigma_2 = Ey\frac{d^2 v_2}{dx^2} \tag{6.7-2}$$

Example (a)   Fixed-Fixed Beam

A fixed-fixed rectangular beam has a temperature distribution

$$T = \frac{T_o y}{h} \sin\frac{\pi x}{L} \tag{1}$$

BOUNDARY LOAD SUPERPOSITION

As shown in Fig. 6.7-1, the boundary restraints at the right end of the beam

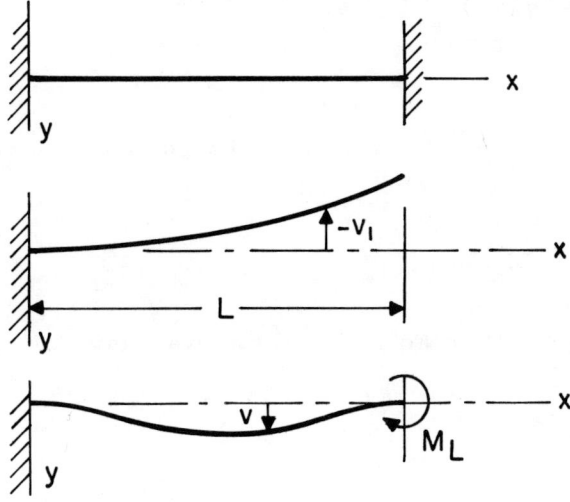

Fig. 6.7-1

are removed and an indeterminate moment, M_L, is applied at that boundary. An indeterminate shear load is not required since the symmetry of the temperature distribution, and consequently the symmetry of deformation indicates that there will be no such load. The mean temperature and thermal moment are

$$\bar{T} = 0 \qquad M_t = \frac{T_o I}{h} \sin \frac{\pi x}{L} \qquad (2)$$

The free thermal deformation equation is

$$\frac{d^2 v_1}{dx^2} = -\frac{\alpha}{I} M_t = -\frac{\alpha T_o}{h} \sin \frac{\pi x}{L} \qquad (3)$$

The boundary conditions for the statically determinate cantilever beam are $v_1(o) = v_1'(o) = 0$. Integration of Eq. (3) yields

$$v_1 = \frac{\alpha T_o L}{\pi h} \left(\frac{L}{\pi} \sin \frac{\pi x}{L} - x \right) \qquad (4)$$

At $x = L$ the slope of the cantilever beam is

$$\left(\frac{dv_1}{dx}\right)_L = -\frac{2\alpha T_o L}{\pi h} \tag{5}$$

The end rotation in a cantilever beam produced by an end moment M_L is

$$\left(\frac{dv_2}{dx}\right)_L = \frac{M_L L}{EI} \tag{6}$$

Equating the slopes, from Eqs. (5) and (6), we obtain

$$M_L = \frac{2EI\alpha T_o}{\pi h} \tag{7}$$

The deflection due to M_L is

$$v_2 = \frac{M_L x^2}{2EI} = \frac{\alpha T_o x^2}{\pi h} \tag{8}$$

and the net deflection is

$$v = v_1 + v_2 = \frac{\alpha T_o L^2}{\pi h}\left(\frac{1}{\pi}\sin\frac{\pi x}{L} - \frac{x}{L} + \frac{x^2}{L^2}\right) \tag{9}$$

From Eqs. (6.7-2), the unmodified stress is

$$\sigma = -\frac{2E\alpha T_o y}{\pi h} \tag{10}$$

Example (b)   Beam on Three Supports

A rectangular beam on three simple supports has a temperature distribution

$$T = \frac{T_o y}{h}\sin\frac{\pi x}{L} \tag{1}$$

Fig. 6.7-2 shows the central support removed to make the problem

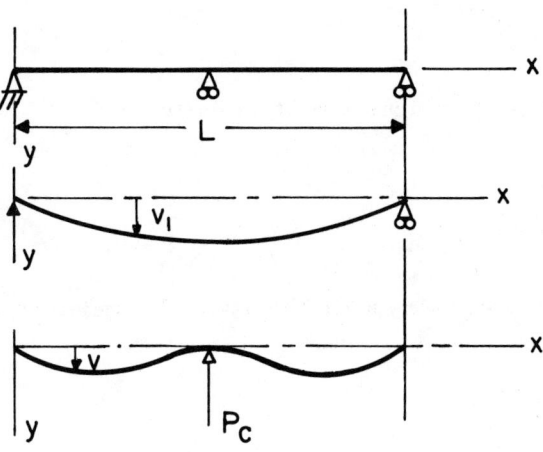

Fig. 6.7-2

statically determinate. An indeterminate boundary load P_c is applied at this point. The mean temperature and thermal moment are

$$\overline{T} = 0 \qquad M_t = \frac{T_o I}{h} \sin \frac{\pi x}{L} \qquad (2)$$

The free deformation thermal beam equation is

$$\frac{d^2 v}{dx^2} = -\frac{\alpha}{I} M_t = -\frac{\alpha T_o}{h} \sin \frac{\pi x}{L} \qquad (3)$$

Integration of the foregoing equation, using the boundary conditions $v_1(0) = v_1(L) = 0$ yields

$$v_1 = \frac{\alpha T_o L^2}{\pi^2 h} \sin \frac{\pi x}{L} \qquad (4)$$

with the central deflection equal to

$$v_{1c} = \frac{\alpha T_o L^2}{\pi h} \tag{5}$$

The load P_c produces a central deflection

$$v_{2c} = \frac{P_c L^3}{48 EI} \tag{6}$$

Equating v_{1c} and v_{2c} we obtain the statically indeterminate load P_c as

$$P_c = \frac{48 EI \alpha T_o}{\pi^2 h L} \tag{7}$$

The deflection due to P_c (from any handbook) is

$$v_2 = -\frac{P_c x}{48 EI}(3L^2 - 4x^2) = -\frac{\alpha T_o x}{\pi^2 h L}(3L^2 - 4x^2) \qquad 0 < x < \frac{L}{2} \tag{8}$$

The net deflection is

$$v = v_1 + v_2 = \frac{\alpha T_o L^2}{\pi^2 h}\left(\sin\frac{\pi x}{L} - \frac{3x}{L} + \frac{4x^3}{L^3}\right) \qquad 0 < x < \frac{L}{2} \tag{9}$$

and the stress in the left span, $0 < x < \frac{L}{2}$, is

$$\sigma = -\frac{24}{\pi^2} E \alpha T_o \left(\frac{x}{L}\right)\left(\frac{y}{h}\right) \tag{10}$$

The deflection and stress are symmetrically distributed in the right span.

In a wide beam, requiring a correction for biaxiality of stress, the correction would be applied only to σ_1, obtained from the first of Eqs. (6.7-2). To this σ_2 from the second equation would be added to obtain the net stress.

6.8 EQUIVALENT LOAD AND MOMENT AREA ANALYSIS

The solution of beam problems by means of equivalent thermal loads permits the direct application of the methods of beam analysis that are commonly used in the solution of isothermal beam problems. The equivalent thermal loads act on the beam in the same manner as real loads, and when real loads are actually present they are lumped with the thermal loads. The solution of the problem, under the action of the combined loads, gives directly the true beam deflection and the displacement stresses, σ'. The net stresses are then obtained from the relationship

$$\sigma = \sigma' - E\alpha T \qquad (6.8\text{-}1)$$

In the discussion which follows, the primed quantities represent displacement stresses and displacement loads. Some of the thermal load parameters that are employed in equivalent thermal beam analysis are: the thermal thrust, N_t; the thermal moment M_t; the displacement thermal thrust, N_t'; the displacement thermal moment, M_t'. They are defined as

$$N_t = \int_A T dA \qquad M_t = \int_A Ty dA \qquad (6.8\text{-}2)$$

$$N_t' = E\alpha N_t = E\alpha \int_A T dA \qquad M_t' = E\alpha M_t = E\alpha \int_A Ty dA \qquad (6.8\text{-}3)$$

with N_t' and M_t' representing equivalent thermal loads.

The beam is assumed to have a temperature distribution $T(x, y)$, with a slow rate of variation of the temperature in the axial direction. Fig. 6.8-1 shows a length of beam on which there are acting both thermal and real loads. The distributed thermal body forces are $-E\alpha \frac{\partial T}{\partial x}$, and the thermal surface

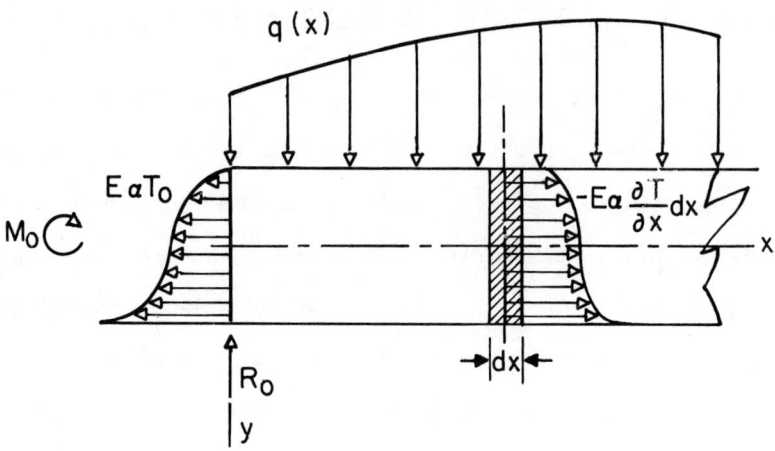

Fig. 6.8-1

body forces are $-E\alpha \frac{\partial T}{\partial x} dx$, and the thermal surface tractions, shown acting on the left side of the beam, are $E\alpha T_o$. Thus

$$X' = -E\alpha \frac{\partial T}{\partial x} \qquad \overline{X}'_o = E\alpha T_o \qquad (6.8-4)$$

Also shown are a real distributed transverse load q(x) and an end moment and shear M_o and R_o.

A body thermal moment m'_t, is caused by the distributed thermal body forces. The body forces, X', from Eq. (6.8-4), are multiplied by their distance from the centroid, y, and integrated over the cross section area. This yields the body thermal moment as

$$m'_t = \int_A X' y dA = -E\alpha \int_A \frac{\partial T}{\partial x} y dA = -E\alpha \frac{\partial}{\partial x} \int_A Ty dA = -E\alpha \frac{\partial M_t}{\partial x} \qquad (6.8-5)$$

The end moment, resulting from the surface tractions is

$$M'_{to} = \int_A \overline{X}' y dA = E\alpha \int_A T_o y dA = E\alpha M_{to} \qquad (6.8-6)$$

The net displacement thermal moment at any beam section, at a distance x from the end, is

$$M'_t = M'_{to} - \int_o^x m'_t \, d\xi = E\alpha M_{to} - \int_o^x -E\alpha \frac{\partial M_t}{\partial \xi} d\xi = E\alpha M_t \qquad (6.8-7)$$

The displacement moment M' is the sum of the displacement thermal moment, $E\alpha M_t$, and the real moment $M = M_o + R_o x + M(q)$. Thus

$$M' = E\alpha M_t + M = E\alpha M_t + M_o + R_o x + M(q) \qquad (6.8-8)$$

with

$$M(q) = -\int_o^x q(\xi) \frac{(x-\xi)^2}{2} d\xi \qquad (6.8-9)$$

The displacement stresses and loads are related to the strains and deformations in accordance with the stress-strain and load-deformation equations that apply to isothermal beams. The displacement moment M' is therefore given as

$$M' = -EI \frac{d^2 v}{dx^2} \qquad (6.8-10)$$

and from Eq. (6.8-8) we have

$$M' = -EI \frac{d^2 v}{dx^2} = M'_t + M_o + R_o x + M(q) \qquad (6.8-11)$$

As in isothermal beam analysis the displacement bending stress is

$$\sigma'_b = \frac{M' y}{I} \qquad (6.8-12)$$

In the case of a statically determinate beam with no real transverse loading, we have $q = R_o = M_o = 0$, and the displacement bending stress becomes

$$\sigma'_b = E\alpha M_t \frac{y}{I} \qquad (6.8-13)$$

The distributed displacement body load, n'_t, is

$$n'_t = \int_A X' dA = -E\alpha \frac{\partial}{\partial x} \int_A T dA = -E\alpha A \frac{dT}{dx} \qquad (6.8\text{-}14)$$

and the surface thermal displacement load is

$$N'_{to} = \int_A \overline{X}' dA = E\alpha \int_A T_o dA = E\alpha A \overline{T}_o \qquad (6.8\text{-}15)$$

The extensional displacement stress at any location x is the sum of the surface displacement load and the distributed body loads, n'_t, divided by the cross section area, A. Thus

$$\sigma'_e = \frac{1}{A}(N'_{to} - \int_0^x n'_t d\xi) = E\alpha \overline{T}_o + \int_0^x E\alpha \frac{d\overline{T}}{d\xi} d\xi = E\alpha \overline{T}(x) \qquad (6.8\text{-}16)$$

The net displacement stress is

$$\sigma' = \sigma'_e + \sigma'_b = E\alpha \overline{T} + \frac{M'y}{I} \qquad (6.8\text{-}17)$$

and the net stress is

$$\sigma = \sigma' - E\alpha T = E\alpha(\overline{T} - T) + \frac{M'y}{I} \qquad (6.8\text{-}18)$$

with M' representing the moment due to the equivalent body forces and surface tractions, and any real loads that may be present. As $M' = EI\frac{d^2v}{dx^2}$ Eq. (6.8-18) can be written as

$$\sigma = E\alpha(\overline{T} - \frac{y}{\alpha}\frac{d^2v}{dx^2} - T) \qquad (6.8\text{-}19)$$

which is the expression for stress derived earlier in Sect. 6.2.

The equivalent load method af analysis permits the direct determination of statically indeterminate boundary loads and stresses in beams without the need of obtaining first the beam deflections. In moment-area methods and conjugate beam methods the beam is loaded with the M/EI diagram. In the present analysis, the real and thermal loads produce a moment M', and the beam is loaded with the M'/EI diagram. The thermal loading component of M'/EI includes the thermal surface traction $E\alpha M_{to}$ and the distributed thermal body moments $-E\alpha \frac{dM_t}{dx}$. The sum of these, as shown by Eq. (6.8-7), is $M'_t = E\alpha M_t(x)$, so that the thermal part of the M'/EI diagram is always

$$\frac{M'_t}{EI} = \frac{E\alpha M_t}{EI} = \frac{\alpha}{I} M_t \qquad (6.8\text{-}20)$$

The other parts of the M'/EI diagram are due to support reactions and real beam loading.

Example (a) Fixed-Pinned Beam

A fixed-pinned rectangular beam has a temperature distribution

$$T = \frac{T_o}{4} \left(\frac{2y}{h} + 1\right)^2 \frac{x}{L} \qquad (1)$$

The mean temperature and thermal moment are

$$\overline{T} = \frac{T_o}{3} \frac{x}{L} \qquad M_t = \frac{T_o I}{h} \cdot \frac{x}{L} \qquad (2)$$

The displacement thermal moment and thermal moment area are

$$M'_t = \frac{E\alpha T_o I}{h} \cdot \frac{x}{L} \qquad \frac{M'_t}{EI} = \frac{\alpha T_o}{h} \cdot \frac{x}{L} \qquad (3)$$

376 EQUIVALENT LOAD AND MOMENT AREA ANALYSIS

Fig. 6.8-2(b) and (c) show the thermal beam loading and thermal moment

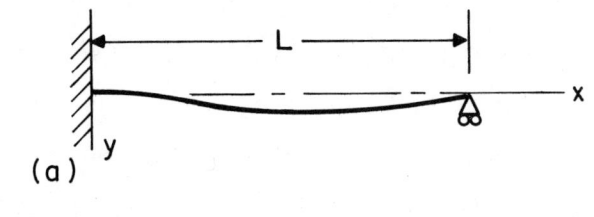

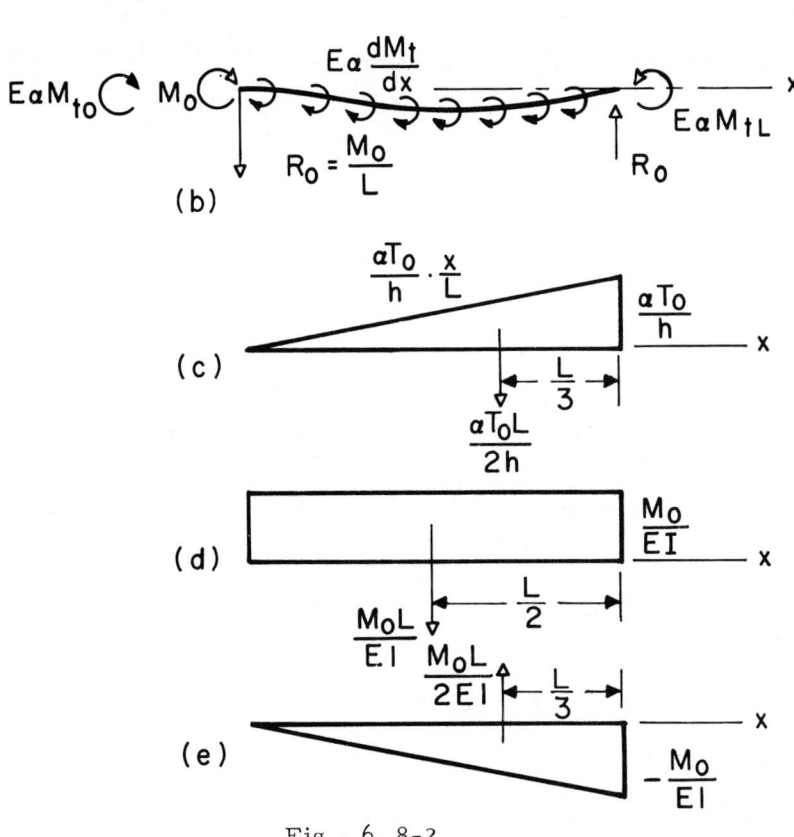

Fig. 6.8-2

area. Fig. 6.8-2(d) and (e), show the moment areas due to the end moment and shear, M_o and M_o/L. As there is no deflection at the right end of the beam and the slope at the left end is zero, the sum of the moments of the areas (c), (d), and (e), about the right end of the beam must be zero. Thus

$$\frac{\alpha T_o L}{2h} \cdot \frac{L}{3} + \frac{M_o L}{EI} \cdot \frac{L}{2} = \frac{M_o L}{2EI} \cdot \frac{L}{3} \tag{4}$$

EQUIVALENT LOAD AND MOMENT AREA ANALYSIS   377

from which

$$M_o = -\frac{E\alpha T_o I}{2h} \qquad R_o = -\frac{E\alpha T_o I}{2hL} \tag{5}$$

Using Eq. (6.8-8) we have

$$M' = \frac{E\alpha T_o I}{h} \cdot \frac{x}{L} - \frac{E\alpha T_o I}{2h} + \frac{E\alpha T_o I}{2h} \cdot \frac{x}{L} = \frac{E\alpha T_o I}{2h}\left(\frac{3x}{L} - 1\right) \tag{6}$$

From Eq. (6.8-12) the displacement bending stress is

$$\sigma_b' = \frac{M'y}{I} = \frac{E\alpha T_o y}{2h}\left(\frac{3x}{L} - 1\right) \tag{7}$$

The net stress is then given by Eq. (6.8-18) as

$$\sigma = E\alpha T_o\left[\frac{x}{3L} + \left(\frac{3x}{L} - 1\right)\frac{y}{2h} - \frac{1}{4}\left(\frac{2y}{h} + 1\right)\frac{x}{L}^2\right] \tag{8}$$

Example (b)   Fixed-Fixed Beam

A fixed-fixed rectangular beam has a temperature distribution

$$T = \frac{T_o y}{h}\sin\frac{\pi x}{L} \tag{1}$$

The mean temperature and temperature moment are

$$\overline{T} = 0 \qquad M_t = \frac{T_o I}{h}\sin\frac{\pi x}{L} \tag{2}$$

and the displacement thermal moment and thermal moment area are

$$M_t' = \frac{E\alpha T_o I}{h}\sin\frac{\pi x}{L} \qquad \frac{M_t'}{EI} = \frac{\alpha T_o}{h}\sin\frac{\pi x}{L} \tag{3}$$

Fig. 6.8-3(b) and (c) shows the thermal beam loading and the associated

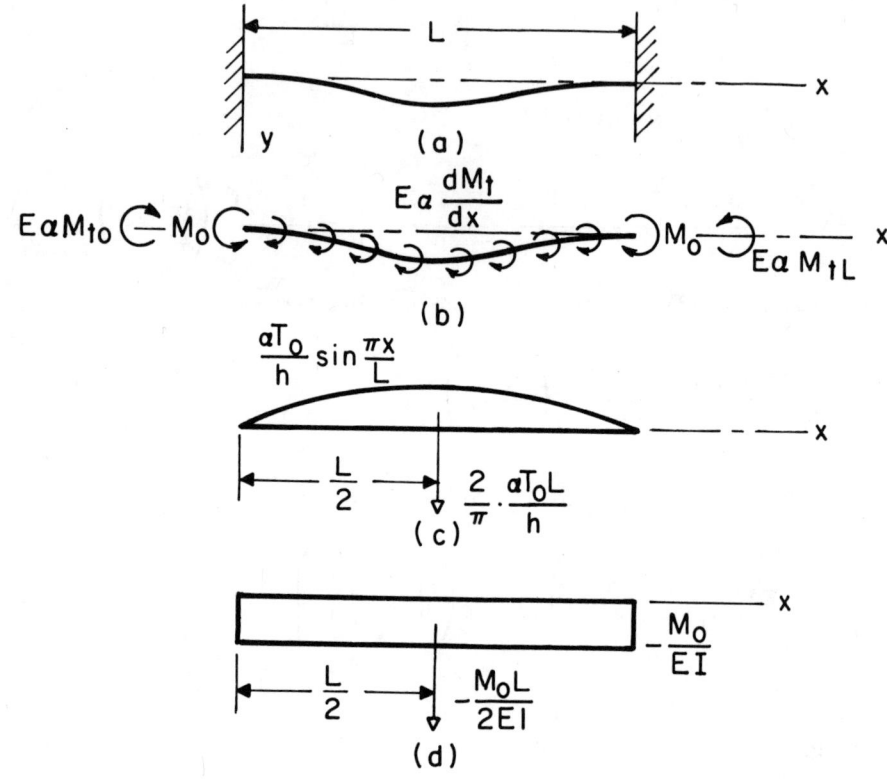

Fig. 6.8-3

moment area. Fig. 6.3-8(d) shows the moment area due to the end moment M_o. The sum of the areas shown in (c) and (d) must be zero since the slopes at both ends of the beam are zero. We note that the area under a sine curve is

$$\text{Area} = \frac{2}{\pi} \cdot \text{amplitude} \cdot \text{length} = \frac{2}{\pi} \cdot \frac{\alpha T_o}{h} \cdot L \tag{4}$$

The addition of areas yields

$$-\frac{M_o L}{EI} + \frac{2}{\pi} \frac{\alpha T_o}{h} \cdot L = 0 \tag{5}$$

from which we obtain M_o as

$$M_o = \frac{2}{\pi} \frac{E I \alpha T_o}{h} \tag{6}$$

The displacement moment is

$$M' = \frac{E \alpha T_o I}{h} \sin \frac{\pi x}{L} - \frac{2}{\pi} \frac{E \alpha T_o I}{h} \tag{7}$$

and the displacement bending stress and net stress are

$$\sigma_b' = \frac{M' y}{I} = \frac{E \alpha T_o y}{h} \left(\sin \frac{\pi x}{L} - \frac{2}{\pi} \right) \tag{8}$$

and

$$\sigma = E \alpha (\overline{T} - T) + \frac{M' y}{I} = \frac{2}{\pi} E \alpha T_o \cdot \frac{y}{h} \tag{9}$$

Example (c) Beam on Three Supports

A rectangular beam on three simple supports has a temperature distribution

$$T = \frac{T_o y}{h} \sin \frac{\pi x}{L} \tag{1}$$

As in the preceding problem the thermal parameters are

$$\overline{T} = 0 \qquad M_t = \frac{T_o I}{h} \sin \frac{\pi x}{L} \tag{2}$$

$$M_t' = \frac{E \alpha T_o I}{h} \sin \frac{\pi x}{L} \qquad \frac{M_t'}{EI} = \frac{\alpha T_o}{h} \sin \frac{\pi x}{L} \tag{3}$$

380 EQUIVALENT LOAD AND MOMENT AREA ANALYSIS

Fig. 6.8-4(b) and (c) show the beam loading and the thermal moment area. Fig. 6.8-4(d) shows the moment area due to the end reaction. As the slope is zero at the center of the span, the moment of the areas shown in (c) and (d), about the left support must add up to zero.

We note that the center of gravity of a half sine curve area lies at a distance $\frac{L}{\pi}$ from the end. Taking moments, we obtain

$$\frac{2}{\pi} \cdot \frac{\alpha T_o}{h} \cdot \frac{L}{2} \cdot \left(\frac{L}{2} - \frac{1}{\pi}\right) = \frac{1}{2} \cdot \frac{L}{2} \cdot \frac{RL}{4EI} \cdot \frac{L}{3} \tag{4}$$

which yields the center reaction, R, as

$$R = \frac{48 E \alpha T_o I}{\pi^2 h L} \tag{5}$$

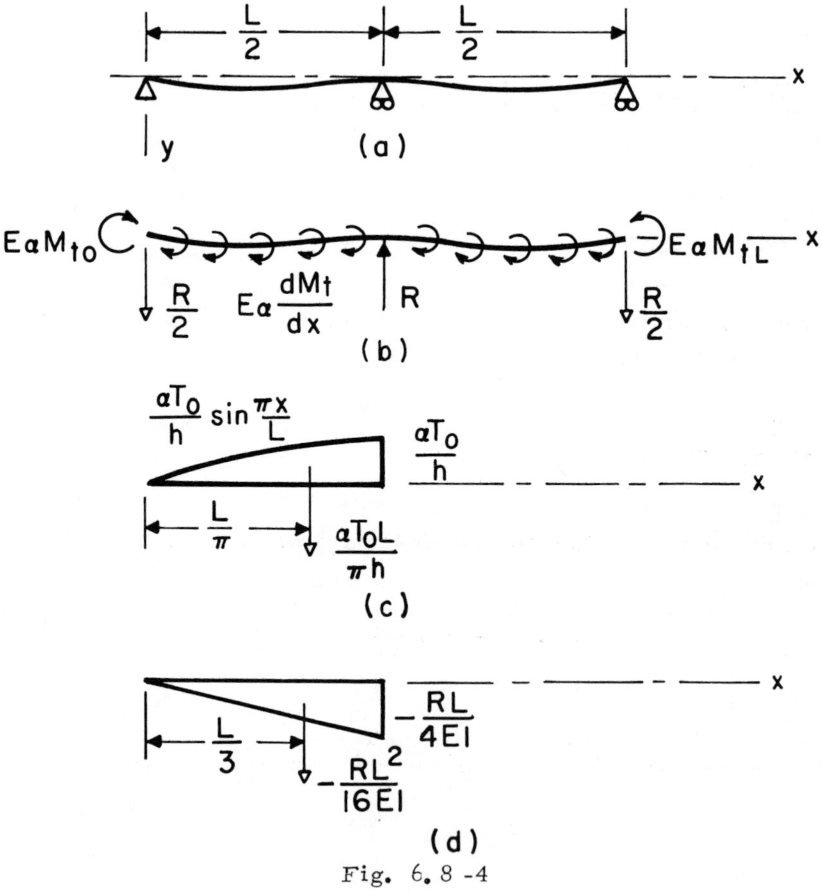

Fig. 6.8-4

The displacement moment is

$$M' = M'_t - \frac{R}{2}x = \frac{E\alpha T_o I}{h}\sin\frac{\pi x}{L} - \frac{24 E\alpha T_o I}{\pi^2 h} \cdot \frac{x}{L} \tag{6}$$

and the displacement bending stress is

$$\sigma'_b = \frac{M'y}{I} = \frac{E\alpha T_o y}{h}\left(\sin\frac{\pi x}{L} - \frac{24}{\pi^2} \cdot \frac{x}{L}\right) \tag{7}$$

The net stress is

$$\sigma = E\alpha(\overline{T} - T) + \frac{M'y}{I} = -\frac{24}{\pi^2} E\alpha T_o \cdot \frac{y}{h} \cdot \frac{x}{L} \tag{8}$$

6.9 COMPLEMENTARY ENERGY SOLUTIONS

The strain energy methods used in isothermal beam analysis can be used in the solution of thermal beam problems. Those that are especially convenient to use in thermal beam analysis are: (1) the complementary energy method; (2) Castigliano's Theorem applied to equivalent thermal beam loading; (3) the thermal reciprocal theorem.

Chapter 5 contains a general discussion of strain energy methods. In this section and in the two sections which follow the application of these methods in this solution of thermal beam problems is demonstrated. We consider first thermal beam solutions based on the principle of minimum complementary energy.

It is assumed that the beam under consideration has a length which is much greater than the depth and that the rate of axial temperature variation is small. The strain energy of shear is then small in comparison to the bending strain energy, and the complementary energy, in accordance with Eq. (5.2-5) is

$$U_1 = \int_0^L \int_A \left(\frac{\sigma^2}{2E} + \alpha T \sigma \right) dA\, dx \tag{6.9-1}$$

The expression for stress in a statically indeterminate thermal beam, in terms of the thermal moment and real moment, is given by Eq. (6.2-10). Substitutions into Eq. (6.9-1) yields

$$U_1 = \int_0^L \int_A \frac{1}{2E} \left[E\alpha\left(\overline{T} + \frac{M_t y}{I} - T\right) + \frac{My}{I} \right]^2 dA \cdot dx \\ + \int_0^L \int_A \alpha T \left[E\alpha\left(\overline{T} + \frac{M_t y}{I} - T\right) + \frac{My}{I} \right] dA \cdot dx \tag{6.9-2}$$

Let R_i and M_i represent indeterminate shears and moments, in terms of which the moment M is expressed. The principle of stationary (minimum) complementary energy requires that

$$\frac{\partial U_1}{\partial R_i} = 0 \qquad \frac{\partial U_1}{\partial M_i} = 0 \tag{6.9-3}$$

Performing this operation on Eq. (6.9-2) we obtain

$$\frac{\partial U_1}{\partial R_i} = \int_0^L \int_A \left[E\alpha\left(\overline{T} + \frac{M_t y}{I}\right) + \frac{My}{I} \right] \frac{y}{I} \cdot \frac{\partial M}{\partial R_i} \cdot dA \cdot dx = 0 \tag{6.9-4}$$

and

$$\frac{\partial U_1}{\partial M_i} = \int_0^L \int_A \left[E\alpha\left(\overline{T} + \frac{M_t y}{I}\right) + \frac{My}{I} \right] \frac{y}{I} \cdot \frac{\partial M}{\partial M_i} \cdot dA \cdot dx = 0 \tag{6.9-5}$$

As $\frac{\partial M}{\partial R_i}$ and $\frac{\partial M}{\partial M_i}$ are functions of x only and as

$$\int_A \overline{T} y \, dA = 0 \qquad \int_A \frac{y^2 dA}{I} = 1 \qquad (6.9-6)$$

the integration over the beam cross section area, of Eqs. (6.9-4) and (6.9-5) yields

$$\int_0^L M \frac{\partial M}{\partial R_i} dx = -E\alpha \int_0^L M_t \frac{\partial M}{\partial R_i} dx \qquad (6.9-7)$$

$$\int_0^L M \frac{\partial M}{\partial M_i} dx = -E\alpha \int_0^L M_t \frac{\partial M}{\partial M_i} dx \qquad (6.9-8)$$

These are the complementary beam equations, which are utilized directly to obtain indeterminate thermal beam reactions and moments.

Example (a) Fixed-Pinned Beam

A fixed-pinned rectangular beam has a temperature distribution

$$T = \frac{T_o}{4} \left(\frac{2y}{h} + 1 \right)^2 \frac{x^2}{L^2} \qquad (1)$$

The beam has only one indeterminate reaction. It is taken as M_o, as shown in Fig. 6.9-1. The real moment and thermal moment are

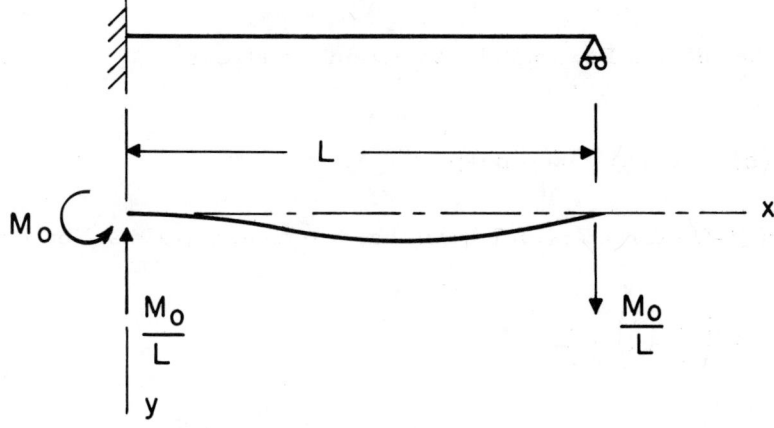

Fig. 6.9-1

$$M = M_o\left(\frac{x}{L} - 1\right) \qquad M_t = \frac{T_o I}{h} \frac{x^2}{L^2} \tag{2}$$

The partial derivative of the real moment with respect to the indeterminate moment M_o is

$$\frac{\partial M}{\partial M_i} = \frac{\partial M}{\partial M_o} = \frac{x}{L} - 1 \tag{3}$$

This is set into Eq. (6.9-8) which becomes

$$M_o \int_o^L \left(\frac{x}{L} - 1\right)^2 dx = -\frac{E\alpha T_o I}{hL^2} \int_o^L \frac{x^2}{L^2} \left(\frac{x}{L} - 1\right) dx \tag{4}$$

and the performance of the integration yields the end moment M_o as

$$M_o = \frac{E\alpha T_o I}{4h} \tag{5}$$

The moment in the beam, in accordance with Eq. (2) is

$$M = \frac{E\alpha T_o I}{4h} \left(\frac{x}{L} - 1\right) \tag{6}$$

and it can be set into Eq. (6.2-10) to obtain the stress.

Example (b) Fixed-Fixed Beam

A fixed-fixed rectangular beam has a temperature distribution

$$T = \frac{T_o}{4} \left(\frac{2y}{h} + 1\right)^2 \frac{x^2}{L^2} \tag{1}$$

This beam has two indeterminate reactions, M_o and R_o as shown in

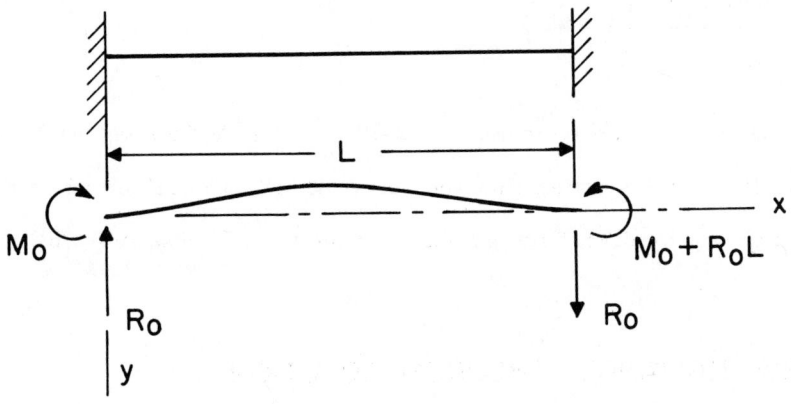

Fig. 6.9-2

Fig. 6.9-2. The real moment and thermal moment are

$$M = M_o + R_o x \qquad M_t = \frac{T_o I}{h} \frac{x^2}{L^2} \qquad (2)$$

The partial derivatives of the real moment with respect to M_o and R_o are

$$\frac{\partial M}{\partial M_o} = 1 \qquad \frac{\partial M}{\partial R_o} = x \qquad (3)$$

Substitution into Eqs. (6.9-7) and (6.9-8) yields respectively

$$\int_o^L (R_o x + M_o)(1) dx = -\frac{E\alpha T_o I}{h} \int_o^L \left(\frac{x^2}{L^2}\right)(1) dx \qquad (4)$$

$$\int_o^L (R_o x + M_o)(x) dx = -\frac{E\alpha T_o I}{h} \int_o^L \left(\frac{x^2}{L^2}\right)(x) dx \qquad (5)$$

Carrying out the integration and solving simultaneously for M_o and R_o yields

$$M_o = \frac{E\alpha T_o I}{6h} \qquad R_o = -\frac{E\alpha T_o I}{hL} \qquad (6)$$

These values are set into Eq. (2) and the moment is found to be

$$M = \frac{E\alpha T_o I}{h}\left(\frac{1}{6} - \frac{x}{L}\right) \qquad (7)$$

The stress is obtained from Eq. (6.2-10), using M, above, and M_t from Eq. (2). It can be verified that the same result is obtained as in the solution of this problem by direct integration, in Sect. 6.5, Example (h).

6.10 CASTIGLIANO'S THEOREM SOLUTIONS

Castigliano's theorem may be applied to a beam loaded with equivalent thermal loads, or a combination of equivalent thermal loads and real loads, to determine beam displacements and indeterminate boundary reactions. Since equivalent thermal loads produce true body displacements, Castigliano's theorem will give the true beam deflections and rotations when the thermal loads are treated as though they were real loads.

We define the bending strain energy due to the combined thermal and real loads, as U_t. If P_i and M_i represent either real loads or dummy loads at some beam location, then, in accordance with Castigliano's theorem, the deflection and (or) rotation at these points is given by the partial derivative of the thermal strain energy (inclusive of real loads) with respect to the actual or dummy loads P_i and M_i. The deflections and rotations are

$$\delta_i = \frac{\partial U_t}{\partial P_i} \qquad \theta_i = \frac{\partial U_t}{\partial M_i} \qquad (6.10-1)$$

In order to find statically indeterminate shears and moments the partial derivatives with respect to the indeterminate reactions, say P_i and M_i are set equal to zero. Thus when there is no deflection and no rotation,

Eqs. (6.10-1) become

$$\frac{\partial U_t}{\partial P_i} = 0 \qquad \frac{\partial U_t}{\partial M_i} = 0 \qquad (6.10-2)$$

The displacement bending moment M' is defined, as before, as the sum of the thermal and real moments. That is

$$M' = M'_t + M = E\alpha M_t + M \qquad (6.10-3)$$

with the moments due to dummy loads included in M. The bending strain energy due to the thermal and real (plus dummy) loads is therefore

$$U_t = \frac{1}{2} \int_0^L \frac{(M')^2}{EI} \, dx \qquad (6.10-4)$$

Eq. (6.10-4) is set into Eqs. (6.10-1). These yield δ_i and θ_i as

$$\delta_i = \int_0^L \frac{M'}{EI} \cdot \frac{\partial M'}{\partial P_i} \, dx \qquad \theta_i = \int_0^L \frac{M'}{EI} \cdot \frac{\partial M'}{\partial M_i} \, dx \qquad (6.10-5)$$

To compute the statically indeterminate loads we set Eq. (6.10-4) into Eqs. (6.10-2) and obtain the expressions

$$\int_0^L \frac{M'}{EI} \cdot \frac{\partial M'}{\partial P_i} dx = 0 \qquad \int_0^L \frac{M'}{EI} \cdot \frac{\partial M'}{\partial M_i} dx = 0 \qquad (6.10-6)$$

The expression for M', Eq. (6.10-3), is set into Eqs. (6.10-5). The result, for a statically indeterminate beam, is

$$\delta_i = \int_0^L \left(\frac{M}{EI} + \frac{a}{I} M_t\right) \frac{\partial M}{\partial P_i} \, dx \qquad \theta_i = \int_0^L \left(\frac{M}{EI} + \frac{a}{I} M_t\right) \frac{\partial M}{\partial M_i} \, dx \qquad (6.10-7)$$

In the case of a statically determinate beam in which M = 0, the foregoing expressions become

$$\delta_i = \int_0^L \frac{\alpha}{I} M_t \cdot \frac{\partial M}{\partial P_i} dx \qquad \theta_i = \int_0^L \frac{\alpha}{I} M_t \cdot \frac{\partial M}{\partial M_i} dx \qquad (6.10-8)$$

Dummy loads, which are set equal to zero after the partial derivatives are taken, may be used in Eqs. (6.10-7) and (6.10-8) to determine deflections and rotations at desired locations. It is noted that when a dummy load is employed to find a deflection or rotation in a statically indeterminate beam, the moments due to the dummy load should be calculated for the statically determinate beam. This is discussed in Sect. 5.3.

To obtain the indeterminate shears and moments, $M' = E\alpha M_t + M$, is set into Eqs. (6.10-6). Assuming that EI is not a function of x, we obtain

$$\int_0^L M \frac{\partial M}{\partial P_i} dx = -E\alpha \int_0^L M_t \frac{\partial M}{\partial P_i} dx \qquad (6.10-9)$$

and

$$\int_0^L M \frac{\partial M}{\partial M_i} dx = -E\alpha \int_0^L M_t \frac{\partial M}{\partial M_i} dx \qquad (6.10-10)$$

Note that these equations coincide with Eqs. (6.9-7) and (6.9-8) derived by means of the stationary complementary energy principle.

Example (a) Pinned-Pinned Beam

We wish to obtain the rotation of the left end of the simply supported rectangular beam, shown in Fig. 6.10-1, which has a temperature distribution

$$T = \frac{T_o y}{h} \sin \frac{\pi x}{L} \qquad (1)$$

Fig. 6.10-1 shows the simple supported beam with a dummy moment,

CASTIGLIANO'S THEOREM SOLUTIONS

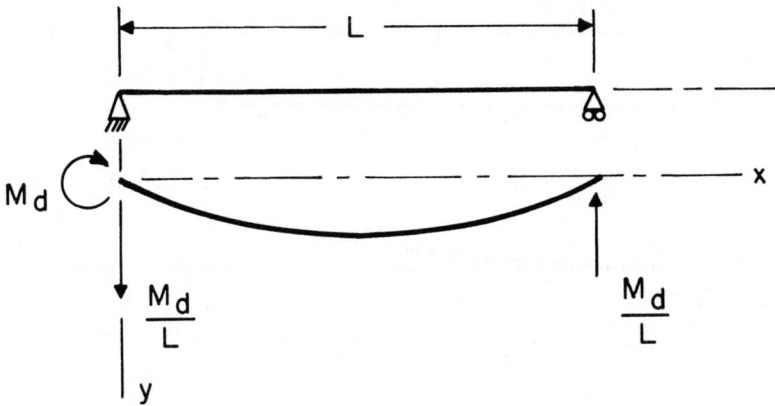

Fig. 6.10-1

M_d, at the left end. The thermal moment and moment due to M_d are

$$M_t = \frac{T_o I}{h} \sin \frac{\pi x}{L} \qquad M = M_d \left(1 - \frac{x}{L}\right) \qquad (2)$$

From the second of Eqs. (6.10-8), with $M_i = M_d$, we have

$$\theta_o = \frac{\alpha T_o}{h} \int_o^L \left(\sin \frac{\pi x}{L}\right)\left(1 - \frac{x}{L}\right) dx = \frac{\alpha T_o L}{\pi h} \qquad (3)$$

Example (b)   End Deflection of a Cantilever Beam

We wish to obtain the thermal deflection of the right end of the rectangular cantilever beam shown in Fig. 6.10-2. The temperature distribution is

$$T = \frac{T_o}{4}\left(\frac{2y}{h} + 1\right)^2 \frac{x^2}{L^2} \qquad (1)$$

Fig. 6.10-2 shows a dummy load P_d at the beam end. The thermal moment

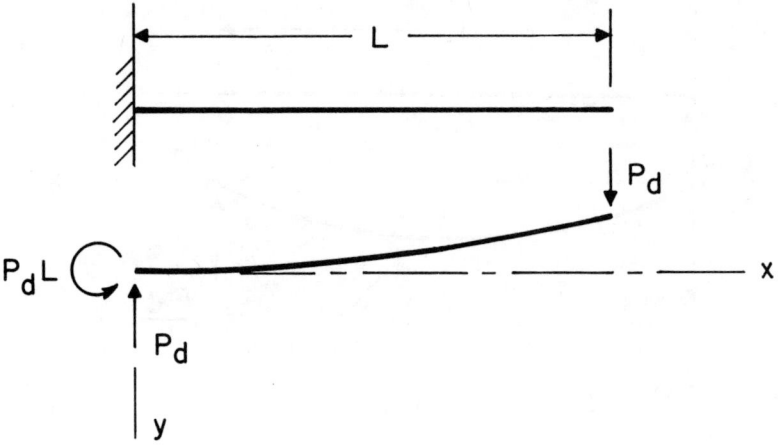

Fig. 6.10-2

M_t and the moment M due to the dummy load P_d are

$$M_t = \frac{T_o I}{h} \cdot \frac{x^2}{L^2} \qquad M = P_d(x-L) \tag{2}$$

Using the first of Eqs. (6.10-8) we have

$$\delta_L = \frac{\alpha T_o}{h} \int_o^L \left(\frac{x^2}{L^2}\right)(x-L)dx = -\frac{\alpha T_o L^2}{12h} \tag{3}$$

Example (c) Deflection of Fixed-Fixed Beam

We wish to obtain the deflection of the center of a fixed-fixed rectangular beam whose temperature distribution is specified as

$$T = \frac{T_o y}{h} \sin \frac{\pi x}{L} \tag{1}$$

Fig. 6.10-3(a) shows the beam with indeterminate end moments M_o.

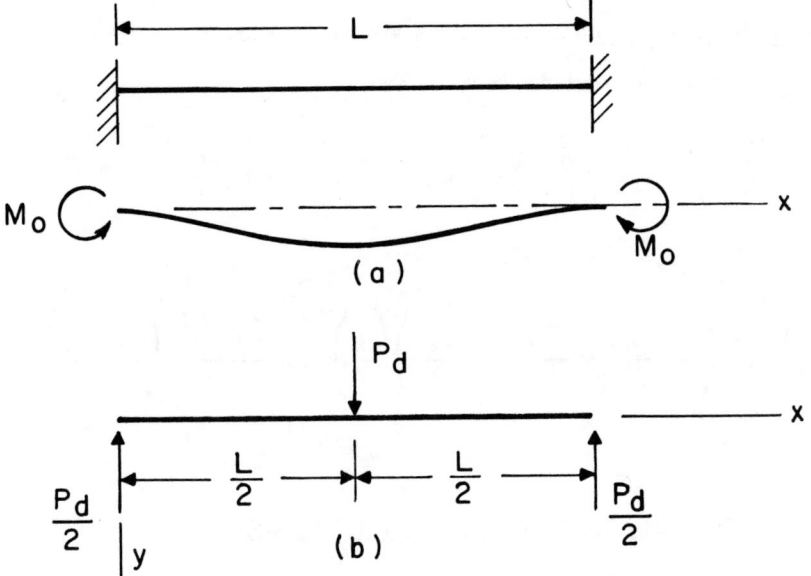

Fig. 6.10-3

We first obtain M_o, as follows: the thermal moment, the real moment, and the real moment derivative are

$$M_t = \frac{T_o I}{h} \sin \frac{\pi x}{L} \qquad M = -M_o \qquad \frac{\partial M}{\partial M_o} = -1 \qquad (2)$$

From Eq. (6.10-10) with $M_i = M_o$, we have

$$\int_0^L (-M_o)(-1) = -\frac{E\alpha T_o I}{h} \int_0^L (\sin \frac{\pi x}{L})(-1)dx \qquad (3)$$

which yields M_o as

$$M_o = \frac{2E\alpha T_o I}{\pi h} \qquad (4)$$

To find the center deflection a dummy load P_d is placed at the center of the statically determinate beam, as shown in Fig. 6.10-3(b). The moment due to P_d and the moment derivative are

$$M = \frac{P_d x}{2} \qquad \frac{\partial M}{\partial P_d} = \frac{x}{2} \qquad 0 < x < \frac{L}{2} \tag{5}$$

From the first of Eqs. (6.10-7), we have

$$\delta_c = 2 \int_0^{\frac{L}{2}} \left(-\frac{2\alpha T_o}{\pi h} + \frac{\alpha T_o}{h} \sin \frac{\pi x}{L} \right) \left(\frac{x}{2} \right) dx = \frac{\alpha T_o L^2}{\pi h} \left(-\frac{1}{4} + \frac{1}{\pi} \right) \tag{6}$$

6.11 RECIPROCAL THEOREM SOLUTIONS

The thermal reciprocal theorem was discussed in Section 5.4. As shown there, the reciprocal theorem affords a convenient means to compute deflections and volume changes, particularly in those cases where the solution of the isothermal auxiliary load problem is known. The thermal reciprocal theorem gives the thermally induced displacements explicitly as the volume integral of the product of the temperature distribution and the sum of the principal stresses. In the case of beams it is utilized to find deflections and rotations at designated points, and it can be used to find the deflection curve of the beam.

The reader is referred to Section 5.4. The quantity θ^* is designated as the sum of the principal stresses, and in a beam, θ^* is equal to the auxiliary bending stress, σ^*, generated by auxiliary loads P_o^* or M_o^*, applied at the beam location where the deflection or rotation is desired. Assuming a constant α, Eqs. (5.4-11) and (5.4-12) give the beam deflection, δ, or beam rotation, φ, as

$$\delta = \frac{\alpha}{P_o^*} \int_V T\Theta^* dV = \alpha \int_V T\frac{\Theta^*}{P_o^*} dV = \alpha \int_V T\frac{\partial \sigma^*}{\partial P_o^*} dV \qquad (6.11\text{-}1)$$

$$\varphi = \frac{\alpha}{M_o^*} \int_V T\Theta^* dV = \alpha \int_V T\frac{\Theta^*}{M_o^*} dV = \alpha \int_V T\frac{\partial \sigma^*}{\partial M_o^*} dV \qquad (6.11\text{-}2)$$

In the isothermal beam loaded with the auxiliary loads, the auxiliary stress and the stress derivatives are

$$\sigma^* = \frac{M^* y}{I} \qquad \frac{\partial \sigma^*}{\partial P_o^*} = \frac{\partial M^*}{\partial P_o^*} \frac{y}{I} \qquad \frac{\partial \sigma^*}{\partial M_o^*} = \frac{\partial M^*}{\partial M_o^*} \frac{y}{I} \qquad (6.11\text{-}3)$$

Substitution of the foregoing stress derivatives into Eqs. (6.11-1) and (6.11-2), and assuming I constant, we have

$$\delta = \alpha \int_V T\frac{\partial M^*}{\partial P_o^*} \cdot \frac{y}{I} dV = \frac{\alpha}{I} \int_o^L \frac{\partial M^*}{\partial P_o^*} dx \int_A Ty dA = \frac{\alpha}{I} \int_o^L M_t \frac{\partial M^*}{\partial P_o^*} dx \qquad (6.11\text{-}4)$$

$$\varphi = \alpha \int_V T\frac{\partial M^*}{\partial M_o^*} \frac{y}{I} dV = \frac{\alpha}{I} \int_o^L \frac{\partial M^*}{\partial M_o^*} dx \int_A Ty dA = \frac{\alpha}{I} \int_o^L M_t \frac{\partial M^*}{\partial M_o^*} dx \qquad (6.11\text{-}5)$$

It is important to note that M^*, in a statically indeterminate beam, is the distributed moment in the beam due to the auxiliary loads P_o^* or M_o^*, applied to the <u>statically indeterminate</u> beam. This is at variance with the procedure used in solutions by means of Castigliano's theorem, in which the dummy loads are applied to the <u>statically determinate</u> beam. Eqs. (6.11-4) and (6.11-5) are comparable to the expressions for deflection and rotation in a statically indeterminate beam, derived by means of Castigliano's theorem, i.e., Eq. (6.10-7), but the significance of the terms M and M^* are not the same. For statically determinate beams, however, the reciprocal theorem and Castigliano's theorem give the identical equations for deflection and rotation. That is, Eqs. (6.11-4) and (6.11-5) when applied to statically determinate beams are identical with Eqs. (6.10-8).

Example (a) End Rotation of a Pinned-Pinned Beam

The temperature distribution in a rectangular beam is specified as

$$T = \frac{T_o}{4}\left(\frac{2y}{h} + 1\right)^2 \frac{x^2}{L^2} \tag{1}$$

Fig. 6.11-1 shows the auxiliary load system for reciprocal theorem analysis.

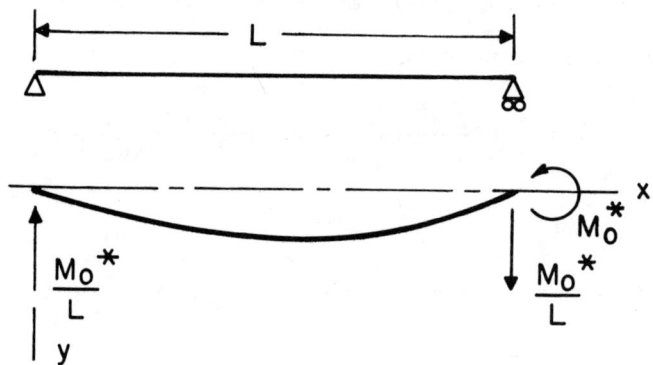

Fig. 6.11-1

An auxiliary moment M_o^* is placed at the right end in order to find the rotation at that point. The thermal moment, the auxiliary moment, and the derivative are

$$M_t = \frac{T_o I}{h} \cdot \frac{x^2}{L^2} \qquad M^* = \frac{M_o^* x}{L} \qquad \frac{\partial M^*}{\partial M_o^*} = \frac{x}{L} \tag{2}$$

Eq. (6.11-5) is used to find the end slope φ_L. Substitution of Eqs. (2) into Eq. (6.11-5) yields

$$\varphi_L = \frac{\alpha T_o}{h} \int_o^L \left(\frac{x^2}{L^2}\right)\left(\frac{x}{L}\right) dx = \frac{\alpha T_o L}{4h} \tag{3}$$

Example (b) Center Deflection of a Fixed-Fixed Beam

The rectangular beam has a temperature distribution

$$T = \frac{T_o y}{h} \sin \frac{\pi x}{L} \tag{1}$$

Fig. 6.11-2 shows the auxiliary load system with the auxiliary load P_o^* placed at the center of the beam, in order to obtain the beam deflection at that point.

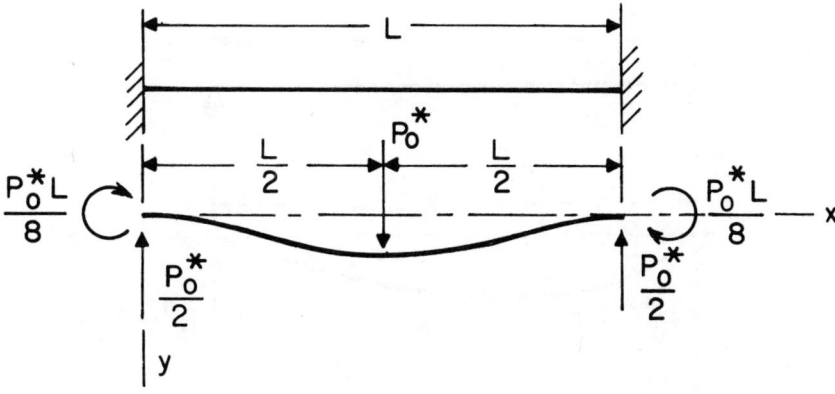

Fig. 6.11-2

Note that P_o^* is acting on the <u>statically indeterminate</u> beam. The thermal moment, the auxiliary moment, and its derivative, for the left half of the span, are

$$M_t = \frac{T_o I}{h} \sin \frac{\pi x}{L} \qquad M^* = P_o^*\left(\frac{x}{2} - \frac{L}{8}\right) \qquad \frac{\partial M^*}{\partial P_o^*} = \frac{x}{2} - \frac{L}{8} \tag{2}$$

Substitutions into Eq. (6.11-4) yields

$$\delta_c = \frac{2\alpha T_o}{h} \int_o^{L/2} \left(\sin\frac{\pi x}{L}\right)\left(\frac{x}{2} - \frac{L}{8}\right) dx = \frac{\alpha T_o L^2}{\pi h}\left(\frac{1}{\pi} - \frac{1}{4}\right) \qquad (3)$$

Example (c) Deflection Curve of Pinned-Pinned Beam

The temperature distribution in the simply supported rectangular beam is specified as

$$T = \frac{T_o}{4}\left(\frac{2y}{h} + 1\right)^2 \frac{x^2}{L^2} \qquad (1)$$

Fig. 6.11-3 shows the auxiliary load system with the auxiliary load P_o^* placed at a variable distance x from the left end. The axial coordinate for

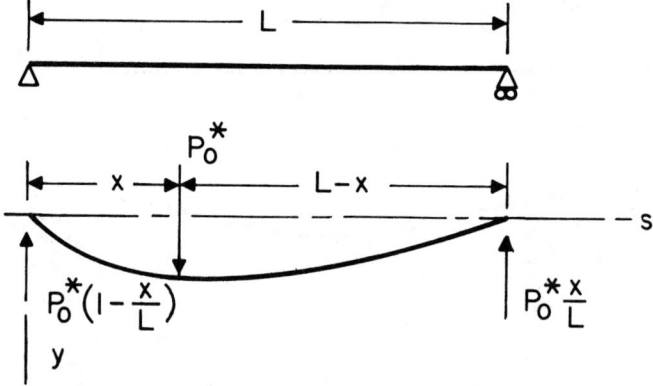

Fig. 6.11-3

integration purposes is s. The thermal moment and auxiliary moments are

$$M_t = \frac{T_o I}{h} \cdot \frac{s^2}{L^2} \qquad M^*_{0<s<x} = P_o^*\left(s - \frac{sx}{L}\right) \qquad M^*_{x<s<L} = P_o^*\left(x - \frac{sx}{L}\right) \qquad (2)$$

Substitution into Eq. (6.11-4) yields

$$v = \frac{\alpha T_o}{L^2 h} \left[\int_o^x (s^2)(s-\frac{sx}{L})ds + \int_x^L (s^2)(x-\frac{sx}{L}) \, ds \right] \quad (3)$$

or

$$v = \frac{\alpha T_o L^2}{12h} \left(\frac{x}{L} - \frac{x^4}{L^4} \right) \quad (4)$$

This example as well as Examples (a) and (b) demonstrate that a minimum amount of computation is required to obtain deflections and rotations by means of the reciprocal theorem. It should be recognized that in each of the examples it was not necessary to compute the stress distribution due to the auxiliary loads, as these were well-known solutions. In cases where the stress distribution due to the auxiliary loads is not known, and must be computed, the use of the reciprocal theorem could require more extensive computations, and would not necessarily involve less labor than some other method.

6.12 CONTINUOUS BEAMS

The solution of a problem of a beam on many supports is often carried out with the use of the three-moment equation. The latter is a relationship between the moments at three adjacent beam supports. Assume that a continuous beam, on n equidistant supports, such as that shown in Fig. 6.12-1, is subjected to an axial and transverse temperature variation $T = T(x, y)$, with the axial temperature gradients assumed to be small.

398 CONTINUOUS BEAMS

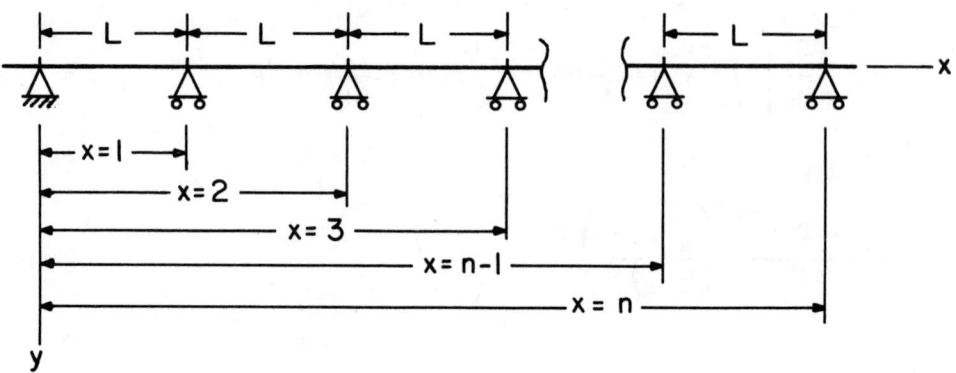

Fig. 6.12-1

In the coordinate system shown in Fig. 6.12-1, x can be only a whole number.

Consider a length of beam over three supports. It is broken up into two simply supported sections, as shown in Fig. 6.12-2, with internal moments M_{x-1}, M_x, and M_{x+1} at the support points.

From Eq. (6.11-5), using the same procedure as in Example (a) of Sect. 6.11, the end rotations

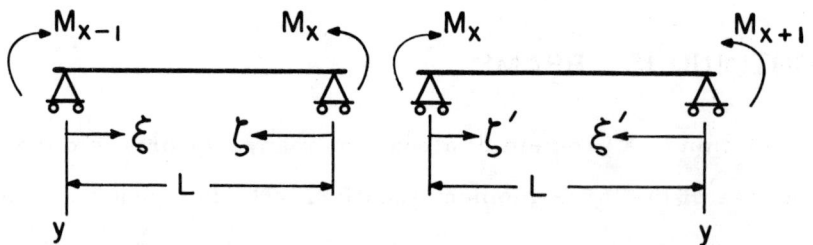

Fig. 6.12-2

at the center support in Fig. 6.12-2, due to the temperature distribution, are

$$\theta_{xt} = -\frac{\alpha}{IL} \int_0^L M_t(\xi) \cdot \xi \, d\xi \qquad \text{(left span)} \qquad (6.12-1)$$

CONTINUOUS BEAMS

$$\theta'_{xt} = +\frac{\alpha}{IL} \int_0^L M_t(\xi') \cdot \xi' \, d\xi' \qquad \text{(right span)} \qquad (6.12\text{-}2)$$

The end rotation at the center support in Fig. 6.12-2, due to the moments M_{x-1}, M_x, and M_{x+1}, are

$$\theta_{xm} = -\frac{L}{6EI}(2M_x + M_{x-1}) \qquad (6.12\text{-}3)$$

$$\theta'_{xm} = \frac{L}{6EI}(2M_x + M_{x+1}) \qquad (6.12\text{-}4)$$

As the angular rotations at the junction of the two spans must coincide we have $\theta_{xt} + \theta_{xm} = \theta'_{xt} + \theta'_{xm}$, or

$$M_{x-1} + 4M_x + M_{x+1} = -\frac{6E\alpha}{L^2}\left[\int_0^L M_t(\xi) \cdot \xi \, d\xi + \int_0^L M_t(\xi') \cdot \xi' \, d\xi'\right] \qquad (6.12\text{-}5)$$

This is the three-moment equation for a thermal beam.

At the left end of the beam we have, for the case of a pin support, $M_{x=0} = 0$. When the beam is clamped at the left end the sum of the slopes due to the thermal moment and real moment must be zero. Eq. (6.12-1) then becomes

$$\theta_{(x=0)t} = \frac{\alpha}{IL} \int_0^L M_t(\zeta) \cdot \zeta \, d\zeta \qquad (6.12\text{-}6)$$

and Eq. (6.12-3) becomes

$$\theta_{(x=0)m} = \frac{L}{6EI}(M_{x=1} + 2M_{x=0}) \qquad (6.12\text{-}7)$$

Since $\theta_{(x=0)t} + \theta_{(x=0)m} = 0$ we have

$$2M_{x=0} + M_{x=1} = -\frac{6E\alpha}{L^2} \int_0^L M_t(\zeta) \cdot \zeta \, d\zeta \qquad (6.12\text{-}8)$$

Eq. (6.12-8) is the boundary condition for a continuous beam clamped at $x = 0$. Similarly we obtain, for a beam clamped at the right end, $x = n$, the boundary relationship

$$2M_{x=n} + M_{x=n-1} = -\frac{6E\alpha}{L^2} \int_0^L M_t(\zeta') \cdot \zeta' \, d\zeta' \qquad (6.12-9)$$

When the temperature is not a function of the axial coordinate, these boundary conditions become

$$2M_{x=0} + M_{x=1} = -3E\alpha M_t \qquad \text{(left end clamped)} \qquad (6.12-10)$$

$$2M_{x=n} + M_{x=n-1} = -3E\alpha M_t \qquad \text{(right end clamped)} \qquad (6.12-11)$$

For a beam which is clamped at both ends and has only a single central support, Eqs. (6.12-5), (6.12-10) and (6.12-11) become

$$M_0 + 4M_1 + M_2 = -6E\alpha M_t \qquad (6.12-12)$$

$$2M_0 + M_1 = -3E\alpha M_t \qquad (6.12-13)$$

$$2M_2 + M_1 = -3E\alpha M_t \qquad (6.12-14)$$

The solution is

$$M_0 = M_1 = M_2 = -E\alpha M_t \qquad (6.12-15)$$

It is clear that any continuous beam with a transverse temperature distribution, which is clamped at both ends and restrained from axial expansion, will have a stress distribution $\sigma = -E\alpha T$, and there will be no deflection.

Transverse Temperature Distribution

For the case where the temperature is a function of the transverse coordinate only the general three-moment equation, Eq. (6.12-5), becomes

CONTINUOUS BEAMS

$$M_{x-1} + 4 M_x + M_{x+1} = -6 E\alpha M_t \qquad (6.12\text{-}16)$$

As the complementary solution of the difference equation we let

$$M_x = C e^{\gamma x} \qquad (6.12\text{-}17)$$

Substitution into the homogeneous equation

$$M_{x-1} + 4 M_x + M_{x+1} = 0 \qquad (6.12\text{-}18)$$

yields

$$C\left(e^{\gamma(x-1)} + 4 e^{\gamma x} + e^{\gamma(x+1)}\right) = 0 \qquad (6.12\text{-}19)$$

from which

$$e_1^{\gamma} = -2 + \sqrt{3} \qquad e_2^{\gamma} = -2 - \sqrt{3} \qquad (6.12\text{-}20)$$

The complementary or general solution is then

$$M_{xg} = C_1 (-2 + \sqrt{3})^x + C_2 (-2 - \sqrt{3})^x \qquad (6.12\text{-}21)$$

For a beam of infinite extent C_2 would have to be zero as otherwise the moment would become infinitely large.

The particular solution is taken as $M_{xp} = K$, a constant. It follows that $M_{x-1,p}$ and $M_{x+1,p}$ must also be equal to K. From Eq. (6.12-16) we have

$$M_{xp} = - E\alpha M_t \qquad (6.12\text{-}22)$$

The complete solution is therefore

$$M_x = M_{xg} + M_{xp} = -E\alpha M_t + C_1(-2 + \sqrt{3})^x + C_2(-2 - \sqrt{3})^x \qquad (6.12\text{-}23)$$

Three Simple Supports

For a beam on three simple supports the boundary conditions are $M_{x-1} = M_{x+1} = 0$. Eq. (6.12-23) then yields $C_1 = C_2 = -\dfrac{E\alpha M_t}{4}$, and the equation becomes

$$M_x = -E\alpha M_t \left[1 + \frac{(-2+\sqrt{3})^x}{4} + \frac{(-2-\sqrt{3})^x}{4} \right] \qquad (6.12\text{-}24)$$

The moment over the center support, $x = 0$, is then $M_o = -\dfrac{3}{2} E\alpha M_t$.

Long Beam on Many Supports

In a very long beam C_2 is zero, and if we let position x-1 represent the end support where the moment is zero, we obtain $C_1 = (-2 + \sqrt{3})\, E\alpha M_t$, and the equation for the long continuous beam on simple supports becomes

$$M_x = -E\alpha M_t \left[1 - (-2 + \sqrt{3})^{x+1} \right] \qquad (6.12\text{-}25)$$

It is observed that at $x = 0$, which is one span length from the end of the beam, the moment over the support is

$$M_o = -E\alpha M_t (3 - \sqrt{3}) = 1.268\, E\alpha M_t \qquad (6.12\text{-}26)$$

As we proceed to the second span and those following it, the support moments oscillate and approach $E\alpha M_t$. Far from the end (four or five span lengths) the moment is constant and equal to $-E\alpha M_t$.

Finite Beam on Simple Supports

In a continuous finite beam having an odd number of simple supports, with M_o at the center support, the moments $M_{-\frac{n}{2}}$ and $M_{+\frac{n}{2}}$, at the extreme ends, are zero. From Eq. (6.12-23) we obtain, for the case of simple end supports,

$$C_1 = C_2 = \frac{(-2+\sqrt{3})^{n/2} - (-2-\sqrt{3})^{n/2}}{(-2+\sqrt{3})^n - (-2-\sqrt{3})^n} E\alpha M_t \qquad (6.12-27)$$

and the moment M_x is

$$M_x = -E\alpha M_t \left\{ 1 - \frac{(-2+\sqrt{3})^{n/2} - (-2-\sqrt{3})^{n/2}}{(-2+\sqrt{3})^n - (-2-\sqrt{3})^n} \left[-2+\sqrt{3})^x + (-2-\sqrt{3})^x \right] \right\}$$
$$(6.12-28)$$

Note than when n is large, the moment at the center of the beam, $x = 0$, is equal to $-E\alpha M_t$.

The deflection curves of the infinite beam and of the finite beam are obtained with the help of the reciprocal theorem. Each of the simply supported spans in Fig. 6.12-2 deforms as a result of the transverse temperature distribution, the end moments, M_x, and the associated reactions. The transverse temperature distribution alone produces a deflection, at the center of any span, which is equal to

$$v_t = \frac{\alpha}{I} \frac{M_t L^2}{8} \qquad (6.12-29)$$

As we move away from the ends the real moment, M_x, at the support points approaches $-E\alpha M_t$. These end moments produce a deflection

$$v_m = -\frac{\alpha}{I} \frac{M_t L^2}{8} \qquad (6.12-30)$$

at the center of a span. The net center deflection, in spans which are not close to the end spans, is therefore $v_t + v_m$ which is equal to zero.

Deflection Curve

In a simply supported continuous beam with a transverse temperature distribution, the maximum deflection occurs in the end spans. The moments at the support points, in a continuous beam over many supports, is given by Eq. (6.12-25). The end moment is zero and the moment over the support which is one span removed from the end is $M_o = -E\alpha M_t (3 - \sqrt{3})$ as given by Eq. (6.12-26). This moment acts on the beam as shown in Fig. 6.12-3.

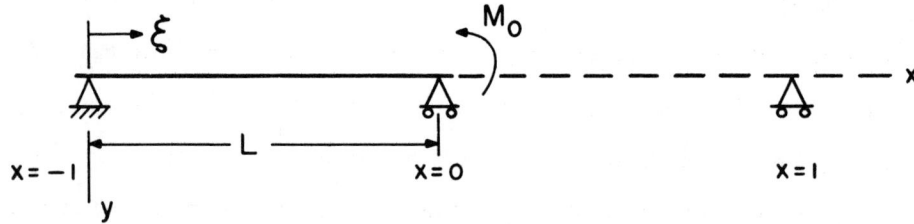

Fig. 6.12-3

To obtain the deflection curve of the end span, which is due to M_o and M_t, we integrate the beam deformation equation

$$\frac{d^2 v}{d\xi^2} = -\frac{\alpha}{I} M_t - \frac{M_o \xi}{EIL} \tag{6.12-31}$$

Integration yields

$$v = \frac{\alpha M_t L^2}{6I} \left[\sqrt{3}\, \frac{\xi}{L} - 3\left(\frac{\xi}{L}\right)^2 + (3 - \sqrt{3})\left(\frac{\xi}{L}\right)^3 \right] \quad 0 < \xi < L \tag{6.12-32}$$

The deflection curves of the other spans may be obtained in a similar manner. We can deduce from Eq. (6.12-25) that the beam deflection falls off very rapidly as we proceed along the beam. For example, at the second support,

$x = 1$, Eq. (6.12-25) shows that $M_{x=1} = -0.928 \, E\alpha M_t$. However, the moment M' that is effective in producing deformation is only $0.072 \, E\alpha M_t$.

One conclusion that we draw is that a continuous beam with a transverse temperature distribution may be assumed to be an infinite beam when it has more than about five supports, without incurring a significant error.

6.13 ELASTIC FOUNDATIONS

Problems of thermal beams on elastic foundations differ from problems of beams on a finite number of supports. The difference is that the equilibrium equations for the distributed elastic support problem is in the form of a differential equation, while the equilibrium equations for a beam on a number of discrete supports are directly integrable. The solutions of problems of beams on elastic foundations, particularly those involving beams of finite length, are therefore analytically more complex.

An element of beam of length dx, loaded with the internal shear loads and moments, having an external distributed load and elastic foundation reaction, is shown in Fig. 6.13-1. The foundation modulus is k (force per unit

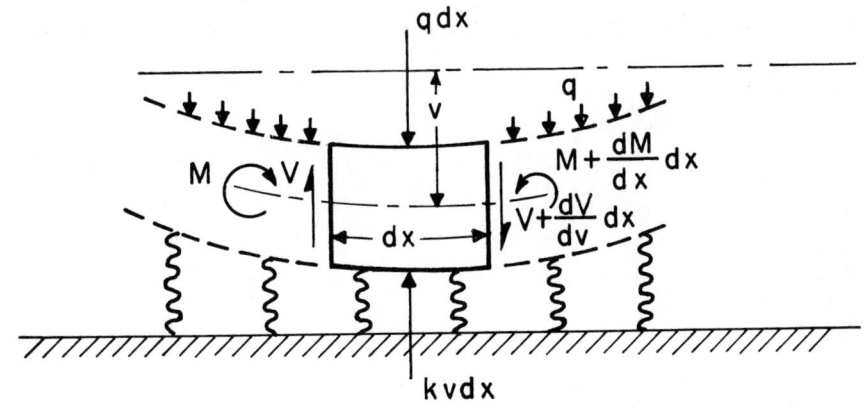

Fig. 6.13-1

length). The foundation reaction load, which is proportional to the beam deflection, v, is therefore kv. The equilibrium of the loads and moments acting on the beam element is expressed as

$$-\frac{dV}{dx} = q - kv \qquad (6.13\text{-}1)$$

and

$$V = \frac{dM}{dx} \qquad (6.13\text{-}2)$$

The moment M, in accordance with Eq. (6.2-8) is

$$M = -EI\frac{d^2v}{dx^2} - E\alpha M_t \qquad (6.13\text{-}3)$$

Substitution of V, from Eq. (6.13-2), into Eq. (6.13-1) yields

$$-\frac{d^2M}{dx^2} + kv = q \qquad (6.13\text{-}4)$$

and setting M. from Eq. (6.13-3), into the foregoing, we obtain

$$\frac{d^4v}{dx^4} + \frac{k}{EI}v = -\frac{\alpha}{I}\frac{d^2M_t}{dx^2} + \frac{q}{EI} \qquad (6.13\text{-}5)$$

The factor k/EI in this equation is defined in terms of a reciprocal relaxation length, β, as

$$\frac{k}{EI} = 4\beta^4 \qquad (6.13\text{-}6)$$

or

$$\beta = \left(\frac{k}{4EI}\right)^{1/4} \qquad (6.13\text{-}7)$$

Eq. (6.13-5) then becomes

$$\frac{d^4v}{dx^4} + 4\beta^4 v = -\frac{\alpha}{I}\frac{d^2 M_t}{dx^2} + \frac{q}{EI} \tag{6.13-8}$$

This is the differential equation for a thermal beam on an elastic foundation. Its solution is

$$v = e^{-\beta x}(A\cos\beta x + B\sin\beta x) + e^{\beta x}(C\cos\beta x + D\sin\beta x) + v_p \tag{6.13-9}$$

with v_p the particular solution.

Four boundary conditions are needed to evaluate the constants, A, B, C, and D. The boundary conditions will generally be in the form of restrained rotations or deflections, or as points of zero moment or zero shear. The latter two conditions are expressed by Eqs. (6.3-7) and (6.3-8) as

$$\frac{d^2 v}{dx^2} = -\frac{\alpha}{I} M_t \qquad (M = 0) \tag{6.13-10}$$

$$\frac{d^3 v}{dx^3} = -\frac{\alpha}{I} \frac{dM_t}{dx} \qquad (V = 0) \tag{6.13-11}$$

Homogeneous Solution

When a transverse applied load is not present, and when $\frac{d^2 M_t}{dx^2} = 0$, the beam differential equation becomes

$$\frac{d^4 v}{dx^4} + 4\beta^4 v = 0 \tag{6.13-12}$$

The solution of this equation is

$$v = e^{-\beta x}(A\cos\beta x + B\sin\beta x) + e^{\beta x}(C\cos\beta x + D\sin\beta x) \tag{6.13-13}$$

Long Beam

In a short beam, the constants A, B, C, and D, are evaluated in accordance with the boundary conditions at the ends of the span. In a long beam the constants C and D must be zero and the solution takes the form

$$v = e^{-\beta x}(A\cos\beta x + B\sin\beta x) \qquad (6.13-14)$$

In order to determine whether a beam is long or short, the quantity βL, which is the ratio of beam length to relaxation length, should be computed. When

$$\beta L > 8 \qquad (6.13-15)$$

the beam can be treated as an infinite or semi-infinite beam, for which the solution, Eq. (6.13-14) is applicable. For $\beta L < 8$ the beam span should be taken as finite, and Eq. (6.13-13) is the appropriate solution.

Semi-Infinite Beam

In the case of a semi-infinite beam on an elastic foundation, shown in Fig. 6.13-2, with $d^2 M_t/dx^2 = 0$, the constants A and B in Eq. (6.13-16) are

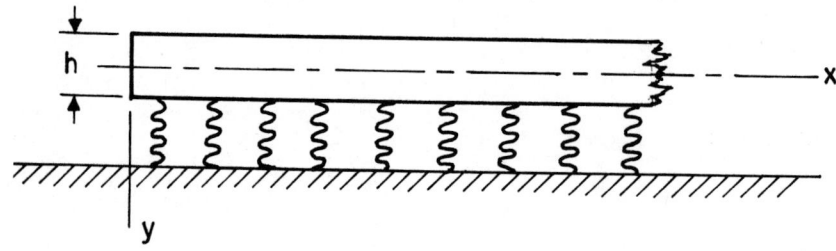

Fig. 6.13-2

determined with the use of the boundary conditions of zero shear and zero moment at $x = 0$, i.e.,

$$\left(\frac{d^2v}{dx^2}\right)_0 = -\frac{\alpha}{I} M_{to} \tag{6.13-16}$$

$$\left(\frac{d^3v}{dx^3}\right)_0 = -\frac{\alpha}{I}\left(\frac{dM_t}{dx}\right)_0 \tag{6.13-17}$$

The beam deflection and the first, second, and third deflection derivatives are then

$$v = \frac{\alpha e^{-\beta x}}{2\beta^3 I}\left[\beta M_{to}(\sin\beta x - \cos\beta x) - \left(\frac{dM_t}{dx}\right)_0 \cos\beta x\right] \tag{6.13-18}$$

$$\frac{dv}{dx} = \frac{\alpha e^{-\beta x}}{2\beta^2 I}\left[2\beta M_{to}\cos\beta x + \left(\frac{dM_t}{dx}\right)_0 (\cos\beta x + \sin\beta x)\right] \tag{6.13-19}$$

$$\frac{d^2v}{dx^2} = -\frac{\alpha e^{-\beta x}}{\beta I}\left[\beta M_{to}(\cos\beta x + \sin\beta x) + \left(\frac{dM_t}{dx}\right)_0 \sin\beta x\right] \tag{6.13-20}$$

$$\frac{d^3v}{dx^3} = \frac{\alpha e^{-\beta x}}{I}\left[2\beta M_{to}\sin\beta x + \left(\frac{dM_t}{dx}\right)_0 (\sin\beta x - \cos\beta x)\right] \tag{6.13-21}$$

The expression for beam curvature, Eq. (6.13-20) is set into Eq. (6.2-9), and we obtain the stress

$$\sigma = E\alpha\left\{\overline{T} - T + \frac{M_{to}y}{I}e^{-\beta x}\left[(\cos\beta x + \sin\beta x) + \frac{1}{\beta M_{to}}\left(\frac{dM_t}{dx}\right)_0 \sin\beta x\right]\right\} \tag{6.13-22}$$

This expression for the beam stress is valid as long as the elastic foundation does not restrain the beam from axial expansion. If it is completely restrained from axial expansion then the term $\overline{T}$ should be set equal to zero.

When the temperature distribution is a function of the transverse coordinate only, $T = T(y)$, then $\frac{dM_t}{dx} = \left(\frac{dM_t}{dx}\right)_0 = 0$, and the maximum deflection, obtained at $x = 0$, is

$$v_o = -\frac{\alpha M_{to}}{2\beta^2 I} = -\frac{\alpha M_t}{2\beta^2 I} \tag{6.13-23}$$

The stress near the end of the beam ($x \approx 0$), from Eq. (6.13-24) is

$$\sigma_o = E\alpha\left(\overline{T} + \frac{M_t y}{I} - T\right) \tag{6.13-24}$$

It is clear that the stress referred to above is not precisely at the extreme end of the beam since at that point the normal stress is obviously zero. Rather it occurs at a distance from the end equal to, more or less, the depth of the beam.

Eq. (6.13-24) is the same as the equation obtained for the stress in an unrestrained beam with a transverse temperature distribution. At locations which are far from the end of the beam, Eq. (6.13-22) shows the stress to be

$$\sigma = E\alpha(\overline{T} - T) \tag{6.13-25}$$

which is the stress in a beam that is completely restrained from bending. This latter stress is the maximum stress in the beam.

Example (a) Free-Free Long ($\beta L > 8$) Rectangular Beam

The temperature distribution in the beam is specified as

$$T = \frac{T_o}{4}\left(\frac{2y}{h} + 1\right)^2\left(1 - \frac{x^2}{L^2}\right) \tag{1}$$

with x the distance measured along the length of the beam. The beam is supported on an elastic foundation as in Fig. 6.13-2. We wish to find the deflection curve of the beam.

The temperature moment and its first and second derivatives are

$$M_t = \frac{T_o I}{h}\left(1 - \frac{x^2}{L^2}\right) \qquad \frac{dM_t}{dx} = -\frac{2T_o I}{hL}\left(\frac{x}{L}\right) \qquad \frac{d^2 M_t}{dx^2} = -\frac{2T_o I}{hL^2} \tag{2}$$

At $x = 0$ we have the boundary conditions for zero shear and zero moment

$$\left(\frac{d^2 v}{dx^2}\right)_0 = -\frac{\alpha}{I} M_{to} = -\frac{\alpha T_o}{h} \qquad \left(\frac{d^3 v}{dx^3}\right)_0 = -\frac{\alpha}{I}\left(\frac{dM_t}{dx}\right)_0 = 0 \tag{3}$$

The pertinent differential equation is

$$\frac{d^4 v}{dx^4} + 4\beta^4 v = -\frac{\alpha}{I}\frac{d^2 M_t}{dx^2} = \frac{2\alpha T_o}{hL^2} \tag{4}$$

and the particular solution is

$$v_p = \frac{\alpha T_o}{2\beta^4 h L^2} \tag{5}$$

Since $\beta L > 8$, the general solution is as given by Eq. (6.13-14), and the complete solution, subject to the boundary conditions (3) is

$$v = \frac{\alpha T_o}{2\beta^4 h L^2} + \frac{\alpha T_o e^{-\beta x}}{2\beta^2 h}(\sin\beta x - \cos\beta x) \tag{6}$$

The first term in Eq. (6) represents the uniform deflection that is produced by a constant value of $d^2 M_t/dx^2$. We can see from Eq. (4) that a constant second derivative of a thermal moment is equivalent to subjecting the beam to a uniformly distributed load. The second term is due to the end thermal moment $M_{to} = T_o I/h$. Since βL is large we find that at $x = 0$ the first term can be omitted since it is smaller than the second term by a factor $1/(\beta L)^2$. Thus at $x = 0$ we have

$$v_o = -\frac{\alpha T_o}{2\beta^2 h} \qquad (7)$$

at a distance far from the end such that $e^{\beta x} \gg (\beta L)^2$, the deflection is given by the first term of Eq. (6). It is

$$v = \frac{\alpha T_o}{2\beta^2 h} \cdot \frac{1}{(\beta L)^2} \qquad e^{\beta x} \gg (\beta L)^2 \qquad (8)$$

The stress is obtained by taking the second derivative of Eq. (6) and setting it into Eq. (6.2-9).

6.14 LAYERED BEAMS

Layered elements in the form of beams are often found in thermostatic devices. The knowledge of the motion of these elements, in response to temperature change, is necessary when designing bimetallic instruments or controls. We examine the behavior of a bimetallic beam subjected to a transverse temperature variation. It is assumed that the beam is of arbitrary cross section but symmetrical about the y (vertical) axis.

Bimetallic Beam

The bimetallic beam constituents have depths h_1 and h_2 and the distance between the centroidal axes is H, as shown in Fig. 6.14-1.

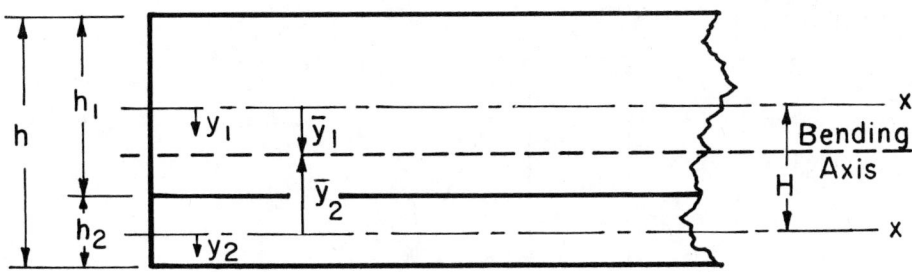

Fig. 6.14-1

Separate coordinates y_1 and y_2 are used, which are measured from the centroidal axes of constituents 1 and 2, respectively. The temperatures, $T_1 = T_1(x, y_1)$ and $T_2 = T_2(x, y_2)$, are taken as slowly varying functions of the axial coordinate, x, and functions of the transverse coordinates y_1 and y_2. The strain is a linear function of the transverse coordinate and is expressed in terms of the centroidal fibre strains, ϵ_{10} and ϵ_{20}, of the constituents, and the flexural strains y_1/R and y_2/R measured from the respective centroidal fibres. Thus

$$\epsilon_1 = \epsilon_{10} + \frac{y_1}{R} = \frac{\sigma_1}{E_1} + \alpha_1 T_1 \qquad (6.14-1)$$

$$\epsilon_2 = \epsilon_{20} + \frac{y_2}{R} = \frac{\sigma_2}{E_2} + \alpha_2 T_2 \qquad (6.14-2)$$

Fig. 6.14-2 shows the geometry of the beam deformation. The axial displacement of a differential element, δu, can be taken as an extension of the centroidal fibre of constituent 1, δu_{10}, plus a rotation about a point on this fibre. From the figure we note the following relationships

LAYERED BEAMS

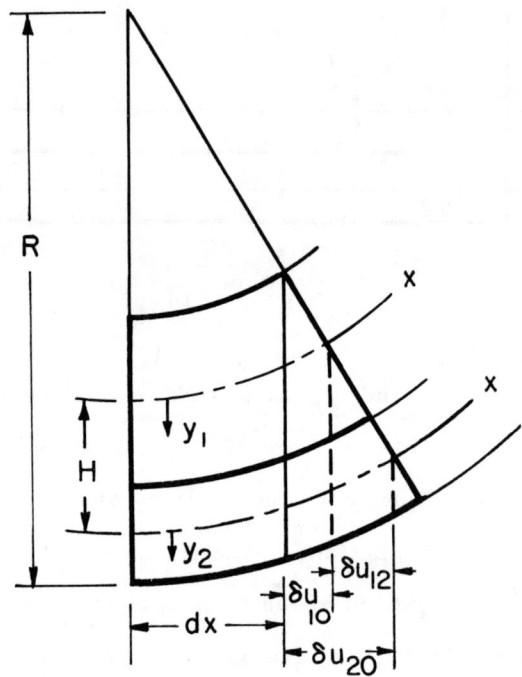

Fig. 6.14-2

$$\epsilon_{10} = \frac{\delta u_{10}}{dx} \qquad \epsilon_{20} = \frac{\delta u_{20}}{dx} \qquad (6.14-3)$$

and

$$\frac{\delta u_{12}}{dx} \pm \frac{\delta u_{20} - \delta u_{10}}{dx} = \frac{H}{R}$$

From the foregoing we obtain the difference in the displacements of the centroidal fibres, δ_{12}, in a distance dx, as

$$\epsilon_{12} = \epsilon_{20} - \epsilon_{10} = \frac{H}{R} \qquad (6.14-4)$$

LAYERED BEAMS

Substitution of ϵ_{20} from the last equation into Eq. (6.14-2) gives

$$\epsilon_2 = \epsilon_{10} + \frac{H}{R} + \frac{y_2}{R} = \frac{\sigma_2}{E_2} + \alpha_2 T_2 \tag{6.14-5}$$

Eqs. (6.14-1) and (6.14-5) are multiplied by differential cross section areas dA_1 and dA_2, respectively, and integrated over the cross section areas A_1 and A_2. It is noted that the area integrals of $y_1 dA_1$ and $y_2 dA_2$ are zero, so that the integration yields

$$E_1 \epsilon_{10} A_1 = P_1 + E_1 \alpha_1 \overline{T}_1 A_1 \tag{6.14-6}$$

$$\frac{E_2 H A_2}{R} + E_2 \epsilon_{10} A_2 = P_2 + E_2 \alpha_2 \overline{T}_2 A_2 \tag{6.14-7}$$

P_1 and P_2 represent the integrated stresses over the constituent areas, A_1 and A_2, and as there are no external axial loads

$$P_1 = -P_2 \tag{6.14-8}$$

Eqs. (6.14-6) and (6.14-7) are added, and we obtain

$$\epsilon_{10} = \frac{E_1 \alpha_1 \overline{T}_1 A_1 + E_2 \alpha_2 \overline{T}_2 A_2}{E_1 A_1 + E_2 A_2} - \frac{H}{R} \frac{E_2 A_2}{E_1 A_1 + E_2 A_2} \tag{6.14-9}$$

From Eq. (6.14-4) we have $\epsilon_{20} = \epsilon_{10} + H/R$, or

$$\epsilon_{20} = \frac{E_1 \alpha_1 \overline{T}_1 A_1 + E_2 \alpha_2 \overline{T}_2 A_2}{E_1 A_1 + E_2 A_2} + \frac{H}{R} \frac{E_1 A_1}{E_1 A_1 + E_2 A_2} \tag{6.14-10}$$

The bending axis lies in the plane of the isothermal beam ($T_1 = T_2 = 0$) where the stress and strain are zero. The stress σ, and the temperature distribution T_1 are set equal to zero in Eq. (6.14-1). This locates the bending axis, $\bar{y}_1$. From Eq. (6.14-1)

$$\frac{\bar{y}_1}{R} = -\epsilon_{10} \qquad \bar{y}_1 = -\epsilon_{10} R \qquad (6.14\text{-}11)$$

Substitution of ϵ_{10}, from Eq. (6.14-9) (with $\bar{T}_1 = \bar{T}_2 = 0$) yields

$$\bar{y}_1 = H \frac{E_2 A_2}{E_1 A_1 + E_2 A_2} \qquad (6.14\text{-}12)$$

The location of the bending axis can also be expressed in terms of the y_2 coordinate as

$$\bar{y}_2 = H \frac{E_1 A_1}{E_1 A_1 + E_2 A_2} \qquad (6.14\text{-}13)$$

Note that

$$\bar{y}_1 - \bar{y}_2 = H \qquad |\bar{y}_1| + |\bar{y}_2| = H \qquad (6.14\text{-}14)$$

The centroidal fibre strains of the two constituents, given by Eqs. (6.14-9) and (6.14-10) can be written as

$$\epsilon_{10} = \overline{\alpha T} - \frac{\bar{y}_1}{R} \qquad (6.14\text{-}15)$$

$$\epsilon_{20} = \overline{\alpha T} - \frac{\bar{y}_2}{R} \qquad (6.14\text{-}16)$$

where $\overline{\alpha T}$ is defined as

$$\overline{\alpha T} = \frac{E_1 \alpha_1 \bar{T}_1 A_1 + E_2 \alpha_2 \bar{T}_2 A_2}{E_1 A_1 + E_2 A_2} \qquad (6.14\text{-}17)$$

The mean load in each of the constituents, $P_1 = -P_2$, is obtained from Eq. (6.14-6) as

$$P_1 = E_1 A_1 (\epsilon_{10} - \alpha_1 \overline{T}_1) = E_1 A_1 \left[\frac{E_2 A_2 (\alpha_2 \overline{T}_2 - \alpha_1 \overline{T}_1)}{E_1 A_1 + E_2 A_2} - \frac{\overline{y}_1}{R} \right] \qquad (6.14\text{-}18)$$

from which

$$P_1 H = E_1 A_1 \overline{y}_1 (\alpha_2 \overline{T}_2 - \alpha_1 \overline{T}_1 - \frac{H}{R}) \qquad (6.14\text{-}19)$$

Eqs. (6.14-1) and (6.14-5) are now multiplied by $y_1 dA_1$ and $y_2 dA_2$ respectively, and integrated. The result is

$$\frac{E_1 I_1}{R} = M_1 + E_1 \alpha_1 M_{t1} \qquad (6.14\text{-}20)$$

$$\frac{E_2 I_2}{R} = M_2 + E_2 \alpha_2 M_{t2} \qquad (6.14\text{-}21)$$

where

$$I_1 = \int_{A_1} y_1^2 dA_1 \qquad I_2 = \int_{A_2} y_2^2 dA_2 \qquad (6.14\text{-}22)$$

$$M_1 = \int_{A_1} \sigma_1 y_1 dA_1 \qquad M_2 = \int_{A_2} \sigma_2 y_2 dA_2 \qquad (6.14\text{-}23)$$

$$M_{t1} = \int_{A_1} T_1 y_1 dA_1 \qquad M_{t2} = \int_{A_2} T_2 y_2 dA_2 \qquad (6.14\text{-}24)$$

The addition of Eqs. (6.14-20) and (6.14-21) gives

$$M_1 + M_2 = \frac{1}{R}(E_1 I_1 + E_2 I_2) - (E_1 \alpha_1 M_{t1} + E_2 \alpha_2 M_{t2}) \qquad (6.14\text{-}25)$$

Fig. 6.14-3 shows the internal loads P_1, M_1, and M_2, and the external loads M_o and R_o which produce a moment $M = M_o + R_o x$ at the location x.

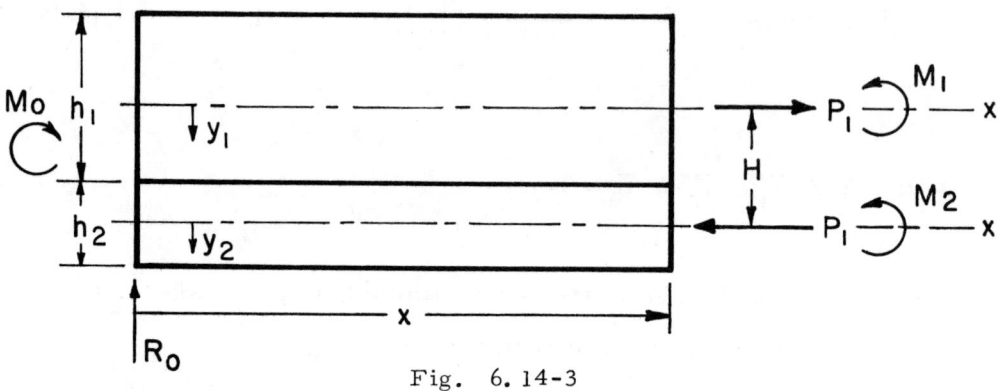

Fig. 6.14-3

The loads P_1 form a couple which is in equilibtium with the moments M, M_1, and M_2. Equilibrium requires that

$$P_1 H = M_1 + M_2 - M \qquad (6.14-26)$$

$P_1 H$ is given by Eq. (6.14-19), and is substituted in Eq. (6.14-26), which is then set into (6.14-25). This yields the curvature 1/R as

$$\frac{1}{R} = -\frac{d^2 v}{dx^2} = \frac{E_1 A_1 \bar{y}_1 (\alpha_2 \bar{T}_2 - \alpha_1 \bar{T}_1) + E_1 \alpha_1 M_{t1} + E_2 \alpha_2 M_{t2} + M}{E_1 I_1 + E_2 I_2 + E_1 A_1 \bar{y}_1 H} \qquad (6.14-27)$$

This is the equation for the thermal deformation of a bimetallic beam. The temperature can vary with the axial coordinate as well as with the transverse coordinate. Eq. (6.14-27) gives the local curvature when $\bar{T}_1$, $\bar{T}_2$, M_t, M_{t2} and M are functions of x. The deflection curve of the bimetallic beam is obtained by integration of Eq. (6.14-27) subject to the appropriate boundary restraints.

The stresses in the beams are obtained from Eq. (6.14-1) and (6.14-2). They are

$$\sigma_1 = E_1(\epsilon_{10} - \alpha_1 T_1 + \frac{y_1}{R}) = E_1(\overline{\alpha T} - \alpha_1 T_1 + \frac{y_1 - \overline{y}_1}{R}) \qquad (6.14-28)$$

$$\sigma_2 = E_2(\epsilon_{20} - \alpha_2 T_2 + \frac{y_2}{R}) = E_2(\overline{\alpha T} - \alpha_2 T_2 + \frac{y_2 - \overline{y}_2}{R}) \qquad (6.14-29)$$

When there is no bending $\frac{1}{R} = 0$ and the expressions become

$$\sigma_1 = E_1(\overline{\alpha T} - \alpha_1 T_1)$$

$$\sigma_2 = E_2(\overline{\alpha T} - \alpha_2 T_2) \qquad (6.14-30)$$

which are the expressions found earlier for the case of sandwich strips. When the beam is homogeneous, as when $h_2 = 0$, Eq. (6.14-28) becomes

$$\sigma_1 = E_1 \alpha_1 (\overline{T}_1 - \frac{y_1}{\alpha} \frac{d^2 v}{dx^2} - T) \qquad (6.14-31)$$

which is the equation for stress derived earlier for the case of homogeneous beams. The expressions for stress, Eqs. (6.14-28) and (6.14-29) were derived on the basis of a uniaxial stress condition. This assumption is not valid at the interface, where the stress is biaxial. Whereas in homogeneous beams the increased stress due to biaxiality depends upon the width to depth ratio, in a bimetallic beam biaxiality is always present at the interfacial plane. The stress in an unrestrained bimetallic beam is therefore more conservatively given as

$$\sigma_1 = \frac{E_1}{1-\nu_1} (\overline{\alpha T}' - \alpha_1 T_1 + \frac{y_1 - \overline{y}_1}{R}) \qquad (6.14-32)$$

$$\sigma_2 = \frac{E_2}{1-\nu_2} (\overline{\alpha T}' - \alpha_2 T_2 + \frac{y_2 - \overline{y}_2}{R}) \qquad (6.14-33)$$

In Eqs. (6.14-32) and (6.14-33) the quantity $\overline{\alpha T}'$ is

$$\overline{\alpha T}' = \frac{\dfrac{E_1 \alpha_1 \overline{T}_1 A_1}{1-\nu_1} + \dfrac{E_2 \alpha_2 \overline{T}_2 A_2}{1-\nu_2}}{\dfrac{E_1 A_1}{1-\nu_1} + \dfrac{E_2 A_2}{1-\nu_2}} \qquad (6.14-34)$$

Bending of Layered Concentric Circular Rods and Tubes

We consider next the problem of circular layered elements such as the clad rod and the dual tube shown in Fig. 6.14-4, subjected to a transverse temperature distribution. The bending axis is the diameter which is

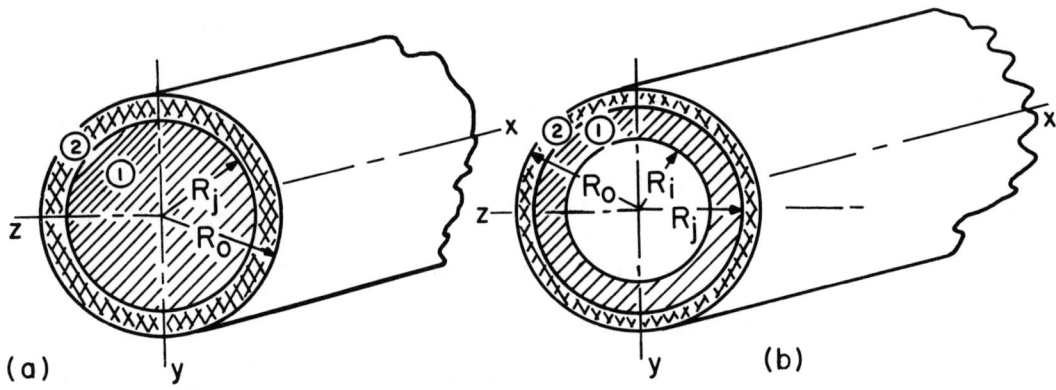

Fig. 6.14-4

perpendicular to the direction of temperature variation.

We assume that the assembly, with constituents (1) and (2) having different material properties, is subjected to a transverse temperature distribution $T = T(y)$, so that the bending axis is the z axis. The transverse strain variation is assumed to be linear, and we assume a uniaxial state of stress. The linear strain variation and stress-strain relationships are expressed as

$$\epsilon = \epsilon_o + \frac{y}{R} = \frac{\sigma_1}{E_1} + \alpha_1 T = \frac{\sigma_2}{E_2} + \alpha_2 T \qquad (6.14\text{-}35)$$

There is no axial load present. Therefore

$$\int_{A_1} \sigma_1 \, dA_1 + \int_{A_2} \sigma_2 \, dA_2 = 0 \qquad (6.14\text{-}36)$$

From Eqs. (6.14-35) and (6.14-36) we have

$$\epsilon_o = \frac{\overline{\sigma}_1}{E_1} + \alpha_1 \overline{T}_1 = \frac{\overline{\sigma}_2}{E_2} + \alpha_2 \overline{T}_2 \qquad (6.14\text{-}37)$$

and

$$\overline{\sigma}_1 A_1 + \overline{\sigma}_2 A_2 = 0 \qquad (6.14\text{-}38)$$

From the two foregoing equations we obtain the axial centroidal strain as

$$\epsilon_o = \frac{E_1 \alpha_1 \overline{T}_1 A_1 + E_2 \alpha_2 \overline{T}_2 A_2}{E_1 A_1 + E_2 A_2} = \overline{\alpha T} \qquad (6.14\text{-}39)$$

We now multiply Eq. (6.14-35) by $y dA_1$ and $y dA_2$ and integrate over the respective cross sections. Thus

$$\int_{-h_1/2}^{h_1/2} \epsilon_o y \, dA_1 + \int_{-h_1/2}^{h_1/2} \frac{y^2}{R} \, dA_1 = \int_{-h_1/2}^{h_1/2} \frac{\sigma_1 y}{E_1} \, dA_1 + \int_{-h_1/2}^{h_1/2} \alpha_1 T y \, dA_1 \qquad (6.14\text{-}40)$$

$$\int_{-h_2/2}^{h_2/2} \epsilon_o y \, dA_2 + \int_{-h_2/2}^{h_2/2} \frac{y^2}{R} \, dA_2 = \int_{-h_2/2}^{h_2/2} \frac{\sigma_2 y}{E_2} \, dA_2 + \int_{-h_2/2}^{h_2/2} \alpha_2 T y \, dA_2 \qquad (6.14\text{-}41)$$

where $h_1/2$ is equal to R_j, and $h_2/2$ is equal to R_o.

The first term in each of the foregoing equations is zero, the second terms are the respective moments of inertia, and the last terms are $\alpha_1 M_{t1}$ and $\alpha_2 M_{t2}$ respectively. The third term in each of the equations represents the internal moment divided by the modulus of elasticity. We designate these as M_1/E_1 and M_2/E_2. Eqs. (6.14-40) and (6.14-41) become

$$M_1 = \frac{E_1 I_1}{R} - E_1 \alpha_1 M_{t1} \tag{6.14-42}$$

$$M_2 = \frac{E_2 I_2}{R} - E_2 \alpha_2 M_{t2} \tag{6.14-43}$$

Since the sum $M_1 + M_2$ represents the moment, M, at a given location, x, the addition of the foregoing equations yields

$$M = \frac{1}{R}(E_1 I_1 + E_2 I_2) - E_1 \alpha_1 M_{t1} - E_2 \alpha_2 M_{t2} \tag{6.14-44}$$

As $1/R = -d^2v/dx^2$, the equation can be written as

$$\frac{d^2v}{dx^2} = -\frac{E_1 \alpha_1 M_{t1} + E_2 \alpha_2 M_{t2}}{E_1 I_1 + E_2 I_2} - \frac{M}{E_1 I_1 + E_2 I_2} \tag{6.14-45}$$

Eq. (6.14-45) is the thermal beam deformation equation for a layered circular beam.

In a statically determinate beam, it is not necessary to solve Eq. (6.14-45) in order to obtain the stresses. In such a beam the moment M is zero and we have

$$\frac{1}{R} = \frac{E_1 \alpha_1 M_{t1} + E_2 \alpha_2 M_{t2}}{E_1 I_1 + E_2 I_2} \tag{6.14-46}$$

and from Eq. (6.14-35) the stresses σ_1 and σ_2 are

$$\sigma_1 = E_1 \left(\overline{\alpha T} + y \, \frac{E_1 \alpha_1 M_{t1} + E_2 \alpha_2 M_{t2}}{E_1 I_1 + E_2 I_2} - \alpha_1 T \right) \tag{6.14-47}$$

$$\sigma_2 = E_2 \left(\overline{\alpha T} + y \, \frac{E_1 \alpha_1 M_{t1} + E_2 \alpha_2 M_{t2}}{E_1 I_1 + E_2 I_2} - \alpha_2 T \right) \tag{6.14-48}$$

In the derivation of the foregoing expressions the assumption of a uni-axial state of stress was made, in which only normal stresses in the axial direction were present. It is clear that tangential stresses may be present whose order of magnitude is the same as the axial stresses. We should therefore properly include the stress modification obtained by replacing E_1 by E_1^* and E_2 by E_2^*, with

$$E_1^* = \frac{E_1}{1-\nu_1} \qquad E_2^* = \frac{E_2}{1-\nu_2} \tag{6.14-49}$$

It should be recognized that Eq. (6.14-45), and Eqs. (6.14-47) and (6.14-48) apply to all types of layered beams having a symmetrical material distribution. They apply for example to layered rectangular beams with a symmetrical material distribution, i.e., sandwich beams.

Example (a)   Bimetallic Rectangular Beam at a Uniform Temperature

A bimetallic rectangular beam, such as shown in Fig. 6.14-1, is heated to a uniform temperature T. The constituents are assumed to be of equal thickness and the moduli of elasticity are assumed to be the same. We wish to find the beam curvature and stresses in the constituents. We have

$$h_1 = h_2 = \frac{h}{2} \qquad\qquad E_1 = E_2 = E \qquad\qquad (1)$$

The centroidal fibre location, from Eqs. (6.14-13) and (6.14-14) is

$$\bar{y}_1 = \frac{h}{4} \qquad\qquad \bar{y}_2 = -\frac{h}{4} \qquad\qquad (2)$$

The curvature, from Eq. (6.14-27) is

$$\frac{1}{R} = \frac{h_1 \bar{y}_1 T(\alpha_2 - \alpha_1)}{I_1 + I_2 + h_1 \bar{y}_1 H} = \frac{\frac{h}{2} \cdot \frac{h}{4} \cdot T(\alpha_2 - \alpha_1)}{2 \frac{(h/2)^3}{12} + \frac{h}{2} \cdot \frac{h}{4} \cdot \frac{h}{2}} = \frac{3}{2} \frac{(\alpha_2 - \alpha_1) T}{h} \qquad (3)$$

and the stresses, from Eqs. (6.14-28) and (6.14-29) are

$$\sigma_1 = \frac{3}{2} E (\alpha_2 - \alpha_1) T \left(\frac{y_1}{h} + \frac{1}{12}\right) \qquad\qquad (4)$$

$$\sigma_2 = \frac{3}{2} E (\alpha_2 - \alpha_1) T \left(\frac{y_2}{h} - \frac{1}{12}\right) \qquad\qquad (5)$$

The maximum stress occurs at the interfacial plane and is

$$\sigma_{max} = \pm \frac{E(\alpha_2 - \alpha_1) T}{2} \qquad\qquad (6)$$

All of the foregoing stresses are based on uniaxial analysis. To account for biaxiality the stresses should be multiplied by a factor $1/(1-\nu)$.

Example (b) Clad Circular Rod with Transverse Temperature Distribution

The rod cross section shown in Fig. 6.14-4(a) represents nuclear fuel rod. It is simply supported so that it can bend freely. The temperature distribution in the fuel rod due to internal heat distribution and surface cooling is specified as

$$T = T_a \left(1 - \frac{r^2}{R_o^2}\right) + T_b \frac{y}{R_o} \qquad\qquad (1)$$

The temperature distribution is thus the combination of a paraboloid of temperature and a linear temperature gradient. It should be recognized that a linear temperature distribution in the clad rod will produce stresses, since the rod is not homogeneous.

The mean temperatures and temperature moments are

$$\overline{T}_1 A_1 = \int_0^{R_j} T_a \left(1 - \frac{r^2}{R_o^2}\right) 2\pi r\, dr = T_a \pi \left(R_j^2 - \frac{R_j^4}{2R_o^2}\right) \quad (2)$$

$$\overline{T}_2 A_2 = \int_{R_j}^{R_o} T_a \left(1 - \frac{r^2}{R_o^2}\right) 2\pi r\, dr = T_a \pi \left(R_o^2 - R_j^2 - \frac{R_o^4 - R_j^4}{2R_o^2}\right) \quad (3)$$

$$M_{t_1} = \int_{A_1} T_b \frac{y}{R_o} \cdot y\, dA_1 = \frac{T_b}{R_o} I_1 = T_b \frac{\pi R_j^4}{4 R_o} \quad (4)$$

$$M_{t_2} = \int_{A_2} T_b \frac{y}{R_o} \cdot y\, dA_2 = \frac{T_b}{R_o} I_2 = T_b \frac{\pi (R_o^4 - R_j^4)}{4 R_o} \quad (5)$$

From Eq. (6.14-39), the extensional strain is

$$\epsilon_o = \overline{\alpha T} = \frac{T_a \left[E_1 \alpha_1 \left(R_j^2 - \frac{R_j^4}{2 R_o^2}\right) + E_2 \alpha_2 \left(R_o^2 - R_j^2 - \frac{R_o^4 - R_j^4}{2 R_o^2}\right)\right]}{E_1 R_j^2 + E_2 (R_o^2 - R_j^2)} \quad (6)$$

Eq. (6.14-46) gives the curvature as

$$\frac{1}{R} = \frac{T_b}{R_o} \cdot \frac{E_1 \alpha_1 R_j^4 + E_2 \alpha_2 (R_o^4 - R_j^4)}{E_1 R_j^4 + E_2 (R_o^4 - R_j^4)} \quad (7)$$

The axial stresses are obtained from Eq. (6.14-35), and they are modified, in accordance with Eqs. (6.14-49) to account for the presence of stresses in the other coordinate directions. Thus

$$\sigma_1 = \frac{E_1}{1-\nu}\left(\overline{\alpha T} + \frac{y}{R} - \alpha_1 T\right) \qquad (8)$$

$$\sigma_2 = \frac{E_2}{1-\nu}\left(\overline{\alpha T} + \frac{y}{R} - \alpha_2 T\right) \qquad (9)$$

T is given by Eq. (1), $\overline{\alpha T}$ by Eq. (6), and $1/R$ by Eq. (7).

If the fuel rod cladding had the same E and α as the core then Eq. (6) would become

$$\epsilon_o = \overline{\alpha T} = \frac{\alpha T_a}{2} \qquad (10)$$

and Eq. (7) would become

$$\frac{1}{R} = \frac{\alpha T_b}{R_o} \qquad (11)$$

The modified axial stress in the rod would then be

$$\sigma = \frac{E}{1-\nu}\left[\frac{\alpha T_a}{2} + \frac{\alpha T_b y}{R_o} - \alpha T_a\left(1 - \frac{r^2}{R_o^2}\right) - \alpha T_b \frac{y}{R_o}\right] \qquad (12)$$

or, upon simplifying

$$\sigma = \frac{E \alpha T_a}{1-\nu}\left(\frac{r^2}{R_o^2} - \frac{1}{2}\right) \qquad (13)$$

It is observed that there is no stress due to the linear temperature variation $\alpha T_b y/R_o$, in the homogeneous rod. The terms due to the linear temperature distribution cancel out in Eq. (12).

6.15 PURE BENDING OF PLATES

When the temperature in homogeneous or layered plates is a function of the transverse (normal to the plate) coordinate only, and the plate is free to deform, the transverse temperature distribution, $T = T(z)$ will cause the plate to deform into a spherical surface of constant curvature. The analysis of the stresses and the accompanying bending deformation is relatively simple in this type of problem since we have a condition in which the stress normal to the plate is equal to zero ($\sigma_z = 0$), and the plane stress in each plate lamina is uniform and isotropic. That is, the stress does not change from point to point in a plate lamina (constant z), and at each location x and y within the lamina, the stress is the same in all coordinate directions. As a result no shear stresses are generated, and the plate undergoes pure bending with

$$\sigma(z) = \sigma_x(z) = \sigma_y(z) \qquad \tau = \tau_{xy} = 0 \qquad (6.15\text{-}1)$$

In problems of pure bending of plates it is possible to employ the results of one-dimensional beam analysis by replacing the modulus of elasticity, E, of the one-dimensional beam analysis with a modified modulus of elasticity $E'' = E/1-\nu$ in accordance with the discussion in Sect. 1.7. This modification is not restricted to rectangular plates. It applies to circular plates or plates of any other shape, and also to layered plates with either symmetrical or unsymmetrical material distribution. The requirement is simply that the temperature be a function of the thickness coordinate only and that there be no external restraints to prevent deformation.

Homogeneous Plates

The analysis of the pure thermal flexure of a plate is carried out as for the beam in Sect. 6.2. E is replaced by $E/(1-\nu)$ and in lieu of Eq. (6.2-3) we have

$$\epsilon = \epsilon_e + \frac{z}{R} = \frac{\sigma(1-\nu)}{E} + \alpha T \qquad (6.15\text{-}2)$$

Instead of Eq. (6.2-7) we now have, with $M = 0$,

$$\frac{1}{R} = -\frac{d^2w}{ds^2} = \frac{\alpha}{I} M_t = \frac{12\alpha M_t}{h^3} \tag{6.15-3}$$

with s any coordinate in the plane of the plate, and M_t defined as

$$M_t = \int_{-h/2}^{h/2} Tz\, dz$$

h, being the plate thickness.

The stress in the plate, which is free to deform, is the same as that given by Eqs. (6.2-9) and (6.2-10) with $M = 0$ and $E/1-\nu)$ replacing E. A further restriction for pure plate bending is that $T = T(z)$ is invariant with the plane plate coordinates, say x and y. The stress equation for the unrestrained plate is thus

$$\sigma = \frac{E\alpha}{1-\nu} (\overline{T} + \frac{z}{\alpha R} - T) \tag{6.15-4}$$

or

$$\sigma = \frac{E\alpha}{1-\nu} (\overline{T} + \frac{M_t z}{I} - T) \tag{6.15-5}$$

with

$$\overline{T} = \frac{1}{h} \int_{-h/2}^{h/2} T\, dz \qquad I = \frac{h^3}{12}$$

Clamped Plate

In the case of plates which are clamped all around but are not restrained from extension deformation, the stress will be

$$\sigma = \frac{E\alpha}{1-\nu} (\overline{T} - T) \tag{6.15-6}$$

and there will be no bending deformation. A plate which is held rigidly at its boundaries so that both extensional and rotational deformation are prevented, will develop a stress equal to

$$\sigma = -\frac{E\alpha T}{1-\nu} \qquad (6.15-7)$$

The validity of Eqs. (6.15-6) and (6.15-7) are readily established when one considers the effect of the restraints on the edge thermal surface tractions.

When there are mixed boundary conditions or when the temperature is a function of both the transverse and plane coordinates, the problem is more complex. One method of solving a problem in which $T = T(z)$, and the plate is neither free all around nor clamped all around, is to compute first the edge rotations assuming an unrestrained plate, and then determine the edge loads that are required to satisfy the boundary conditions. The stresses due to these edge loads are additive to the unrestrained deformation stresses.

Layered Plates

The deformation and the state of stress in an unrestrained layered plate with a transverse temperature distribution can be obtained directly from the one-dimensional solution of the comparable layered beam problem. This problem is discussed in Sect. 6.14. The equivalent thermal loads that act on the layered plate are the edge thermal moments and mean loads due to the surface tractions. It is clear that these will produce a uniform extension plus a uniform curvature. They will also generate uniform displacement stresses in each plate lamina; and as the temperature is taken as a function of the thickness coordinate only, the net thermal stresses will also be uniform in each plate lamina.

As in the homogeneous plate, the directional uniformity of stress permits us to employ a modified modulus of elasticity, $E'' = E/(1 - \nu)$, to transform the one-dimensional layered beam problem to a two-dimensional layered plate problem. The analysis of the layered beam, given in Sect. 6.14, applies to the layered plate, with the proviso that wherever E_1 appears it is replaced by $E_1/(1-\nu_1)$, and wherever E_2 appears it is replaced by $E_2/(1-\nu_2)$.

With this modification, the uniform curvature of the plate, taken from Eq. (6.14-27), with $M = 0$, is

$$\frac{1}{R} = \frac{\dfrac{E_1 h_1 \bar{z}_1 (\alpha_2 \bar{T}_2 - \alpha_1 \bar{T}_1)}{1 - \nu_1} + \dfrac{E_1 \alpha_1 M_{t1}}{1 - \nu_1} + \dfrac{E_2 \alpha_2 M_{t2}}{1 - \nu_2}}{\dfrac{E_1 I_1}{1 - \nu_1} + \dfrac{E_2 I_2}{1 - \nu_2} + \dfrac{E_1 h_1 \bar{z}_1 h}{2(1 - \nu_1)}} \qquad (6.15\text{-}8)$$

In most materials $\nu_1 = \nu_2 = .3$. When the Poisson's ratios are equal, the curvature of the layered plate and of the layered beam becomes the same. In Eq. (6.15-8), h_1 is the thickness of one layer and h_2 the thickness of the other layer. The distances of the bending axis from the central planes of layers 1 and 2 (see Fig. 6.14-1) are

$$\bar{z}_1 = \frac{h}{2} \frac{E_2 h_2}{E_1 h_1 + E_2 h_2} \qquad \bar{z}_2 = -\frac{h}{2} \frac{E_1 h_1}{E_1 h_1 + E_2 h_2} \qquad (6.15\text{-}9)$$

with $h = h_1 + h_2$. The mean temperatures of layer 1 and layer 2 are $\bar{T}_1$ and $\bar{T}_2$, and the temperature moments

$$M_{t1} = \int_{-h_1/2}^{h_1/2} T_1 z_1 \, dz, \qquad M_{t2} = \int_{-h_2/2}^{h_2/2} T_2 z_2 \, dz_2 \qquad (6.15\text{-}10)$$

and are taken about their respective central planes. The moments of inertia I_1 and I_2 are

$$I_1 = \frac{h_1^3}{12} \qquad I_2 = \frac{h_2^3}{12} \qquad (6.15\text{-}11)$$

The stresses in the unrestrained layered plate are obtained directly from Eqs. (6.14-28) and (6.14-29). They are

$$\sigma_1 = \frac{E_1}{1-\nu_1}\left(\overline{\alpha T}' - \alpha_1 T_1 + \frac{z_1 - \overline{z}_1}{R}\right) \qquad (6.15\text{-}12)$$

$$\sigma_2 = \frac{E_2}{1-\nu_2}\left(\overline{\alpha T}' - \alpha_2 T_2 + \frac{z_2 - \overline{z}_2}{R}\right) \qquad (6.15\text{-}13)$$

The mean extensional strain $\overline{\alpha T}'$ is defined, in accordance with Eq. (6.14-17), as

$$\overline{\alpha T}' = \frac{\dfrac{E_1 \alpha_1 \overline{T}_1 h_1}{1-\nu_1} + \dfrac{E_2 \alpha_2 \overline{T}_2 h_2}{1-\nu_2}}{\dfrac{E_1 h_1}{1-\nu_1} + \dfrac{E_2 h_2}{1-\nu_2}} \qquad (6.15\text{-}14)$$

The coordinates z_1 and z_2 are measured from the respective central planes of layers 1 and 2 in the same manner as y_1 and y_2 are measured in Fig. 6.14-1.

When rotation is prevented at the edges of the plate, but extension is not restrained, Eqs. (6.15-12) and (6.15-13) become

$$\sigma_1 = \frac{E_1}{1-\nu_1}\left(\overline{\alpha T}' - \alpha_1 T_1\right) \qquad (6.15\text{-}15)$$

$$\sigma_2 = \frac{E_2}{1-\nu_2}\left(\overline{\alpha T}' - \alpha_2 T_2\right) \qquad (6.15\text{-}16)$$

On the other hand, if the plate is rigidly clamped at its boundaries, and both rotation and extension are prevented, then the foregoing equations become

$$\sigma_1 = - \frac{E_1 \alpha_1 T_1}{1 - \nu_1} \qquad (6.15\text{-}17)$$

$$\sigma_2 = - \frac{E_2 \alpha_2 T_2}{1 - \nu_2}$$

These results are readily obtained by examining the effect of thermal surface tractions at the edges of a clamped plate.

Example (a) Homogeneous Plate with Transverse Temperature Variation

A plate which is simply supported at its four corners is subjected to a temperature distribution

$$T = T_o e^{-\frac{z}{h}} \qquad (1)$$

The mean temperature and the temperature moment are

$$\bar{T} = \frac{1}{h} \int_{-h/2}^{h/2} T\, dz = 2 T_o \sinh \tfrac{1}{2} = 1.044\, T_o \qquad (2)$$

$$M_t = \int_{-h/2}^{h/2} Tz\, dz = T_o h^2 (2 \sinh \tfrac{1}{2} - \cosh \tfrac{1}{2}) = -.084\, h^2 T_o \qquad (3)$$

From Eq. (6.15-3) the curvature of the plate is

$$\frac{1}{R} = \frac{12\, \alpha\, M_t}{h^3} = -1.008\, \frac{\alpha T_o}{h} \qquad (4)$$

The stress distribution in the plate, in accordance with Eq. (6.15-4), is

$$\sigma = \frac{E\alpha}{1-\nu} (\bar{T} + \frac{z}{\alpha R} - T) = \frac{E\alpha T_o}{1-\nu} (1.044 - 1.008\, \frac{z}{h} - e^{-\frac{z}{h}}) \qquad (5)$$

PURE BENDING OF PLATES 433

If the plate were rigidly clamped on all edges, the curvature would be zero and the stress would be

$$\sigma = -\frac{E\alpha T}{1-\nu} = -\frac{E\alpha T_o}{1-\nu} e^{-\frac{z}{h}} \tag{6}$$

Example (b) Layered Plate with Transverse Temperature Variation

A layered plate consists of two materials as shown in Fig. 6.15-1.

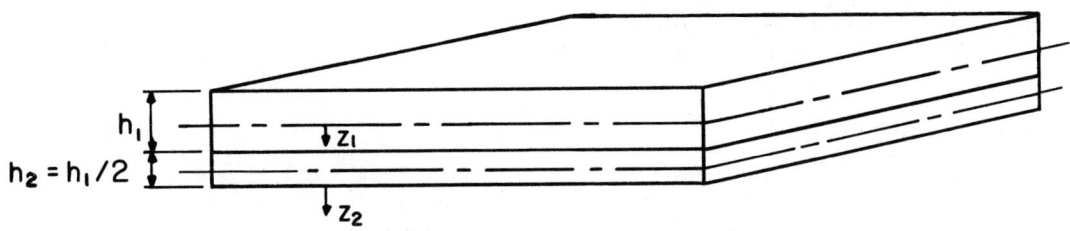

Fig. 6.15-1

The plate, which is free to deform, is subjected to a temperature distribution

$$T_1 = T_o - \frac{T_o}{4}\left(1 - \frac{2 z_1^2}{h_1}\right)$$
$$T_2 = T_o \tag{1}$$

This would be the kind of temperature distribution that would be obtained if the lower plate did not contain heat sources and were insulated at the bottom, and the upper plate contained a uniformly distributed heat source.

The mean temperatures and temperature moments are

$$\bar{T}_1 = \frac{2}{3} T_o \qquad M_{t1} = \frac{T_o h_1^2}{12} = \frac{T_o I_1}{h_1} \qquad (2)$$

$$\bar{T}_2 = T_o \qquad M_{t2} = 0 \qquad (3)$$

The location of the bending axis is

$$\bar{z}_1 = \frac{3 E_2 h_1}{8 E_1 + 4 E_2} \qquad \bar{z}_2 = -\frac{3 E_1 h_1}{4 E_1 + 2 E_2} \qquad (4)$$

If we assume that $\nu_1 = \nu_2 = \nu$ we obtain the mean extensional strain as

$$\frac{\bar{\epsilon}}{\alpha T} = \frac{4 E_1 \alpha_1 + 3 E_2 \alpha_2}{6 E_1 + 3 E_2} T_o \qquad (5)$$

and the curvature as

$$\frac{1}{R} = \frac{T_o}{h_1} \frac{(3\alpha_2 - 2\alpha_1)\left(\frac{E_1 E_2}{2 E_1 + E_2}\right) + \frac{E_1 \alpha_1}{3}}{\frac{E_1}{3} + \frac{E_2}{4} + \frac{9}{4}\left(\frac{E_1 E_2}{2 E_1 + E_2}\right)} \qquad (6)$$

Substitution of these values in Eqs. (6.15-12) and (6.15-13) yields the stresses.

PROBLEMS

6-1 A rectangular beam of depth h has a temperature distribution

$$T = \frac{T_o}{2}\left(1 + \frac{8 y^3}{h^3}\right)$$

with y measured from the beam center line. Find the thermal stress distribution and beam curvature for the following cases:

(a) a cantilever beam.

(b) a simply supported beam.

Without integration of beam deformation equation find the deflection of the end of the cantilever beam and the deflection at the center of the simply supported beam. Take the length of the beam as L.

6-2 Assume that the beam, with a temperature distribution as given in Prob. 6-1, is hinged rather than simply supported, so that it cannot expand axially. Fig. 6-2 shows such a support arrangement. Find the thermal stress distribution in the hinged beam. What is the approximate increase in deflection due to the axial thrust?

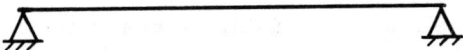

Fig. 6-2

6-3 The beams in Figs. 6-3(a) and 6-3(b) have horizontally guided supports and clamped supports respectively.

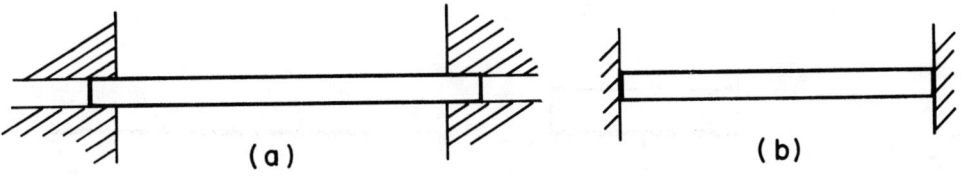

Fig. 6-3

The beams are subjected to the same temperature distribution as given in Prob. 6-1. Find the stress distributions. Do not integrate the thermal beam deformation equation.

PROBLEMS

6-4 A simply supported rectangular beam and a cantilever rectangular beam have a temperature distribution

$$T = \frac{T_o}{2}\left(1 + \frac{32\,y^5}{h^5}\right)\cos\frac{\pi x}{2L}$$

with h the beam depth and y measured downward from the beam center line. L is the beam length and x is measured from the left end. Determine directly the stress distribution and the curvature variation for the simply supported and cantilever beams. State whether the beam deflections are obtainable geometrically as in Prob. 6-1.

6-5 Compute the shear stress distribution in the beams of Prob. 6-4. Find the location and the magnitude of the peak shear stress (away from the ends) and determine the ratio of this stress to the maximum normal stress in the beams.

6-6 The wide-flange beam shown in Fig. 6-6 has a temperature distribution

$$T = \frac{T_o}{2}\left(1 + \frac{8\,y^3}{h^3}\right)\sin\frac{\pi x}{L}$$

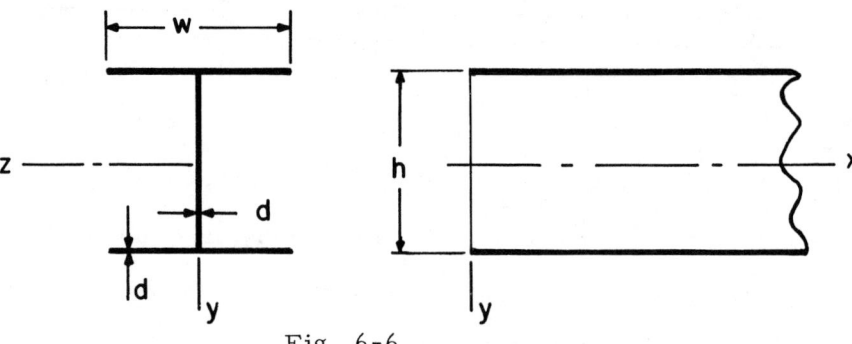

Fig. 6-6

Compute the normal and shear stress variation at the junction of the web and flange. Find the locations and the magnitudes of the peak normal stresses and the peak shear stresses.

6-7 The rectangular beam shown in Fig. 6.1-1 is unrestrained and is subjected to a temperature distribution

$$T = T_o \sinh \frac{\pi y}{h} \cos \frac{\pi z}{h} \sin \frac{\pi x}{L}$$

Determine the normal axial stresses and shear stresses in the beam. Are normal cross section stresses σ_y and σ_z generated in the beam?

6-8 Repeat Prob. 6-7, using a temperature distribution

$$T = T_o \sin \frac{\pi y}{h} \sin \frac{\pi z}{h} \sin \frac{\pi x}{L}$$

Find the normal axial stresses in the beam. Determine whether cross section normal stresses σ_y and σ_z are present. If so, estimate conservatively the required modification of the uniaxial stress, σ_x.

6-9 Integrate the thermal beam deformation equation to obtain the normal stresses, shear stresses and deflection of the rectangular beam shown in Fig. 6-9.

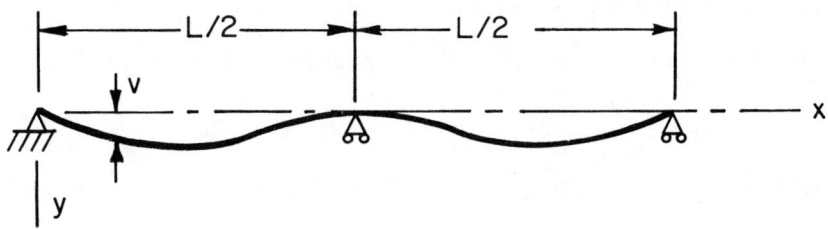

Fig. 6-9

The temperature distribution is

$$T = T_o e^{y/h}$$

with h the depth of the beam and y measured from the center line.

6-10 Integrate the thermal beam deformation equation, and equate slopes at $x = L/3$, to obtain the normal and shear stresses, and the deflection of the rectangular beam shown in Fig. 6-10. It is subjected to the same temperature distribution as given in Prob. 6-9.

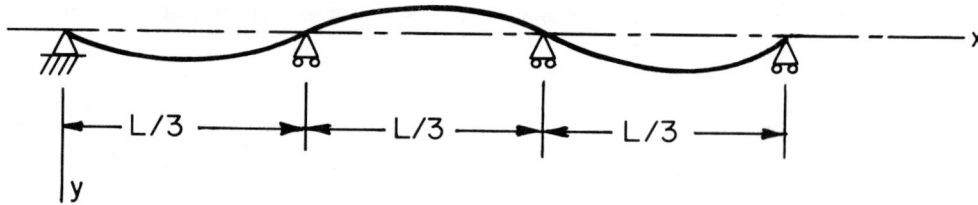

Fig. 6-10

6-11 A beam having a circular cross section of radius R is subjected to a temperature distribution

$$T = \frac{T_o}{16} \left(\frac{y}{R} + 1 \right)^4$$

The beam is supported as shown in Fig. 6.5-3. Find the normal axial stresses and shear stresses, and the deflection. Estimate the increase in the normal stress due to the presence of the stresses σ_y and σ_z.

6-12 A rectangular beam of depth h and length L is clamped at both ends, so that the applicable boundary conditions are $v(o) = v(L) = v'(o) = v'(L) = 0$. It is subjected to a temperature distribution

$$T = \frac{T_o}{16} \left(\frac{2y}{h} + 1\right)^2 e^{x/L}$$

Determine the deflection curve and the bending and shear stresses. Does the beam deflect upward or downward?

6-13 A rectangular beam of depth h and length L is supported as a cantilever with end conditions $v(o) = v'(0) = 0$. It is subjected to a temperature distribution

$$T = \frac{T_o}{16} \left(\frac{2y}{h} + 1\right)^4 \left(\frac{x}{L}\right)^n$$

Determine the normal and shear stress distribution.

6-14 A circular rod of radius R has fixed-pinned supports, as in Fig. 6.5-3, with boundary conditions $v(o) = v'(o) = 0$ and $v(L) = 0$. Remove the support at $x = L$, and by using superposition analysis determine the stresses and deflection of the fixed-pinned rod having a temperature distribution

$$T = T_o \left(1 + \frac{y}{R}\right) \cosh \frac{x}{L}$$

Second, remove the rotational restraint at $x = 0$ so that the beam is simply supported, and by superposition determine the stresses and deflection of the fixed-pinned beam.

6-15 Perform an analysis of Example (b) of Section 6.7 by determining first the deflection and stresses in two disconnected simply supported beams on spans of length $L/2$. By superposition of a moment M_o at the center of the span ($x = L/2$) to restore continuity, find the net deflection and stresses in the beam.

6-16 The rectangular beam shown in Fig. 6-16 consists of two spans of

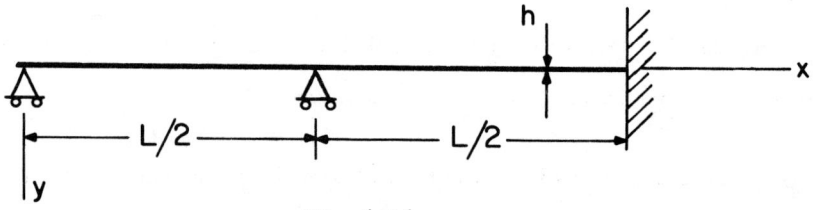

Fig. 6-16

length $L/2$. The left span $0 < x < \frac{L}{2}$ is subjected to a temperature distribution

$$T = \frac{T_o}{2} \left(1 + \frac{2y}{h}\right)$$

The right span $\frac{L}{2} < x < L$ is at $T = 0$. Disconnect the two span lengths at $x = \frac{L}{2}$ so that the left span becomes a pinned-pinned beam and the right span a pinned-fixed beam. By superposition of a moment M_o at the center support, to obtain continuity over this support, compute the stresses and the deformation of the beam.

6-17 The left half, $0 < x < \frac{L}{2}$, of a simply supported rectangular beam of depth h and length L, has a temperature distribution

$$T = \frac{T_o}{2} \left(1 + \frac{2y}{h}\right)$$

By the use of equivalent thermal loads and conjugate beam analysis find the slope at each end of the beam. What are the stresses in each section of the beam?

6-18 In a fixed-pinned rectangular beam of depth h and length L the portion defined by $\frac{L}{4} < x < \frac{L}{2}$ is heated to a temperature

$$T = T_o e^{-y/h}$$

PROBLEMS

Using equivalent thermal loads and moment-area analysis, find the stresses in the beam. The beam is fixed at $x = 0$ and pinned at $x = L$.

6-19 A cantilever beam in the form of a circular rod of radius R is supported at $x = 0$. The middle third of the rod, $\frac{L}{3} < x < \frac{2L}{3}$, is heated to a temperature

$$T = \frac{T_o}{2}\left(1 + \frac{2y}{h}\right)\left(\frac{x}{L} - \frac{1}{3}\right)$$

with y measured from the center line of the beam, and $h = 2R$. The remainder of the rod is at $T = 0$. Using equivalent thermal loads and the moment area method, determine the deflection at the end of the rod, $x = L$.

6-20 A rectangular beam on three simple supports has its central portion $\frac{L}{4} < x < \frac{3L}{4}$ heated to a temperature

$$T = \frac{T_o}{4}\left(1 + \frac{2y}{h}\right)^2$$

as shown in Fig. 6-20. The remainder of the beam is at $T = 0$. Using equivalent thermal loads and moment area analysis, find the normal stresses in the beam.

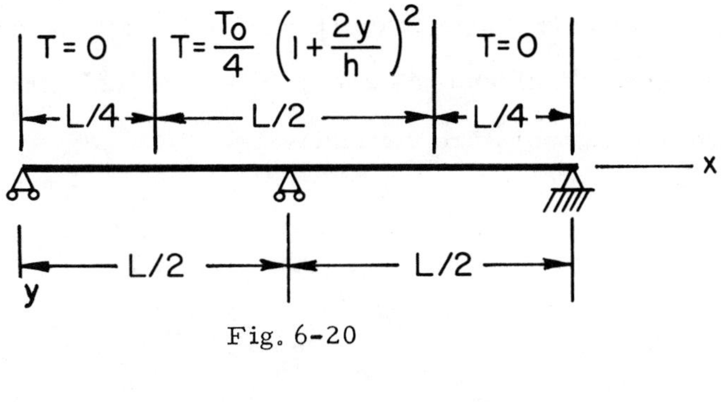

Fig. 6-20

6-21 A stainless steel ($E = 28 \times 10^6$ psi, $\alpha = 9.5 \times 10^{-6}/^\circ F$) nuclear fuel plate assembly has a cross section as shown in Fig. 6-21.

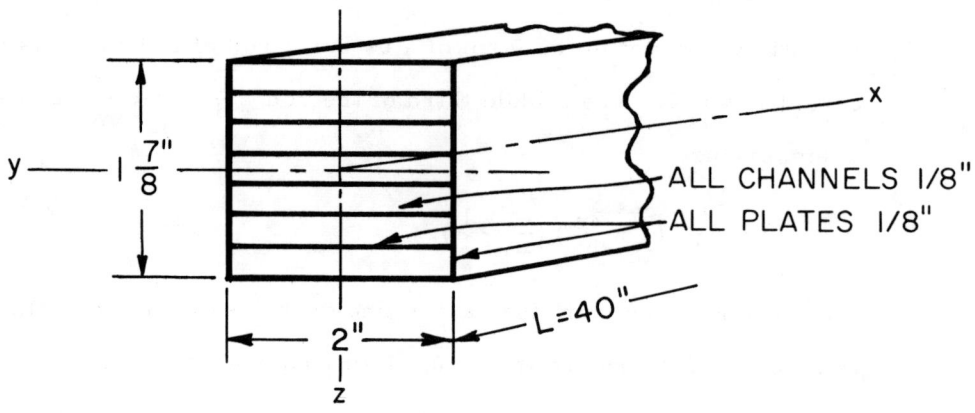

Fig. 6-21

The assembly is rigidly held (fixed) at $x = 0$ and simply supported at $x = L$. The temperature distribution is

$$T = 100 (2 + y) \sin \frac{\pi x}{40}$$

with x and y measured in inches and T in $^\circ F$. Treat the assembly as a beam, and use equivalent thermal loads and moment area analysis to obtain the stress distribution at $x = 0$, the rigidly held end of the assembly.

6-22 A carbon steel wide flange beam ($E = 30 \times 10^6$ psi, $\alpha = 6.5 \times 10^{-6}/^\circ F$) is rigidly held at both ends as shown in Fig. 6-22. The upper flange is subjected to a temperature distribution

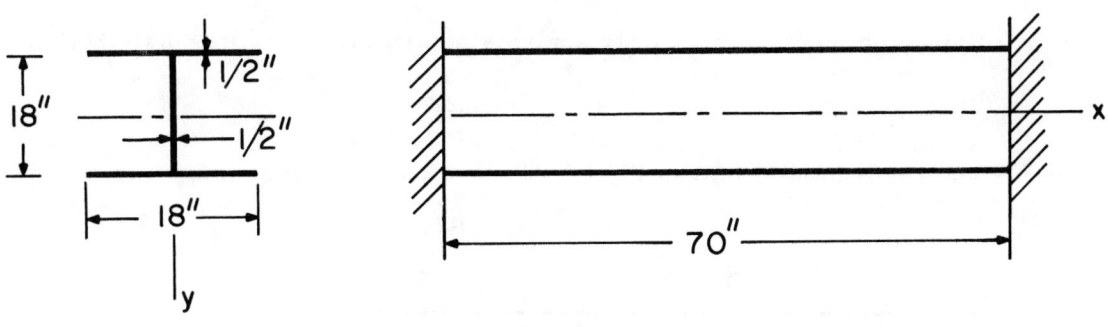

Fig. 6-22

$$T = 250 \sin \frac{\pi x}{70}$$

with T in °F and x in inches. The rest of the beam is at T = 0. Use equivalent thermal loads and moment area analysis to find the end moments and the normal stress distribution. What are the maximum shear stresses at the junction of web and flange?

6-23 The square frame shown in Fig. 6-23 is made up of I-beams, and is loaded uniformly with a linear load q #/in. It has a temperature distribution on each span

$$T = \frac{T_o}{2} \left(1 + \frac{2y}{h}\right) \sin \frac{\pi x}{L}$$

with y measured inward from the beam center line and x from any corner of the frame.

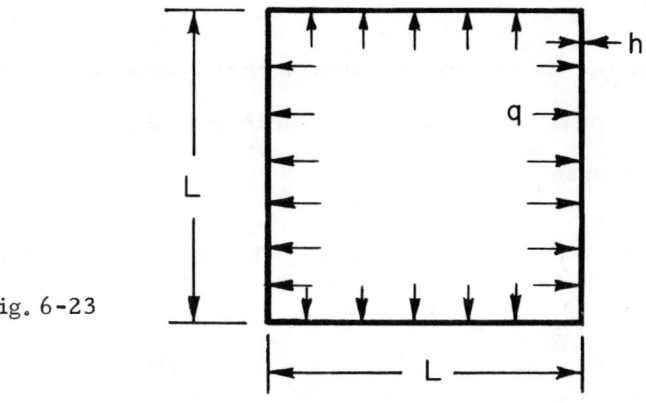

Fig. 6-23

Using complementary energy analysis determine the stress distribution in the frame

6-24 The rectangular beam in Fig. 6-24 rests on four simple supports. The

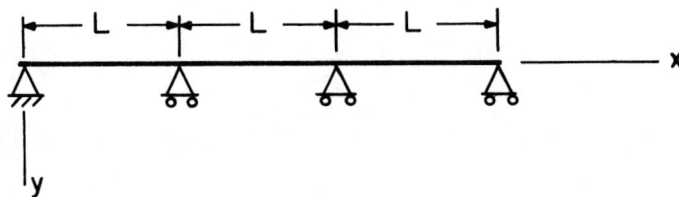

Fig. 6-24

temperature distribution is

$$T = \frac{T_o}{4}\left(1 + \frac{2y}{h}\right)^2 \sin \frac{\pi x}{3L}$$

By means of complementary energy analysis determine the stress distribution in the beam. Hint: The indeterminate reactions at the two inside supports are the same; $R_1 = R_2 = R$, and the strain energy is symmetrically distributed so that only the section $0 < x < \frac{3L}{2}$ need be considered.

6-25 By means of complementary energy analysis determine the stress distribution in the beam shown in Fig. 6-16, for the temperature distribution stated in Prob. 6-16.

6-26 Determine, by means of complementary energy analysis, the moment at the fixed end of the beam described in Prob. 6-14, for the temperature distribution given in that problem.

6-27 The rectangular beam shown in Fig. 6-27 is subjected to a temperature

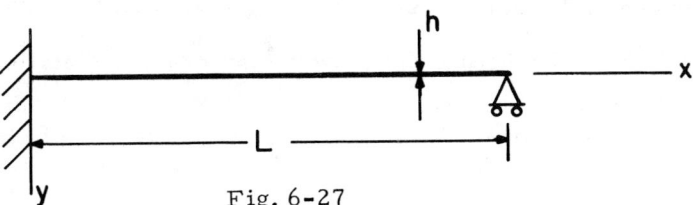

Fig. 6-27

distribution

$$T = T_o \sinh \frac{y}{h}$$

Use Castigliano's theorem to determine the moment at the fixed end and the deflection at the center of the beam. Employ the following three procedures:

(a) Determine the moment M_o at the fixed end. Then put a dummy load at the center of the beam and determine the deflection, assuming a simply supported beam with the known value of M_o acting at $x = 0$.

(b) Place a moment M_o at $x = 0$ and a dummy load P_d at $x = \frac{L}{2}$ and determine concurrently the value of M_o and the deflection under P_d.

(c) Determine first the end moment M_o, as in (a). Then put a dummy load, P_d, at $x = \frac{L}{2}$ and determine the deflection by assuming that P_d acts on the fixed-pinned statically indeterminate beam.

6-28 Using Castigliano's theorem, compute concurrently, for the beam in Prob. 6-16, the following: (a) the slope at $x = 0$; (b) the support reaction at $x = \frac{L}{2}$; (c) the moment at the fixed end, $x = L$.

6-29 Determine, by means of Castigliano's theorem, the deflection at the span mid points of the frame shown in Fig. 6-23. The temperature distribution is as given in Prob. 6-23.

6-30 Obtain the deflection at the center of a pinned-fixed rectangular beam, of depth h and length L, having a temperature distribution

$$T = \frac{T_o y}{h} \sin \frac{\pi x}{L}$$

with x measured from the fixed end. Perform analysis by Castigliano's theorem.

6-31 Obtain the deflection at the center of a fixed-fixed rectangular beam, of depth h and length L, having a temperature distribution

$$T = \frac{T_o y}{h} (1 - \sin \frac{\pi x}{L})$$

Use Castigliano's theorem to obtain solution, and compare with results given in Example (j) of Sect. 6.5.

6-32 Find by means of Castigliano's theorem the end rotations, at $x = 0$ and $x = 3L$, of the beam shown in Fig. 6-24. The temperature distribution is as given in Prob. 6-24.

6-33 Use Castigliano's theorem to obtain the slope and deflection at the free end, $x = L$, of a circular beam in the form of a rod of radius R, supported as shown in Fig. 6-33.

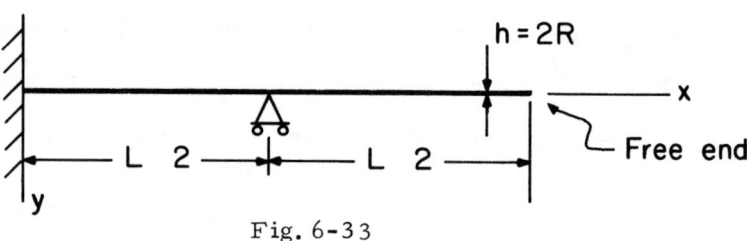

Fig. 6-33

The temperature distribution is

$$T = T_o \frac{y}{h} \cosh\left(\frac{x}{L} - \frac{1}{2}\right)$$

6-34 Obtain, by means of the reciprocal theorem, the deflection and rotation at $x = L$ of the cantilever beam shown in Fig. 6.10-2. The temperature distribution is as given in Example (b) of Sect. 6.10.

6-35 From an engineering handbook we obtain the deflection curve of a pinned-fixed beam with a load P^* at the center of the span as

$$v = \frac{P^* x}{96 EI} (3L^2 - 5x^2) \qquad 0 < x < \frac{L}{2}$$

$$v = \frac{P^*}{96 EI} (x - L)^2 (11x - 2L) \qquad \frac{L}{2} < x < L$$

with x measured from the pinned end. Determine, by means of the reciprocal theorem the deflection at the center of the span, in a beam with a rectangular cross section. The temperature distribution is

$$T = T_o \sinh \frac{y}{h}$$

Compare results with those obtained in Prob. 6-27.

6-36 A rectangular fixed-fixed beam of depth h and length L is subjected to a temperature distribution

$$T = T_o e^{-y/h} \frac{x}{L}$$

Using the reciprocal theorem find the mean deflection of the beam. Hint: the deflection curve of a fixed-fixed beam loaded with a uniform load q^* is

$$v = \frac{q^* x^2}{24 EI} (L - x)^2$$

6-37 Use the reciprocal theorem to obtain the deflection curve of the fixed-fixed rectangular beam shown in Fig. 6-37.

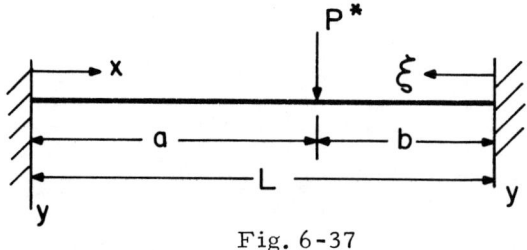

Fig. 6-37

The temperature distribution is

$$T = T_o \sinh \frac{x}{h}$$

Note: The deflection curve due to the auxiliary load P^* is

$$v = \frac{P^* b^2 x^2}{6EIL^3} (3aL - 3ax - bx) \qquad 0 < x < a$$

$$v = \frac{P^* a^2 \xi^2}{6EIL^3} (3bL - 3b\xi - a\xi) \qquad 0 < \xi < b$$

6-38 A rectangular beam of depth h and length L rests on three equally spaced simple supports as shown in Fig. 6.7-2. The temperature distribution in the beam is

$$T = T_o \frac{y}{h} \sin \frac{\pi x}{L}$$

Use Eq. (6.12-5) to obtain the moment over the center support. Check result against Example (b) of Sect. 6.7. Repeat problem for the case of a three-span beam with clamped ends.

6-39 Derive general expressions for a beam resting over: (a), four equally spaced simple supports; (b), five equally spaced supports; (c), six equally spaced supports. The temperature distribution is $T = T(y)$ so that Eq. (6.12-16) applies.

PROBLEMS

6-40 A beam with a transverse temperature distribution $T = T(y)$ rests on seven equally spaced simple supports (6 spans). Obtain the moments over the supports assuming first, that the beam is of infinite extent (Eq. (6.12-25)) and second, that the beam is finite, Eq. (6.12-28)). Compare results by plotting moment diagrams. Because of symmetry only three spans need to be analyzed.

6-41 A free-free beam of moderate length, $\beta L = \frac{\pi}{2}$, rests on an elastic foundation. Using the temperature distribution of Example (a), Sect. 6.13, determine the deflection curve and the stress distribution.

6-42 A long circular nuclear fuel element of radius R has pin end supports and intermediate spring supports which have the character of an elastic foundation. Derive the expression for the deflection and stress distribution in the element for the case of $T = T(y)$.

6-43 A long cylindrical element in the form of a circular tube has an outer radius R_o. The two extremities are clamped and the rest of the tube is supported by closely spaced springs which act as an elastic foundation. The temperature distribution is

$$T = \frac{T_o y}{h} \sin \frac{\pi x}{L}$$

with $h = 2R_o$. Obtain expressions for the deflection and stress distribution in the tube. Indicate the approximate stress distribution near the ends and far from the ends.

6-44 A temperature indicator is in the form of a cantilever bimetallic strip, as shown in Fig. 6-44. The scale at the free end,

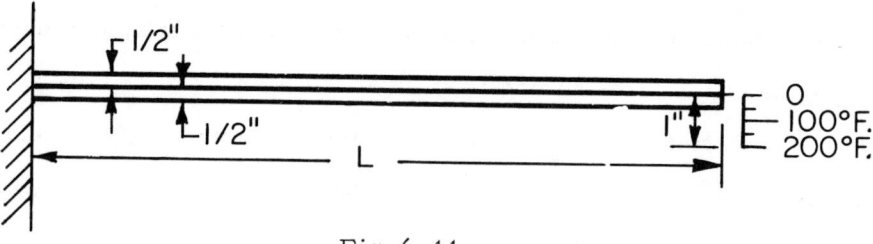

Fig. 6-44

graduated from $0°$ to $200°$ F measures 1". The moduli of elasticity of the upper and lower bimetallic constituents, each $\frac{1}{2}$" thick, are equal and the coefficients of expansion are $1 \times 10^{-6}/°F$ and $9 \times 10^{-6}/°F$ respectively. Find the required length, L, to produce 1" deflection in a temperature change of $200°F$.

6-45 A bimetallic strip consists of two components of dissimilar materials, as in Fig. 6.14-1. It is clamped at the left end, $x = 0$, and pinned at the right end, $x = L$. Find the deflection curve and stress distribution when the temperature variation is

$$T_1 = \frac{T_i}{2}(1 + \frac{2y_1}{h_1}) \sin \frac{\pi x}{L}$$

$$T_2 = \left[T_i + \frac{T_o}{2}(1 + \frac{2y_2}{h_2}) \sin \frac{\pi x}{L}\right]$$

Use subscripts 1 and 2 for the upper and lower constituents.

6-46 A reinforced rectangular concrete beam is unrestrained so that it can deform freely. The beam is heated uniformly to a temperature $T = 500°F$. The coefficient of expansion and modulus of elasticity of the reinforcing bars are $\alpha = 6.5 \times 10^{-6}/°F$ and $E = 30 \times 10^6$ psi, and the corresponding values for the concrete are $\alpha = 5.5 \times 10^{-6}/°F$

and $E = 2 \times 10^6$ psi. The dimensions are shown in Fig. 6-46. Find the stress distribution over the cross section. Note: Disregard the concrete lying below the steel.

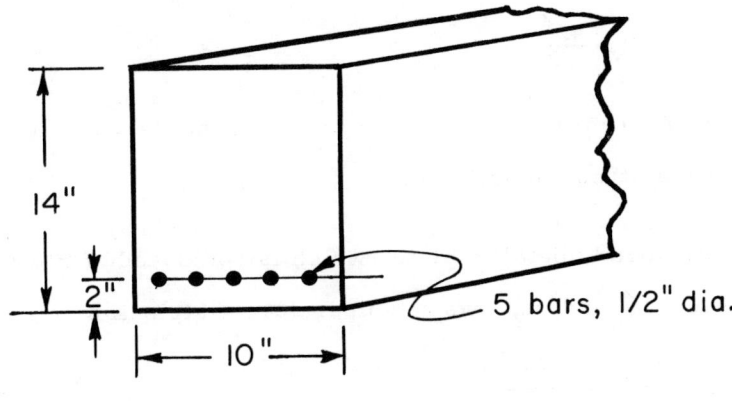

Fig. 6-46

6-47 A clad rod such as that shown in Fig. 6.14-4(a) is rigidly supported (fixed-fixed with axial displacement permitted) and subjected to a temperature distribution

$$T = (T_o + T_1 \frac{y}{h}) \cosh(\frac{x}{L} - \frac{1}{2})$$

with $h = 2R_o$. Derive the expressions for the deflection and stress distribution, assuming the cladding and core to be of dissimilar materials.

6-48 A double-walled heat exchanger tube, such as that shown in Fig. 6.14-4(b) is rigidly supported at its two ends (with axial displacement prevented). The temperature distribution is

$$T = (T_a + T_b \frac{y}{R_o}) \sin \frac{\pi x}{L}$$

Derive expressions for the deflection and stress distribution, assuming dissimilar properties for the two materials.

6-49 An unrestrained homogeneous flat plate, of thickness h, has a temperature distribution which is a function of the thickness coordinate z. It is

$$T = T_o \left(1 - \frac{4z^2}{h^2}\right) + T_1 \frac{z}{h}$$

with z measured from the central plane. Find the curvature of the plate and the stress distribution.

6-50 An unrestrained bimetallic plate, which is composed of two layers of thicknesses h_1 and h_2, has a temperature distribution

$$T_1 = \frac{T_i}{2}\left(1 + \frac{2z_1}{h_1}\right)$$

$$T_2 = T_i + \frac{T_o}{2}\left(1 + \frac{2z_2}{h_2}\right)$$

as a result of lateral heat conduction. The distances z_1 and z_2 are measured downward from the mid-planes of the upper and lower plates, respectively. Assume dissimilar materials, and find the plate curvature and stress distribution in each of the constituents.

6-51 A carbon steel pressure vessel wall is 7" thick and is clad internally with a $\frac{1"}{4}$ layer of stainless steel. The vessel wall may be taken to be a flat plate which is free to expand but whose edges are restrained from rotating. Assume a temperature distribution

$$T_{cs} = 50(1 + 2z_1) \qquad °F$$

$$T_{ss} = 100 + 100(1 + 8z_2) \qquad °F$$

with z_1 measured radially inward from central plane of carbon steel shell and z_2 measured radially inward from central plane of stainless cladding. Assume the following properties: $\nu = 0.3$ for both materials, and

$$E_{cs} = 30 \times 10^6 \text{ psi} \qquad \alpha_{cs} = 6.5 \times 10^6 /°F$$

$$E_{ss} = 27 \times 10^6 \text{ psi} \qquad \alpha_{ss} = 9.5 \times 10^6 /°F$$

Compute the stress distribution in the pressure vessel wall. If the vessel were completely encased in a heavy concrete structure, so that it could not expand radially, what would be the stresses in the vessel wall?

Index

A

Abrupt temperature reversals:
    in rectangular bar, 166
    in rectangular plate, 166
Airy stress function:
    biharmonic equation, 129, 131
    boundary conditions, 130
    cylindrical coordinates, 131
    definition, 127
Analogy for plane thermal stress, 164
Angular rotation components, 4
Assemblies with axial symmetry, 185
Attached parallel elements:
    normal axial stresses, 44, 47
    shear stresses in end attachments, 48
Auxiliary loads, 306
Axially symmetrical assemblies, 185

B

Bars:
    abrupt temperature reversal, 52, 166
    correction of stress, 49-54
    I-beam, 65
    in parallel, 44, 47
    in series, 89
    layered, 61
    parabolic transverse temperature, 164
    plane harmonic cross section temperature, 60
    restrained;
        by classical analysis, 27, 57
        by complementary energy, 286
        by equivalent load analysis, 25, 30
        by Rayleigh-Ritz method, 321
    rigidly enclosed, with axial temperature, 31
    shear stresses, 63
    sinusoidal transverse temperature, 161
    stress-free cross section, 150
    stress modification due to biaxiality, 49
    three-dimensional temperature, 59
    transverse stresses, 26
    triangular transverse temperature, 162
    unrestrained;
        by Castigliano's theorem, 299
        by equivalent load analysis, 24, 26, 29, 55, 56
        by reciprocal theorem, 310
    wide-flange beams, 65
    with non-uniform cross section, 100
Beams:
    basic relationships, 337
    boundary conditions, 341
    by boundary load superposition, 365
    by Castigliano's theorem, 386
    by complementary energy, 381
    by conjugate beam analysis, 371
    by integration of beam equation, 347
    by moment area analysis, 371

    by Rayleigh-Ritz method, 324
    by reciprocal theorem, 392-396
    cantilever;
        by Castigliano's theorem, 389
        by integration of beam equation, 349
    clamped;
        by boundary load superposition, 366
        by Castigliano's theorem, 390
        by complementary energy, 384
        by integration of beam equation, 351, 356
        by moment area analysis, 377
        by reciprocal theorem, 395
        high end temperature gradients, 361
        high mid-span temperature gradient, 359
    continuous;
        analysis, 397
        deflection curve, 404
        finite beam, 402
        long beam, 402
        on three supports, 402
        three moment equation, 399
    corrections due to;
        axial temperature variation, 335
        axial thrust, 336
        geometry, 334, 335
        property changes, 336
    cross section Laplacian, 150, 333
    deformation equation, 339, 340, 341
    equilibrium, 340
    errors in analysis, 333
    fixed-pinned;
        by complementary energy, 383
        by integration of beam equation, 350, 355
        by moment area analysis, 375
    layered;
        analysis, 412
        bending axis, 416
        bimetalic, 412
        circular, 424
        concentric circular tubes, 420

        deformation equation, 418
        rectangular, 423
    on elastic foundations;
        analysis, 405
        long beam, 408
        semi-infinite beam, 408
    on three supports;
        by boundary load superposition, 368
        by moment area analysis, 379
        by three-moment equation, 402
    shear stresses, 342
    simply-supported;
        bimetalic, 423
        by Castigliano's theorem, 388
        by integration of beam equation, 347, 353
        by Rayleigh-Ritz method, 324
        by reciprocal theorem, 394, 396
    stress-free cross sections, 150, 333
    two-plane bending, 363
Biaxial stresses in bars and rings:
    discussion, 49
    homogeneous bars, 50
    homogeneous rings, 53
    layered bars, 62
    modified modulus, E^*, 51, 52
    restrained bars, 53
    stress modification formulae, 51, 52
    unrestrained bars, 53
Biaxial stresses in beams, 333-335
Boundary conditions:
    classical, 3, 5
    displacement stress formulation, 8, 10
Boundary load superposition, 365

C

Castigliano's theorem:
    beam analysis, 386
    cantilever beam, 389
    clamped beam, 390
    concentric tubes, 300

discussion, 289
displacement strain energy, 290
simply supported beam, 388
statically determinate truss, 301
statically indeterminate structures, 291
statically indeterminate truss, 303
unrestrained bar extension, 299
Circular fin, 223
Circular plate:
   axially symmetrical temperature variation, 227
   generalized plane stress, 227
   layered, 234
   radial and axial temperature variation, 227
   radial temperature variation, 212
   sandwich, 234
   step temperature change, 211, 218
   stress concentration in center hole, 218
   symmetrical transverse temperature, 171
   with radially varying thickness, 223
Classical equations:
   cylindrical coordinates, 4
   discussion, 1
   rectangular coordinates, 2
Compatibility:
   approximation in plate, 148
   error magnitude in plate, 148
Compatibility equations:
   in strain, 39, 142
   in stress, 39
Compatible deformation, 2
Circular cylinder: (see Circular rod)
Circular rings:
   axial temperature variation, 73
   concentric, 84
   layered, 76
   radial stress, 82, 85
   radial temperature variation, 80
   shear stresses, 74, 79
   three-dimensional temperature variation, 87

Circular rod or cylinder:
   analysis, 246
   center stress concentration, 252
   hollow, 284
   radial temperature variation, 246
   reciprocal theorem, 316
   solid, 251
   uniform heat generation, 253
Complementary energy:
   beam analysis, 381
   clamped beam, 384
   concentric tube analysis, 287
   discussion, 280
   fixed-pinned beam, 383
   one-dimensional, 281, 282
   plane, 284
   restrained bar, 286
   statically indeterminate truss, 288
   three-dimensional, 285
Concentric circular cylinders:
   bending, 420
   extensional deformation, 239, 254
   transverse temperature distribution, 424
Concentric circular plates:
   analysis, 204, 223
   interference pressure, 209, 223
   three concentric plates, 223
   two concentric plates, 204
Concentric circular tubes:
   bending, 420
   extensional deformation, 239, 254
   transverse temperature distribution, 424
Concentric cylindrical shells:
   extensional deformation, 177
   many concentric shells, 181
   radial temperature variation, 177
   two concentric shells, 177
Concentric rings:
   analysis, 204
   interference pressure, 209
Concentric spheres:
   at uniform temperature, 256
   with radial temperature variation, 268

Concentric spherical shells:
    extensional deformation, 182
    many concentric shells, 185
    radial temperature variation, 182
    two concentric shells, 185
Concentric tubes:
    Castigliano's theorem, 300
    complementary energy, 287
    normal axial stresses, 44, 47
    reciprocal theorem, 311
    shear stresses in end attachments, 48
Conjugate beam analysis, 371
Cylindrical coordinates:
    classical equation, 4
    displacement stress formulation, 9
Cylindrical shell:
    classical analysis, 175
    end stresses, 177
    equivalent load analysis, 173
    extensional deformation, 173
    radial temperature variation, 173

D

Dilatation, 4, 38
Disc and ring assembly, 186
Displacement strain energy, 290
Displacement stress definition, 5
Displacement stress formulation:
    cylindrical coordinates, 9
    rectangular coordinates, 6
Displacement stress transformation:
    one-dimensional, 14
    three-dimensional, 6, 9
    two-dimensional, 12

E

End normal stresses:
    cylindrical shells, 177
    layered bars, 69
    plate strips, 159
    riveted strips, 71
End shear stresses:
    layered bars, 69
    plate strips, 159
    riveted strips, 71
Equilibrium equations:
    classical;
        cylindrical coordinates, 5
        rectangular coordinates, 3
    displacement stress formulation;
        cylindrical coordinates, 10
        rectangular coordinates, 7
Equilibrium of thermal loads:
    axial, 34
    transverse, 36
Equivalent internal pressure:
    one-dimensional, 15
    three-dimensional, 6, 9
    two-dimensional, 13
Equivalent thermal loads:
    body forces, 7, 10, 13, 14
    discussion, 5, 11, 13
    one-dimensional, 14
    surface tractions, 8, 11, 13, 15
    three-dimensional, 7
    two-dimensional, 13
Extensional deformation of:
    bars, 44
    cylindrical shells, 173
    homogeneous plates, 22, 23, 171
    layered bars, 61
    layered plates, 22, 23, 172
    spherical shells, 181

F

Fatigue, 18, 192
Fin: (see Circular fin)
Free plane strain, 19, 138
Free strain, 2, 16

INDEX 459

G

Generalized plane stress:
    circular plate, 228
    definition, 132

H

Heat sources and sinks, 152
Hot area in large plate:
    linear radial temperature, 220
    uniform temperature, 169

I

Internal pressure:
    one-dimensional, 15
    three-dimensional, 6, 9
    two-dimensional, 13
Isothermal stress, 1

L

Lamé constants:
    definition, 38
    equilibrium equations, 40
Layered bars:
    end shear stresses, 69
    interfacial shear stress, 67
    normal stress, 61
Layered plates:
    circular, 234
    extensional deformation, 172
    sandwich, 234
    symmetrical transverse temperature, 172
Linear surface temperature ramp, 194
Linear temperature variation:
    cylindrical coordinates, 145
    discussion, 144
    in bars, 29, 146
    in beams, 29, 146
    in circular tube, 145
    in plates, 146

M

Maximum stress in parallel elements, 190
Mitchell conditions, 153
Modified elastic constants, 18, 22
Modified equivalent loads, 20
Moment area analysis, 371
Multiply connected plates and cross sections, 152

N

Non-uniform cross section bars:
    discussion, 100
    parabolic area and temperature variation, 101
    sinusoidal area and temperature variation, 103

O

One-dimensional equivalent loads, 13
One-joint bar assembly, 33, 104

P

Parallel and series elements, 92
Parallel bar assembly:
    plane displacement, 44
    plane strain, 47
Parallel element assembly:
    normal axial stresses, 44, 47
    shear stresses in end attachments, 48
Pinned two-bar assembly, 33
Plane displacement, 44
Plane harmonic temperature:

in beam cross sections, 333
stress-free deformations, 142
Plane strain:
biharmonic equation, 140
definition, 11, 136
free, 19, 138
in rectangular bars, 155
zero, 19, 137
Plane strain reformulation, 18, 19
Plane stress:
Airy stress function, 127
boundary conditions, 130
biharmonic equation, 129, 131
cylindrical coordinates, 131
discussion, 127
generalized, 132
in plate strips, 155
stress function, 127
Plastic hinge, 92
Plate bending (pure):
clamped, 428
discussion, 21, 426
homogeneous, 427, 432
layered, 429, 433
spherical, 21, 426
Plates: (see Rectangular plates and Circular plates)
Plate strips:
discussion, 155
saw-tooth temperature variation, 160
sinusoidal temperature variation, 156, 161
step temperature change, 168
Potential energy, 319
Pure bending, 21
Pure extension:
definition, 21
layered plates, 135
plane, 22, 134
plate, 22, 134
shell, 22
Pure plane extension:
layered plates, 135

plates, 134

R

Radial fin: (see Circular fin)
Rayleigh-Ritz method:
beam deflection, 324
discussion, 319
potential energy, 319
restrained rod, 321
Reciprocal theorem:
auxiliary loads, 306
cantilever beam, 312
clamped beam, 395
concentric tube assembly, 311
deflection formula, 309
discussion, 306
hollow cylinder, 316
radial displacement, 318
rotation formula, 309
simply supported beam, 394, 396
statically determinate truss, 314
unrestrained rod, 310
volume change of sphere, 317
volume changes, 309
Rectangular coordinates:
classical equations, 2
displacement stress formulation, 6
Rectangular plates:
abrupt temperature reversals, 166
parabolic temperature distribution, 164
saw-tooth temperature variation, 160
sinusoidal temperature variation, 156, 161
symmetrical transverse temperature, 171
triangular temperature distribution, 162
Ring and disc assembly, 186
Rings: (see Circular rings)
Rivet shear stresses, 71
Riveted sandwich strips, 71
Rod and tube assembly, 188
Rotation components, 4

INDEX

S

Sandwich plate:
    circular, 234
    rectangular, 172
Sectionally heated bar, 95
Self-prestressing, 337
Self-relaxing thermal loads, 192, 337
Self-springing, 337
Series bars with transverse temperature, 97
Series elements:
    discussion, 89, 192
    plastic hinge, 92
    vulnerability, 92
Shear stresses in end welds, 48
Shrink-fit:
    circular cylinders, 239, 254
    circular tubes, 239, 254
    concentric plates, 204
    concentric rings, 204
    datum temperature, 205
    stress-free temperature, 205
Spheres:
    center stress concentration, 267
    hollow, 264
    radial displacement, 318
    radial temperature variation, 261
    reciprocal theorem, 317
    solid, 265
    volume change, 317
    with uniform heat generation, 267
Spherical bending, 21, 427
Spherical shell:
    extensional deformation, 181
    radial temperature variation, 181
Step temperature change:
    in circular plate, 218
    in plate strip, 168
Strain-displacement relationships:
    cylindrical coordinates, 4
    rectangular coordinates, 3
Strain energy:
    beam analysis;
        Castigliano's theorem, 386
        complementary energy, 381
        Rayleigh-Ritz method, 324
        reciprocal theorem, 392
    Castigliano's theorem, 289
    complementary energy, 280
    discussion, 279
    one-dimensional, 281
    Rayleigh-Ritz method, 319
    reciprocal theorem, 306
    strain energy density, 281
    three-dimensional, 285
    two-dimensional, 284
Stress-free cross sections, 150
Stress-free temperature distribution:
    Class 1: statically determinate structures, 143
    Class 2: linear temperature variation, 144
    Class 3: flat plate temperature distribution, 146
    Class 4: plane harmonic cross section temperature distributions, 150
    Class 5: multiply connected plates and cross sections, 152
    discussion, 141
    plane harmonic temperature distributions, 142
    summary, 153
    unique stress distribution, 141
Stress-free trusses:
    Castigliano's theorem, 301
    discussion, 143
    pinned two-bar assembly, 33
Stress function:
    biharmonic equation, 129, 131
    boundary conditions, 130
    cylindrical coordinates, 131
    definition, 127
Stress-strain relationships:
    classical equations, 2, 4
    displacement stress formulation, 7, 10
Summary of thermal loading, 16

Surface thermal shock, 194

T

Thermal body forces, 7, 10, 13, 14
Thermal cycling, 17, 192, 337
Thermal dilatation, 38
Thermal fatigue, 18, 192
Thermal loading table, 16
Thermal shield, 194
Thermal shock, 192
Thermal sleeve, 194
Thermal strain, 2, 17, 38
Thermal surface tractions, 8, 11, 13, 15
Thermoelastic reciprocal theorem, 306
Thick-walled cylinder, 248
Thick-walled tube: (see Thick-walled cyclinder)
Three-moment equation:
   axial and transverse temperature, 399
   transverse temperature distribution, 400
Transverse incompatibility:
   in bars, 26
   in plates, 148
Transverse stresses:
   in bars, 26
   in plates, 148, 149
Triaxial stress correction, 52
Trusses:
   statically determinate;
     Castigliano's theorem, 301
     pinned two-bar assembly, 33
     reciprocal theorem, 314
     stress-free, 143
   statically indeterminate;
     by complementary energy, 288
     Castigliano's theorem, 303
     one-joint bar assembly, 104
Tube and rod assembly, 188
Two-dimensional equivalent loads, 11

U

Uniaxial stress:
   definition, 13
   requirements for, 151

V

Volume changes, 309, 317

Z

Zero plane strain, 19, 137